Keeping It Living

Keeping It Living

Traditions of Plant Use and Cultivation
on the Northwest Coast of North America

Edited by

DOUGLAS DEUR

and

NANCY J. TURNER

UNIVERSITY OF WASHINGTON PRESS *Seattle*

UBC PRESS *Vancouver and Toronto*

Keeping it Living: Traditions of Plant Use and Cultivation on the Northwest Coast of North America has been published with the assistance of a generous grant from the Pendleton and Elisabeth Carey Miller Charitable Foundation.

© 2005 University of Washington Press
Printed in the United States of America
12 11 10 09 08 07 06 05   5 4 3 2 1

University of Washington Press
P.O. Box 50096, Seattle, WA 98145
www.washington.edu/uwpress

Published simultaneously in Canada by

UBC Press
The University of British Columbia
2029 West Mall, Vancouver, B.C. V6T 1Z2
www.ubcpress.ca

Library of Congress Cataloging-in-Publication Data
Keeping it living : traditions of plant use and cultivation on the Northwest Coast of North America / edited by Douglas Deur and Nancy J. Turner.
      p.   cm.
      Includes bibliographical references and index.
      ISBN 0-295-98512-7 (hardback : alk. paper); ISBN 0-295-98565-8 (pbk : alk. paper)
      1. Indians of North America—Ethnobotany—Northwest Coast of North America.
      2. Indians of North America—Agriculture—Northwest Coast of North America.
      3. Indians of North America—Food—Northwest Coast of North America. 4. Plants, Cultivated—Northwest Coast of North America.   5. Plants, Useful—Northwest Coast of North America. I. Deur, Douglas, 1969–   II. Turner, Nancy J., 1947–
      E78.N78K45   2005
      581.6'3'09795—dc22   2004029448

Library and Archives Canada Cataloguing in Publication
      Keeping it living: traditions of plant use and cultivation on the Northwest Coast of North America / edited by Douglas Deur and Nancy J. Turner.

Includes bibliographical references and index.
ISBN 0-7748-1266-4 (bound); ISBN 0-7748-1267-2 (pbk.)

      1. Indians of North America—Ethnobotany—Northwest Coast of North America. 2. Indians of North America—Agriculture—Northwest Coast of North America—History. 3. Indians of North America—Food—Northwest Coast of North America—History. 4. Plants, Cultivated—Northwest Coast of North America—History. 5. Plants, Useful—Northwest Coast of North America—History. I. Deur, Douglas, 1969– II. Turner, Nancy J., 1947–

E78.N78K38 2005   581.6'3'09795   C2005-902364-3

The paper used in this publication is acid-free and 90 percent recycled from at least 50 percent post-consumer waste. It meets the minimum requirements of American National Standard for Information Sciences—Permanence of Paper for Printed Library Materials, ANSI Z39.48-1984.

Cover: Ma-Ma Yockland of the Gwatsinux ("Quatsino") Kwakwaka'wakw of northern Vancouver Island, picking salmonberries. She was raised in the mid-nineteenth century; her forehead was flattened in infancy, emblematic of her high status. Photo taken by B. W. Leeson in 1912 and shown here courtesy University of Pennsylvania Museum, Philadelphia.

The chapter opening drawings are by Jeanne R. Janish (from Hitchcock et al., *Vascular Plants of the Pacific Northwest*, vols. 1–5, University of Washington Press, 1955–69).

Contents

Preface

E. RICHARD ATLEO, *Umeek of Ahousat*

M odern discourse about the relationship between the world (the earth and its plants and animals) and human beings began with the onset of colonization. Until recently this discourse has primarily involved colonizers talking among themselves. The quotation attributed to John Locke, who wrote the following in 1690, is one example.

> God gave the World to men in Common; but since he gave it them for their benefit, and the greatest Conveniences of Life they were capable to draw from it, it cannot be supposed that he meant it should always remain common and uncultivated. He gave it to the use of the Industrious and Rational (and Labour was to be his Title to it). [Quoted in Turner et al., this volume]

In a postmodern age of pluralism, the problem with discourse that excludes a large proportion of the affected world may be obvious. The relationship between humans and world in this type of discourse allows for one rather narrow perspective, as though reality were subject to one point of view, one perspective, one lens upon the universe, one worldview. A worldview, it is argued, is inextricably tied to beliefs about origins, beliefs that help to shape the defining characteristics of a society or civilization. Locke's views and those of his successors regarding the relationship between humans and the world can be traced to the biblical injunction "to subdue the land" and to exercise "dominion" over all nonhuman life in it. This is an origin story. In conjunction with relatively recent western cultural developments, such as science or the notion of surplus for profit rather than need, this origin story has contributed to the devastation of the earth's environment. Strip mining and logging clear-cuts are a logical outcome of an interpretation of biblical injunctions and the writings of early European thinkers such as Locke. So too are conceptual distinctions such as agriculture, horticulture and related terms such as tending, domesticating, cultivating and subsistence. Each of these terms describes subtle distinctions between various acts of subduing and exercising dominion. These distinctions are key in understanding western civilization's relationship with and interpretation of the world.

For those born into this worldview it has, perhaps, until very recently, been assumed that all who do not share in this cultural experience belong to a less-developed category of humanity. This latter view has been manifested in Western science, whose own origin story places the beginning of the universe at 15 billion years ago, sometime following which humanity's ancestors came into being and have been evolving ever since. In this story, the hunter-gatherers evolved into a more advanced, agricultural form of humans. Much current debate about First Nations' use of resources represents a refinement of earlier discourses about the apparent links between hunter-gatherers and agriculturalists, and as such retains and does not question these long-standing evolutionary assumptions. In some cases, the assumption is clear that indigenous activity in relation to the nurturing of food plants is easily placed on the backward, less-developed side of the imaginary evolutionary scale. The more-advanced human societies, in that worldview, practice agriculture and the less-advanced practice "incipient" agriculture. If the scientific bases for these assumptions are robust and indisputable then the discourse has some validity. If, however, the scientific bases for these assumptions are challenged from a different worldview then a very different interpretation becomes available. What follows, then, is one example, from a Nuu-chah-nulth[1] perspective, of a different worldview with very different assumptions about the relationship between earth's life-forms.

From the time of birth, among traditionally oriented Nuu-chah-nulth, the first discourses about beginnings and the nature of all relationships are heard. These discourses take place in mythical time and explain, for example, from a Nuu-chah-nulth perspective, how the great abundance and diversity that was originally in Clayoquot Sound came to be. The following is an excerpt from one such story told in the house of Keesta who was born eighty years after fur trading began and forty years before colonial settlement (ca. 1900) in Clayoquot Sound. Keesta was my great-great-grandfather who survived into the 1950s, a full decade after my own birth. This discourse approximates the oral tradition in which it is found and is presented as an excerpt from a play.

SETTING: Somewhere in Clayoquot Sound
TIME: Near the beginning of time
SCENARIO: A person is busy fashioning two knives in preparation to resist change. The prophetic word has it that someone is coming to change everyone's life. The knife maker is unaware that the person who approaches and begins a friendly conversation is the one who, in English, may be referred to as the Transformer, who speaks first.

"What are you making?"
"Knives."
"Oh! For what purpose?"

"They say someone's coming to change us but nobody's going to change me!" (This is said with great conviction and resolve).

"They are beautiful!!! Let me have a look at those!!!" (The knives are handed over to the Transformer).

"Here," (placing the two knives on each side of the person's head) "from now on these shall be your two ears and you will make the forest your new home."

And that's how deer came to be.

In this discourse, creation is complete in a non-Darwinian sense. The "origin of species" is in, and from, the first people who remain essentially the same throughout the ages down to the present day. The grounds of discourse, or the starting point of discourses, defines the critical historical path and requires only that proper relationships be developed and maintained between all life-forms. This in contradistinction to the injunction that humans 'subdue and have dominion over the earth' in an evolutionary process; rather the injunction is for humans to find a balance and harmony between all life-forms. Since the salmon and human have common origins they are brothers and sisters of creation. Since the assumption of all relationships between all life-forms is a common ancestry, protocols become necessary in the exercise of resource management. If the salmon are not properly respected and recognized they cannot properly respect and recognize their human counterparts of creation. This historical process is neither evolutionary nor developmental in the linear sense. Changes are not from simple to complex, as a more modern worldview would have it, but from complex to complex, from equal to equal, from one life-form to another. Thus, in the beginning, the full diversity of life-forms, the biodiversity of the world, is produced from common origins, which, in turn, set the egalitarian grounds of discourse since all life-forms are from the same family. All organisms have origins, experiences, and concerns that are fundamentally the same as those of humans. The nature of this discourse remained unchanged for millennia and accounts for much of the prevalent view attributed to indigenous perspectives that "*heshook ish tsawalk.*"[2] "Everything is one." This unity of creation is not to be interpreted from a linear perspective but from a cyclical one. Species have not evolved over time but remain essentially the same, repeating themselves in apparently endless cycles.

The unity of creation is not to be interpreted from a purely physical, empirical perspective, which has been found (in the Nuu-chah-nulth tradition) to be an unreliable indicator of true reality. A wolf, a deer, a bear, and a bird may all physically differ one from one another and from a human, but they are all of one species in the spiritual realm. Just as there is a significant difference between the letter and the spirit, so too can there be a significant difference between the physical and the spiritual, where the latter is the substance and the former is the shadowy reflection thereof. The prevailing Western worldview holds that objective truth can only be known empirically, that is, with and from the human senses. The indigenous worldview holds the antithesis, that real truth can only be known in and from the spiritual realm, which is the reality upon which empiricism is based, founded upon, and sourced. These conditions have implications for understanding traditional stewardship of tribal lands and resources, or what might be commonly termed "resource management" today.

If the first premise is that creation is complete and does not require further development in the Darwinian sense, then the development of agriculture and horticulture as these practices were known in the Western world is not necessary. Moreover, if surplus is married to the needs of a people, in a particular place and over the long term, rather than profit, clear-cutting and strip mining become impossible. The potential development of agriculture and horticulture cannot be ruled out in this worldview but their development must, of necessity, be constrained by the values based in the worldview. Amidst the abundant resources of the Northwest Coast of North America, territories were considered complete, requiring only tending and nurturing to provide for the needs of each community. If creation is complete there is no need for development. In this view, in the midst of abundance, "subsistence" is not an appropriate descriptor. The Nuu-chah-nulth word _hawilth_, which is translated as "chief" but can also mean "wealth," would be a more appropriate descriptor. A more telling descriptor is the Nuu-chah-nulth word mentioned in Turner et al.'s paper, _hahuulhi,_ which has the same root as _hawilth_ and refers to the sovereign wealth of the chief. In this sense, wealth is not created by human development but by what the Creator has created, and humans then merely tend, nurture, and respectfully, thankfully, take what is needed. Where people seek balance and harmony as a way of life, the earth and its resources become equal partners in an endless cycle of respectful life. The word "subsistence" makes sense only from a Western worldview, which demands that humans dominate the environment and profit from it.

Postmodern discourse about the relationship of humans to the earth, influenced by earlier European thinkers such as John Locke, must now be modified to incorporate the recent Supreme Court of Canada ruling on the _Delgamuukw_ case. In seeking to define Aboriginal title, the Supreme Court stated that Aboriginal title to land is "more than a fungible commodity" (footnote 2, _Delgamuukw v. the Queen_, paragraph 129 [British Columbia 1997]). This is an implicit recognition of the spiritual relationship between humans and the earth and its life-forms. This spiritual dimension accounts for the common notion that humans are not the real owners of land because everything belongs to the Creator. Notwithstanding the spiritual stewardship notion, human inhabitants possess a legal claim to the land, and the relationship of humans to creation is guided and shaped by an interpretation of origins. John Locke had one interpretation and Keesta another. The present state of the earth indicates that the interpretation of John Locke's views about human relationships to the earth, although it has created vast wealth, has also created vast poverty, in addition to creating huge imbalances in the earth's environment.

Perhaps it is time to begin a discourse between John Locke and Keesta, and the worldviews they represent, to see if each may learn from the other.

Notes

1. The Nuu-chah-nulth, formerly referred to as the "Nootka" in the literature of anthropology, are the indigenous inhabitants of the west coast of Vancouver Island. The author is a hereditary chief of the Ahousaht, a constituent subdivision of the Nuu-chah-nulth.

2. The author employs the orthographic standard used by the Nuu-chah-nulth in their language programs and for official correspondence.

in memoriam

The editors wish to acknowledge Wayne Suttles, who made a major contribution to this volume as well as to the meetings and discussions that preceded it, but who passed away shortly before its publication. A student of Erna Gunther, Melville Jacobs, and others in the Boasian tradition, Suttles provided a bridge between generations of researchers interested in documenting the environmental practices of Northwest Coast peoples. Since the 1940s, he had studied Coast Salish languages and cultures and published a number of prominent books and articles on these topics, many of which were brought together in a single volume, *Coast Salish Essays*. He also served as editor of the Smithsonian Institution's *Handbook of North American Indians*, Volume 7: *Northwest Coast*. We owe an immense debt of gratitude to Wayne for advancing the study of Northwest Coast peoples and their relationships to the landscapes, flora, and fauna of this distinctive region.

Keeping It Living

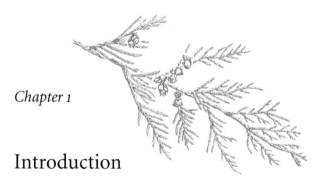

Chapter 1

Introduction

Reconstructing Indigenous Resource Management,
Reconstructing the History of an Idea

DOUGLAS DEUR AND NANCY J. TURNER

From the earliest anthropological research on the Northwest Coast of North America to the present day, there has been little debate as to whether the peoples of this region cultivated plants. Most scholars have accepted that they did not. In fact, Northwest Coast peoples' apparent lack of cultivation and their large, permanent villages of socially stratified foragers provided the anthropological literature with one of its most prominent anomalies. As Alfred Kroeber (1962: 61) and his contemporaries asserted, Northwest Coast societies were exceptional, "a wholly non-planting and non-breeding culture—perhaps the most elaborate such culture in the world." Prevailing wisdom suggested that, as beneficiaries of vast salmon runs, Northwest Coast peoples fed themselves with minimal effort. Plant cultivation, in this view, was unnecessary; the potential for plant cultivation was not apparent to the region's indigenous peoples, and the absence of scarcity extinguished any motive to enhance the natural availability of plants. This anomalous aspect of Northwest Coast subsistence was enthusiastically popularized by Franz Boas and his students, as they sought to rebut evolutionary and environmentalist models of cultural development. In turn, the apparent lack of cultivation on the Northwest Coast became a part of most North American anthropologists' undergraduate training, a prominent observation within many introductory textbooks, and a cornerstone of later theoretical developments within the fields of anthropology, archaeology, geography, and ethnobotany.[1]

Yet, this orthodoxy runs deep, and was taking form well before the arrival of anthropologists on the Northwest Coast in the late nineteenth century, shaped by a sense of this region and its inhabitants that was often superficial, based on brief encounters and biased expectations. Moreover, exploration was inextricably tied to the act of territorial appropriation and, all other cultural biases notwithstanding, its written record shall always be somewhat suspect. Captain James Cook visited the rocky west coast of Vancouver Island in 1778 during the spring season, a time of marine resource harvesting on the outer coast (and, significantly, a season when relatively little intensive plant harvesting occurs) among the resident Nuu-chah-nulth. On the basis of casual observations during this visit, Cook's journals described an indigenous population that was "indolent," "wild & uncouth," and incapable of the most basic civilized pursuits (including agriculture), but blessed by tremendous natural wealth in the form of fish and other marine animals (Cook and King 1784). In turn, inspired by Cook's tales of fur trade wealth, European explorers and fur traders flooded into the region in the late eighteenth and early nineteenth centuries. Each presumed that the entire region was composed of untouched wilderness, on the basis of Cook's accounts as well as their own ethnocentric presuppositions. Thus, some of the earliest explorers, when encountering evidence of indigenous plant management on the Northwest Coast, ascribed the practices to the antecedent diffusion of European influences, despite the rarity of prior cross-cultural interaction.

In 1789, during one of the first Euro-American fur-trading expeditions on the northern Northwest Coast, members of John Meares's crew saw evidence of cultivated plots—probably of tobacco—within Haida villages. In his official log, Meares's assistant, William Douglas (1790: 369), would assert that "In all probability Captain Gray, in the Sloop *Washington,* had fallen in with this tribe, and employed his considerable friendship in forming this garden," though he noted that there was no evidence to support this interpretation. Indeed, Gray does not appear to have visited the village in question, and Meares was probably the first European to pull ashore there. Today, few would ascribe Haida tobacco cultivation to the "considerable friendship" of early fur traders and explorers: the precontact antiquity of Haida tobacco cultivation is widely accepted (Turner and Taylor 1972). Other plant management practices encountered by early explorers were likewise attributed to European influences, if they were documented at all. However, unlike tobacco cultivation, these practices remain largely misunderstood today and are still attributed to European influence, we contend, due in significant part to a priori assumptions about Northwest Coast peoples' lack of plant cultivation among the region's earliest explorers and settlers.

Even during the nineteenth century, as European recolonization of the Northwest Coast proceeded apace, accounts of indigenous plant use remained superficial and tied to the biases and agendas of the colonial project. In the colonizers' view, the people who originally occupied the land used it only minimally, as hunters and random pluckers of shoots and berries. "Of agriculture they are quite ignorant," nineteenth-century writers would assert, "they have no aboriginal plant which they cultivate" (Brown 1873–1876: 50). This claim, in turn, was employed as one of several justifications for the dispossession of First Nations land during the colonial period.[2] In 1868, Gilbert Sproat (who became Land Commissioner in 1876) would characterize First Nations' land-use practices in quite typical terms:

> We often talked about our rights as strangers to take possession of the district. . . . The American woodmen . . . considered that any right in the soil which these natives had as occupiers was partial and imperfect as, with the exception of hunting animals in the forest, plucking wild fruits, and cutting a few trees . . . the natives did not in any civilized sense, occupy the land. [Sproat 1868 (1987): 8]

Such claims would be repeated, often uncritically, within the literatures of anthropology and archaeology. Northwest Coast food gathering would be depicted as undemanding: salmon was the staple food, and, as these fish appeared to be available in limitless abundance, there was simply no need for the peoples of this coast to modify their environment or intensively utilize plants. "Their civilization was built upon an ample supply of goods, inexhaustible, and obtained without excessive expenditure of labour" (Benedict 1934: 174). Boundless fish resources insured that they simply "did not rely heavily on plant foods" (Huelsbeck 1988: 166). The only evidence of plant management was to be found in "small plots of problematical tobacco" (Murdock 1934: 223). "The plants just grew by themselves. The only agriculture was of European introduction" (Drucker 1951: 81). Consistently, early ethnological surveys identified the region's inhabitants as noncultivators, and these secondary sources heavily influenced the thinking of subsequent generations of scholars (Kroeber 1939; Sauer 1936; Spinden 1917). However, as the chapters in this volume demonstrate, it is apparent that the region's natural abundance was overstated, while the role of humans in modifying Northwestern environments was much understated and misunderstood.

In recent decades, there has been a dramatic change in the prevailing view of what constitutes "plant cultivation." Today, most authors question the European biases within past representations of non-Western plant management. "Cultivation," it is now commonly countered, is a problematic term, but clearly should be applied to a broad continuum of plant enhancement practices (see Smith, this volume). Many of these practices are quite different from those characteristic of European agriculture, but all involve the manipulation of both plants *and* their environments as a means of achieving quantitatively and qualitatively enhanced plant production. Plant cultivation was not invented once or twice and then diffused from a small number of "civilized" hearths, as many anthropologists once suggested, nor did its emergence always eclipse efficient and preexisting hunting, fishing, and gathering practices.[3] This broad reevaluation of the very nature of plant cultivation provides us with an opportunity to revisit the puzzling case of the noncultivating Northwest Coast (Deur 2002b, Smith, this volume).

E. S. Curtis, *The Bark Gatherer*, about 1915. The woman is Virginia Tom of Hesquiaht. (British Columbia Archives, D-08330)

The puzzles are many: the region's sedentism, social stratification, complex ceremonialism, and patterns of resource ownership among presumed "foragers" have been depicted as both paradoxical and bewildering to earlier generations of anthropologists. Moreover, Northwest Coast "foragers" also fit many of the preconditions for the development of plant cultivation and agriculture proposed by theorists on agricultural origins. Few regions of the world seem so well suited to the "sedentary fisherfolk" hypothesis of agricultural origins, advanced by Sauer (1952), E. Anderson (1969), and Moseley (1975), or the widespread "affluent foragers" model of agricultural origins as summarized by MacNeish (1992). In each of these models, certain key factors are identified as being likely contributors to the emergence of plant cultivation and ultimately agriculture: resource wealth, sedentary villages of hunter-gatherers, abundant vegetation, and leisure-time opportunities for experimentation with plant enhancement. Few places fit these criteria as well as the Northwest Coast. These modelers consistently point to the Northwest Coast as a curious exception, and have felt compelled to defend their models with claims that the region presents a highly exceptional case. Several explanations have been advanced, in turn, to explain why Northwest Coast peoples did not cultivate plants: the superabundance of salmon eclipsed all other means of subsistence; the region was isolated from paths of American agricultural diffusion and the idea of plant cultivation never occurred to its inhabitants; Mesoamerican crops failed to diffuse across numerous inhospitable climate regions; or the Northwest Coast lacked a pool of suitably cultivatable plants. Yet none of these explanations provides a particularly convincing explanation for the enigma.

Moreover, the development of resource use identified in the archaeological record followed a sequence that coheres with the models of agricultural development outlined above, and seems to have been particularly conducive to the development of agriculture elsewhere in the world (Binford 1968; Boserup 1965; Cohen 1977; Deur 2000, 1999). As sea levels stabilized following the Pleistocene epoch, salmon populations appear to have increased, facilitating growing geographical concentrations of hunter-fisher-gatherers in sedentary villages lining the region's estuaries and major rivers (Fladmark 1975). In the millennia that followed, the elaboration of Northwest Coast social organization and improvements in food processing and storage technologies fostered additional population growth.[4] With time, the large, sedentary villages of extended family houses—of the sort described by Europeans at the time of contact—became widespread throughout the region. As village populations grew, methods of natural-resource management continued to become more innovative and intensive. Archaeological evidence from this period of resource intensification demonstrates a profusion of fish weirs, intertidal stone fish-traps, and fishnets. A corresponding intensification of plant resources would be expected under these circumstances, but few archaeologists have attempted to explore this point. Certainly, archaeological evidence of plant enhancement is elusive, as plant materials are poorly preserved in the archaeological record, but future research may yet yield information that will augment the largely ethnographic content of this volume.

We contend, however, that these puzzles are not as puzzling as they might seem. In our view, Northwest Coast peoples did develop numerous means of enhancing their plant resources, just as they developed numerous means of enhancing their ability to harvest, process, and store animal resources. The source of the puzzlement, we contend, is primarily the misrepresentation of indigenous subsistence practices by explorers, settlers, and early anthropologists, rather than the purportedly anomalous character of the region's indigenous societies. Indeed, we suggest that Northwest Coast peoples were actively cultivating plants, as that term is now defined, and that they were doing so prior to European contact. The story of these cultivation practices and how they came to be overlooked by the peoples of Europe is an enlightening one, and it is among the primary emphases of this book.

The Northwest Coast as a Natural and Cultural Region

First, however, the region must be introduced. The Northwest Coast represents a relatively distinct cultural, geographical, and ecological region, although the exact territory it covers is subject to varying interpretation. As defined by Suttles (1990), it extends along the Pacific Coast of North America, encompassing a narrow strip of coastline bounded on the inland side by the Coast and Cascade mountain ranges, running from the Oregon–California border north to the Copper River delta on the Gulf of Alaska. In general terms, the environment is characterized by a relatively mild climate, temperate rainforests, and rich marine life; it is a land of rugged, densely forested mountainsides and, from the Puget Trough northward, glacier-scoured inlets, islands, and fjords. This concise description is deceptively simple, for the region exhibits considerable localized environmental diversity. Along the coasts of Oregon, Washington, British Columbia, and Southeast Alaska, mountain peaks rise abruptly from sea level to alpine zones well above the timberline, to elevations of over 3,000 meters. Elsewhere, there are gently sloping hills and low coastal bluffs. Sharp temperature and moisture gradients exist between different environments, creating distinct microclimates within shorelines, lowland forests, and montane and alpine areas, even within the same watershed. Marine air moving into the region from across the Pacific Ocean distributes precipitation unevenly, with temperate rainforests on windward slopes, and relatively dry rain-shadow zones on leeward slopes. Within a single traditional First Nation's territory, annual precipitation at one site can be over three meters greater than in another, and each such location possesses distinctive plant and animal communities (Deur 1999). Landslides, floods, glaciers, wind, and fire have produced a diverse patchwork of open meadows and prairies, and clearings that contain a distinctive assortment of herbaceous plants, which were important to indigenous communities and a number of animal species alike. Rivers and streams are abundant and structurally diverse within the region, and there are riparian habitats of multiple scales and levels of biological complexity. Freshwater lakes, marshes, and peat bogs are commonplace, particularly on formerly glaciated terrain. Marine environments vary from open ocean, to ocean shoreline, to protected inlets extending far inland, and the biota of each varies considerably; the shoreline is punctuated by biologically rich tidal flats, lagoons, and alluvial floodplains, coastal dunes, and rocky seastacks and headlands. Wave action, water temperature, and salinity vary considerably within these coastal environments, influencing the distribution of habitats and organisms, while currents pull water, sediments, and organisms from place to place in never-ending cycles (Pojar and MacKinnon 1994). Together, all of these factors yield tremendous variation in habitats and microclimates within the Northwest Coast region, each supporting distinctive forms of plant and animal life. In turn, the geographical distribution of the plants and animals that occupy these environments might also be best described as patchy or uneven. The region's organisms, including the diverse plants, fish, shellfish, mammals, and birds that provided the Northwest Coast peoples' staple foods and materials, will be abundant in one location but entirely absent in many others. Thus, culturally important plants and animals were distributed unevenly, both spatially and temporally. As several sections of this volume demonstrate, this diversity and distribution of ecological opportunities had tremendous implications for the peoples of this region, and influenced the ways that they chose to use, modify, and manage the land (Deur 1999).

In order to succinctly summarize this environmental variability, the region can be usefully subdivided into a few primary ecological zones. Meidinger and Pojar (1991) identify four general "biogeoclimatic zones," defined by vegetation, topography, and climate, within the Northwest Coast region. On the southeast coast of Vancouver Island, in much of the Puget Sound, and in the Willamette Valley, there are areas situated in the rain shadow that lies to the east of the Olympic, Vancouver Island, and Coast mountain ranges. This is a relatively warm, dry zone, the "Coastal Douglas-fir zone." It is characterized by forests dominated by Douglas-fir (*Pseudotsuga menziesii*), and associated tree species (such as grand fir, *Abies grandis*; Pacific madrone, *Arbutus menziesii*; broad-leaved maple, *Acer macrophyllum*; and western redcedar, *Thuja plicata*). In the driest areas one encounters an open parkland with Garry oak (*Quercus garryana*) interspersed with prairies and meadows comprised in large part of grasses and spring-flowering perennials, most notably two blue camas species (*Camassia quamash* and *C. leichtlinii*) and other plants of the lily family, which possess edible bulbs and were of tremendous significance to some indigenous peoples.

At low to mid elevations along the entire remaining coastline, the climate is significantly moister, with annual precipitation levels reaching from 60 to well over 400 cm, but with relatively mild temperatures due to the maritime influence. This is the "Coastal Western Hemlock zone," which at its climax stage is dominated by dense forests of western hemlock (*Tsuga heterophylla*), Pacific silver fir (*Abies amabilis*), western red-cedar, and—close to the shoreline—Sitka spruce (*Picea sitchensis*). At mid to high elevations all along the coast, subalpine forests predominate, with colder winters and more precipitation falling as snowfall rather than rain. This is in the "Mountain Hemlock zone," a zone dominated by cold-tolerant trees such as mountain hemlock (*Tsuga mertensiana*), yellow-cedar (*Chamaecyparis nootkatensis*), and subalpine fir (*Abies lasiocarpa*). Above the tree line in the mountains is a fourth zone, the "Alpine Tundra biogeoclimatic zone," with extended winters that are too cold to support tree cover. Low, perennial plants, including several heath species, predominate here. Within each of these major zones are the many topographically and hydrologically influenced habitat types already noted, each supporting specialized communities of plants and animals, many of these important to Northwest Coast peoples. In each of these zones, the various successional stages of local plant communities, resulting from vegetation disturbance and gradual regrowth, also produce many culturally important species.

The Aboriginal human geography of the Northwest Coast is also considerably more complex than might first be assumed, despite a number of characteristics that set the region's inhabitants apart from other indigenous peoples of North America. While linguistically and culturally diverse, the indigenous peoples of the Northwest Coast shared a distinctive, highly developed woodworking technology especially focusing on the use of western red-cedar. They also shared a marine-oriented economic life, based on fishing, particularly for salmon, hunting of marine and land mammals, shellfish harvesting, and the use of a wide spectrum of plant foods and other plant resources. Most Northwest Coast peoples lived in large villages of multifamily houses during the winter months while, at other times of the year, they traveled in smaller groups to various permanent resource outposts.

The patterns of human settlement and land tenure on the Northwest Coast reflected, in part, the regional biogeography that we have described above. These peoples utilized especially those habitats close to the coast or along the inlets and larger rivers, where most permanent winter villages were located, and many still are today. (Yet despite this maritime focus, Northwest Coast peoples frequented a wide range of specialized habitats, and successfully exploited the successional and climax communities of all of the biogeoclimatic zones described here.) Locating themselves near places where culturally important plants and animals were naturally abundant, Northwest Coast peoples frequently sought to control access to these sites.

Unlike some places in North America, tribal territories commonly had sharply defined and widely recognized borders, particularly in the northern two-thirds of the region defined by Suttles (1990) (see Turner et al., this volume). Within these boundaries, elites controlled many of the marine and terrestrial resources, and these lands and resources were typically inherited along clan or family lines. Often, resource territories were subdivided according to lineages or households. Chiefs made decisions regarding resource production and distribution that affected the larger household and village populations. While elites did tend to own resource sites and their output, they in turn were expected to redistribute their wealth broadly. Wealth redistribution was essential to the validation and legitimation of chiefly powers, and Northwest Coast societies did not condone miserly chiefs. This situation no doubt greatly influenced traditional patterns of resource use. Resource production was required not only to meet the daily dietary needs of Northwest Coast peoples but, in addition, the social, political, and economic structures of these peoples were vitally dependent upon the display and redistribution of wealth acquired through surplus resource harvesting. In order to maintain their position, therefore, Northwest Coast leaders had tremendous incentives to accrue resource wealth and to monitor (or enhance) the productivity of the resources within their finite territories (Maschner 1991).

While the Northwest Coast's forests are among the world's most productive environments, measured as volume of carbon stored per unit of area, this vast amount of biomass is banked primarily in coniferous trees, in the form of inedible cellulose (Alaback and Pojar 1997). The abundance and productivity of vegetation is therefore deceptive, as edible plants are often quite scarce in the region's dense mature forests. Requiring ample sunlight, many berry species do not grow well under the dense forest canopy that blankets much of the region but, rather, thrive and produce berries more abundantly in the more exposed, successional, and edge vegetation found along stream banks or in other, often restricted, moist clearings (Turner et al. 2003). Edible roots, bulbs, and rhizomes were also limited to finite natural environments: camas and other lilies, bracken fern (*Pteridium aquilinum*), and other edible plants only grow well in clearings, while other root foods grow most densely along the margins of salt- or freshwater marshes, creating isolated, focused food-gathering sites that were of tremendous economic, social, and dietary significance. Even the western red-cedar—from which the peoples of the coast made their homes, canoes, bark clothing, and many other durable goods—were irregularly distributed along the coast, and large-diameter cedars often were locally scarce (Hebda and Mathewes 1984). These plant resources also varied in quality from site to site. Likewise, plants were available unevenly over time, with berries, greens, shoots, bark, and other plant materials variously being obtainable only during particular times of the growing season. To be sure, while the peoples of the Northwest Coast were in some manner living in an "abundant" natural environment, they were not living in a place where natural resources were ubiquitous. Their valuation of particular landscapes varied accordingly, and there were often incentives to maximize output and tribal control of particularly convenient or resource-rich sites (Deur 1999).

The diversity and patchiness of Northwest Coastal environments have presented both opportunities and challenges for indigenous peoples, so dependent on unevenly distributed resource sites for their food, clothing, shelter, medicines, and spiritual needs, as well as for the continuous functioning of their society. The region's biogeographic diversity assured that productive fishing areas, hunting sites, shellfish gathering areas, and other animal-harvest sites were often many kilometers apart. Thus, as part of the effort to meet their

E. S. Curtis, *The Tule Gatherer* (Cowichan), 1910 (British Columbia Archives, D-08262)

resource needs, Northwest Coast peoples commonly migrated (and some still migrate) between important resource sites, scheduling their journeys to make the most of the locations and timing of peak resource abundance. The result was a patterned and systematic choreography, taking peoples from one long-standing resource encampment to another within their well-defined territorial holdings. A distinctive settlement pattern emerged from this choreography, mirroring the distributions of important resource sites. Commonly, many people would gather simultaneously at a single site, providing labor for the tasks ahead. Every individual, depending on his or her gender, age, and status, would perform certain specialized resource harvesting and processing duties. Fortunately, important species were frequently concentrated together into clusters of resource-rich sites, where people could gather to harvest and process some combination of plants, fish, shellfish, or game animals. It is important to note that, despite changes in the environment and in indigenous peoples' lifestyles, in many coastal communities, aspects of this traditional pattern persist today.[5]

Estuaries, in particular, have provided exceptionally rich concentrations of multiple resources, and it is not surprising that most large villages were located at estuarine locations. At a single productive estuarine location, for example, groups of families might camp on their lands for extended periods in the summer and fall when various salmon runs were at their prime. While some people fished, others processed and smoked the salmon, while still others dug nearby clams, or hunted waterfowl. Other members of the family also might pick and dry salal berries (*Gaultheria shallon*) and other berries, harvest stems of tule (*Schoenoplectus acutus*) and cattail (*Typha latifolia*) leaves for mats, and stinging nettle (*Urtica dioica*) fiber for cordage and nets, or harvest and process estuarine roots, such as springbank clover (*Trifolium wormskjoldii*) rhizomes, Pacific silverweed (*Potentilla anserina* ssp. *pacifica*) roots and northern rice-root (*Fritillaria camschatcensis*) bulbs, all during the time they were at this location. Elsewhere, at other times, alpine environments and intentionally burned clearings and meadows provided other concentrations of productive resources, perhaps less significant than estuaries, but still of such importance as to warrant regular seasonal visits and temporary residence by members of Northwest Coast societies (Lepofsky et al., this volume).

While the region's ethnographic literature seldom discusses plants as a part of this pattern of resource use, plants undeniably were of great importance, providing an impetus for migrations and specialized harvest areas. In all, approximately 300 species of plants were utilized traditionally by Northwest Coast peoples for food, materials, medicines, and other purposes. This includes about 100 food-plant species, some of which provide more than one type of product. Pacific crabapple (*Pyrus fusca*), for instance, was prized for its fruit, which was gathered in tremendous quantities up and down the coast. However, it was also valued for its tough, resilient wood, good for making implements, and for its bark, which was used for a wide range of medicinal purposes. Many other plants are similarly multipurposed. Thus, the total number of different types of applications of Northwest Coast native plants is sev-

Pacific crabapples (Photo by R. D. Turner)

eral times the number of species. Moreover, in discussions of the role of plants in food systems of Northwest Coast peoples, it is particularly notable that plants provided the majority of materials required to manufacture the items—nets, snares, spears, bows, or weirs, for example—that were used in the procurement of animal resources. Not only were plants nutritionally important in their own right, they also allowed the acquisition, as well as the transport, processing, and storage of the animal foods so significant to the Northwest Coast diet.

Little work has been done to date in quantifying the total amount of plant resources that were used in the past by Northwest Coast peoples. Available data suggest that the quantities of plants—the actual biomass of berries, root vegetables, basket and mat-making materials, and woody materials for construction of canoes, houses, fishing weirs and implements, for example—was enormous. A single large, sewn tule mat, used by Saanich and other coastal peoples, for instance, might require 1,000 tule stalks, and each household might possess ten or more such mats. There are accounts of nineteenth-century Saanich families digging several 50-pound gunnysacks of camas bulbs in one season; given that approximately 100 camas bulbs fill a 1-liter container, and that a gunnysack holds about a bushel of the bulbs in volume, this could translate into well over 10,000 camas bulbs per family per year.[6] If there were 1,000 people in a broad community, with ten people per family, the annual harvest for their locality could be in excess of 1,000,000 bulbs. Ethnographic information suggests similarly large figures for berries. Conservatively assuming that a liter of dried salal berries contains about 2,000 dried berries, and that each 20-liter storage basket held 40,000 dried berries, individual families appear to have commonly consumed from 100,000 to 200,000 salal berries per year. Within a community of 1,000 people, a consumption rate of 10,000,000 to 20,000,000 salal berries per year would be a reasonable estimate. Yet salal berries represent only one of many different kinds of berries gathered in such quantities. As the contributors of this volume demonstrate, certain traditional land-management practices served to augment the natural availability of many of these plants, providing the temporal and spatial concentrations of plant resources necessary to meet such high levels of human consumption.

Reconceptualizing Cultivation

As many researchers have noted in recent years, past generations of scholars tended to equate cultivation with the familiar practices and geometric patterns that characterized European agriculture (Blaut 1993; Butzer 1990; Denevan 1992; Doolittle 1992). In turn, past scholars tended to dismiss many of the unfamiliar and seemingly chaotic anthropogenic plant communities encountered in Africa, Asia, and the Americas as "nonagricultural." European cultivation was the measure by which all other practices were judged, and scholars commonly presumed that cultivation had emanated from within a small number of "civilized" societies. Increasingly, these notions have been discredited. Further, recent researchers have come to accept that foraging, hunting, low-intensity plant cultivation, and intensive agriculture are not mutually exclusive activities. Among peoples called hunter-gatherers, therefore, the emergence of plant cultivation did not always eclipse efficient and preexisting modes of subsistence; often, the development of cultivation merely augmented the outcomes of hunting and gathering, thereby contributing to the overall temporal stability and spatial concentration of food resources. A growing body of archaeological evidence likewise confirms that, in many different times and in many different places, plant cultivation persisted alongside other subsistence strategies for millennia without demanding a transition to agriculture (despite an abundance of evolutionary models that have suggested the contrary).[7] In many cases past and present, hunter-gatherer subsistence strategies and low-intensity plant cultivation can be "overlapping, interdependent, contemporaneous, coequal, and complementary" (Sponsel 1989: 45). Past attempts to categorize all the world's peoples as "hunter-gatherers," "pastoralists," or "cultivators," it seems, represent heuristic efforts of limited value that often conceal as much as they reveal.

The view of what constitutes "cultivation," therefore, has been opened up to critical reevaluation. Gaining distance from European categories, definitions of cultivation, by necessity, have grown more inclusive. They have come to encompass a continuum of practices found throughout the non-Western world, from plant "tending," to plant "cultivation," to plant "domestication." Plant tending, most scholars now agree, involves the minor modification of environments to encourage the growth of naturally occurring plants in situ, while plant cultivation involves a more intensive and extensive pattern of environmental modification. Cultivation, despite continued terminological ambiguities, is now commonly associated with such activities as the seeding or transplanting of propagules (i.e., the parts of plants such as seeds, bulbs, or fragments of rhizome, capable of regenerating into individual new plants), the intentional fertilization or modification of soils, improvements of irrigation or drainage, and the clearing or "weeding" of competing plants. Domestication, by contrast, involves the genetic modification of crops as a result of selective cultivation and propagation of plants with anomalous and desirable traits (Ford 1985; Harlan 1995, 1975). While still fraught with potential ambiguities, these definitions have proven sufficiently flexible to allow the analysis and comparison of plant-management practices found both within and outside of the traditions of the Western world.

Current definitions of cultivation are unified by their focus upon peoples' repeated and intentional manipulation of both plants *and* their environments as a means toward plant resource enhancement. Accordingly, researchers have directed increasing attention to the presence of modified environments, or "agroecosystems," as diagnostic of cultivation practices among aboriginal populations. These agroecosystems represent human-fostered and often genetically simplified environments, maintained by humans to increase the output of valued plants by replicating or enhancing certain naturally occurring conditions. Often in these systems it is not just a single species that is enhanced, but a complex of culturally important species, varying in their growth forms or morphologies, co-existing in complementary associations. Cultivated plants at a single site, then, can include a range of culturally significant trees, shrubs, and herbaceous species, and even epiphytes (plants growing on trees or other plants), which together form a locus of diverse resources. A number of these different species—providing the raw materials for foods, materials, and medicines—are often sought from these culturally modified sites during the same harvest (Ford 1985).

In light of this dramatic change in scholarly opinion regarding non-Western plant cultivation practices, contemporary researchers are compelled to critically reexamine past representations of peoples who, within the ethnographic literature, are traditionally designated as "noncultivators." In particular, it is worthwhile to revisit those ethnographic contexts in which the absence of plant cultivation has proven counterintuitive. Such a critique already has been directed at a number of other North American contexts, including the nearby indigenous societies of California (Blackburn and Anderson 1993), the Columbia Plateau (French 1957; Hunn and Selam 1990; Thoms 1989), and the semi-arid interior Plateau of British Columbia (Peacock 1998; Peacock and Turner 2000). The conventional view of these societies has shifted accordingly. Yet the Northwest Coast of North America has not been effectively subjected to the same critical reevaluation, and textbook accounts of the region still prominently discuss the absence of plant cultivation as one of the region's defining characteristics (Deur 2002a, 2002b). This volume, we hope, will direct overdue critical attention squarely on the case of the Northwest Coast.

Plant Use and Plant Cultivation on the Northwest Coast

In the chapters that follow, we demonstrate that there is considerable evidence supporting the view of Northwest Coast peoples as active managers and cultivators or producers of plants. These populations used a variety of techniques and strategies to maintain and enhance the viability and productivity of plant resources and habitats that were of particular importance to their diets, ceremonies, and economies. Many readers will be familiar with the one notable exception, mentioned previously, to Northwest Coast peoples' reputation as noncultivators: tobacco (*Nicotiana* sp.). At the time of first European contact, several explorers documented their observations of this crop, which was—by that point in history—the one familiar to the western eye, and represented it as the single example of a cultivated plant to be found among Northwest Coast peoples. On the northern Northwest Coast, the Haida, Tlingit, and probably the Tsimshian grew tobacco, of a species or variety apparently no longer in existence but related to *N. quadrivalvis*; it was grown in garden settings and was propagated from seed capsules each year. The tobacco gardens of the Haida were planted in individual, owned plots. Sometimes surrounded with low fences of cedar or crabapple stakes; these plots were also weeded of competing vegetation, tilled, and fertilized in the fall by mixing rotten wood into the soil after the tobacco had been harvested (Turner and Taylor 1972). Meanwhile, on the southern Oregon and northern California coasts, and on the middle Columbia River, a related tobacco, *Nicotiana quadrivalvis* var. *bigelovii*, was grown in similarly cultivated plots. The plant cultivated on the North Coast was not apparently native to the moist outer coast, but was most likely obtained from semiarid lands to the east or south of the region—the acquisition of these original seeds from afar is described in several different Haida narratives (Turner and Taylor 1972). Boas (1966: 23) was the first to popularize Haida tobacco as a puzzling enigma, apparently unrelated to all other plant management in the region, a practice that he suggested might have resulted from "an accidental, solitary influence" from elsewhere.

However, there is good reason to view this practice as part of a much larger pattern of plant tending and cultivation that spanned the whole of the Northwest Coast. If these peoples knew how to cultivate tobacco, why should we presume that they lacked the capacity to apply these same methods to other species? In truth, Northwest Coast peoples were actively modifying many of the roughly 300 plant species that they used as foods, sources of material and medicines, and for spiritual purposes. In different places, and with different plants, these peoples sometimes seeded or transplanted propagules, intentionally modified soils, and weeded out a wide variety of competing plants (Turner and Peacock, this volume). Population- and community-level management practices such as selective harvesting, digging, weeding and pruning, controlled burning, and the rotation of harvesting locales created and maintained ecologically distinct plant communities and enhanced the productivity and diversity of plant resources.

Many plant products were subject to intentional "sustainable harvesting." Some materials—such as planks, stems, or strips of bark—were taken from living, growing trees, shrubs, or herbaceous perennials in ways that did not destroy the plants but allowed them to live and continue producing harvestable materials. Root vegetables were often selectively harvested by size, such that only the older, larger roots or bulbs were taken, and the younger, smaller individuals were left to grow for future harvesting. The replanting and transplanting of plant materials was also widespread. The practice of replanting propagules in situ, and purposely leaving viable growing parts behind during harvesting was also apparently commonplace among people digging camas, rice-root, springbank clover, and other root vegetables. In addition, the transplanting of plants and their propagules from one place to another, to make them more productive or accessible, appears to have been widespread in pre-European times if the ethnographic literatures of the twentieth century are any indication. For example, wapato tubers were transplanted from areas where they were abundant to areas where they were scarce or did not naturally occur (Haeberlin and Gunther 1930: 21; Spurgin 2001; Arvid Charlie, personal communication to NT, 1999). Some ethnographic consultants also describe "weeding" of root patches, to help increase the competitive advantage and increase the

Old Growth western red-cedar CMT (culturally modified tree), showing where a bark sheet was removed many decades previously (Gitga'ata territory, Turtle Point, British Columbia). (Thanks to Marven Robinson and Gitga'at Nation) (Photo by N. Turner)

productivity of desired species. Routine digging and cultivation of the soil, both during the harvesting process or as a separate management activity, was also said to enhance the growth of root vegetables, at once aerating the soil and distributing essential nutrients into the soil matrix, and also creating an uneven surface with pockets in which moisture and nutrients can accumulate (Beckwith 2004). In addition, although documentation has been comparatively scarce, pruning was another cultivation strategy practiced by Northwest Coast peoples. Certain types of berry bushes were cut or broken back periodically; this was said to make them more productive in subsequent years. In other locations, pruning of branches from adjacent trees and shrubs appears to have reduced competition with culturally preferred species. Many of these practices have persisted in some form among Northwest Coast indigenous peoples of the past century. All will be discussed in greater detail in the chapters that follow.

Fire was also an important and widely used means of enhancing productivity and maintaining the open habitats preferred for the growing of camas and some berry plants, as well as for the improvement of hunting areas for elk, deer, and other game species. This technique appears to have been most prevalent in the drier coastal Douglas-fir sites of southeastern Vancouver Island, the Gulf and San Juan Islands, and the prairies of western Washington and Oregon (Boyd 1999; Turner 1999). Still, burning was also practiced as a vegetation-management tool in coastal temperate rainforests and in the subalpine zone (Deur 1999, 2002b; Lepofsky et al., this volume). Ownership of specific plant-harvesting sites by individuals, families, and lineages, and the restriction of non-kin access to these sites, was also a management strategy, widely applied on the Northwest Coast (Turner et al., this volume).

These practices parallel several strategies for the management of animal populations, which have been well documented and, arguably, more widely recognized than plant-management strategies within the ethnographic literature. First Nations peoples tended clam beds, for example, removing sticks and rocks and restricting harvests on beds that were owned; in other cases, they transported salmon smolts or eggs and released them in depleted or previously nonproducing streams (Bouchard and Kennedy 1990; Sproat 1868 [1987]: 148). Edible herring eggs were commonly harvested by placing lattices or "fences" of hemlock boughs in the water, where herring were known to deposit spawn on submerged materials. In other locations, as noted, prime and easily accessible hunting sites were burned to enhance the regrowth of herbs and forbs that served as food for browsing deer and elk, resulting in the concentration of these animals in single, predictable locations. As with plant enhancement methods, these animal management practices concentrated resources close to villages and within each village's territorial control, reduced the risk of fluctuations in salmon populations and other resources, and provided necessary nutritional breadth to the diet of large village populations (Kuhnlein and Turner 1991).[8]

Admittedly, the available evidence for precontact plant management is partial and sometimes problematic. The period following the first recorded contact (1774) has been one of dramatic change for the environments and peoples of the Northwest Coast. By the 1790s, European powers were engaging in regular, brief interactions with some coastal peoples and would soon be constructing fur-trade outposts in key locations up and down the coast. The indigenous peoples of the coast were exposed to foreign ideas and technologies, and would often adopt them with enthusiasm into their own repertoire of cultural practices. Throughout the first 100 years of this exchange, diseases would decimate some First Peoples, decreasing some populations by more than 90 percent. Some portions of the landscape were depopulated and have not, at the time of this writing, returned to their precontact population levels. Population pressure on some natural resources declined, while cultural changes sometimes decreased demand for traditional plant foods and materials.

Not only were Northwest Coast peoples' ethnobotanical practices in rapid flux, but ethnographers seldom addressed these practices as their primary research question. When they did, their accounts were often highly fragmented. Even extremely knowledgeable ethnobotanical consultants were often only exposed to small pieces of their peoples' cumulative plant lore, such as those fragments appropriate to someone of their gender, age, rank, occupation, or generation (Turner 2003b). Some peoples with proprietary traditions of plant use (religious or medicinal uses, for example) have sometimes chosen to not provide the whole story, or they may have omitted the specifics. At other times, oral traditions are not passed between elders and youngsters, or elders' memories can grow understandably fuzzy on the details of plant use with time. Moreover, embedded within past ethnographic descriptions of plant use, we often find data of ambiguous chronology or antiquity. Often we see consultants' comments on "what I remember from my childhood," "what we do now," "what my grandparents told me about their childhood," or "what everyone says we did before white people arrived," all conflated by researchers into a single, undifferentiated account. These problems underscore the precariousness of the entire "salvage ethnography" enterprise, and confound attempts to make sense of historical changes within dynamic cultures. Too often, therefore, attempts to make statements about pre-European ethnobotanical practices reveal a collage of fragmentary ethnographic information, filtered through two centuries of selective recording, social upheaval, and environmental change.

Despite inherent difficulties associated with ethnographic reconstructions of the past, however, the ethnographic evidence presented here cannot be dismissed. References to intensive plant management are available in written and archival sources of diverse antiquity and geographical distribution along the Northwest Coast. (And increasingly, these references are corroborated by archaeological evidence [see Deur and Lepofsky et al., this volume].) Oral traditions and ceremonial traditions are linked to these plant management practices, while plant management technologies are elaborate and are referred to by myriad, etymologically distinct terms within each ethnolinguistic group of the region.

In the past, when ethnographers encountered oral or documentary references to plant cultivation methods on the Northwest Coast, they tended to view them as the product of post-contact developments. Ironically, indigenous uses of animal resources, recorded anthropologically, were widely assumed to be representative of precontact practices; nonetheless, the intensity of *plant* resource use, scholars assumed, developed in only one direction, intensifying as a result of rapid cross-cultural diffusion and technological exchange on the Northwestern frontier. Precontact plant use was seen—a priori—as minimal, so contradictory evidence was summarily interpreted as the result of post-contact influences, which in turn provided additional reinforcement in the ethnographic literature for the prevailing view of precontact past plant use: the reasoning was circular. However, increasing evidence hints that a reverse process may have been at play. In the two centuries following European contact, the peoples of this coast experienced dramatic demographic collapse—with some communities losing more than 90 percent of their population—as well as widespread dislocation and relocation from many principal villages and territories. Simultaneously, many endemic foods were eclipsed by the arrival of new domesticated crops, most notably the potato (*Solanum tuberosum*). For these reasons, endemic cultivation practices in the ethnographic record may well reflect only a *fragment* of the full range or intensity of practices that existed in past times. This point remains open for debate, but there is clearly no valid basis for the assumption that the trend was uniformly toward greater intensity of plant-resource use in the post-contact era. Indeed, the disintensification of resource use—a decline in the intensity and variety of resource-management practices following major social and demographic disruptions—is a development common to many colonial contexts around the globe (Balée 1994; Brookfield 1972; Ford 1985). Likewise, allusions to this process make frequent appearances in the archaeological and ethnographic record of the Northwest Coast.

Overlooking the Gardens in the "Wilderness"

In light of the evidence provided here, we might return to this question: Why was the presence of plant tending and cultivation discounted on the Northwest Coast, even by the most capable anthropologists of the last century? The origin of the region's prominent reputation as a place lacking cultivation can be traced to the nineteenth century, and that perhaps tells us as much about colonial and scholarly agendas of this period as it does about subsistence practices of indigenous peoples. As indicated earlier, written representations of Northwest Coast peoples have been highly selective in their content, and have also been continuously shaped by the agendas of each successive generation of writers (Cannizzo 1983; Said 1993; Webster and Powell 1994; Willems-Braun 1997).

Undeniably, the Northwest Coast plant-management methods that we discuss here defied conventional European notions of cultivation during the first contacts of the late eighteenth and early nineteenth centuries. Many of the first explorers noted the signs of Northwest Coast cultivation practices outlined here, but did not recognize evidence of human cultivation in anthropogenic plant communities lacking rectilinear plantings divided by picket fences, or monocultures of familiar plants.[9] Traveling amidst camas prairies cleared by human burning along the shores of Puget Sound, Captain George Vancouver (1967, orig. 1798) proclaimed,

> I could not possibly believe any uncultivated country had ever been discovered exhibiting so rich a picture. Stately forests . . . pleasingly clothed its eminences and chequered its vallies; presenting in many places, extensive spaces that wore the appearance of having been cleared by art . . . [we] had no reason to imagine this country had ever been indebted for its decoration to the hand of man. [227–29]

Yet it is apparent that the places described by Vancouver were managed landscapes, cleared not by "art" but by people with very specific technologies and objectives. Burned berry patches, weeded and tended root-vegetable plots containing multiple native species: each of these places represented highly modified environments of a sort that was alien to the European world. We contend that cultivated plant communities were observed from the earliest European presence on this coast, but that Europeans routinely recorded these intensively managed landscapes as "natural" features (Menzies 1923: 116–17). Accordingly, impressed by the region's abundant marine and terrestrial life, explorers and colonists concluded that food was at most times and most places naturally plentiful for the region's native inhabitants (Folan 1984). Plant cultivation, they assumed, had been unnecessary in such fecund environments and this had precluded the emergence of indigenous forms of agriculture. Clearly, however, the region's "natural abundance" was, in many respects, not natural but the byproduct of long-term human intervention.

In this light, the Haida tobacco discussed before no longer seems so very "solitary" an example. Tobacco, the one cultivated plant of this coast that *was* familiar to the peoples of Europe, was commonly described as the only evidence of agriculture on this coast, and as a total anomaly (Harris 1997: 219; Murdock 1934: 223; Vancouver 1801).[10] This uneven treatment of tobacco is quite revealing. The methods of tobacco cultivation did not differ substantively from the methods used to cultivate most of the other plants discussed here. Tobacco, like Pacific silverweed, springbank clover, rice-root, camas, and other endemic plants, was maintained on owned plots cleared of competing vegetation, with well-worked and augmented soils. The primary difference is that indigenous peoples regularly planted tobacco by seed (or seedpod), rather than planting it by vegetative cuttings or propagules.[11] Seed propagation was viewed by some nineteenth- and early twentieth-century authors as the litmus test of true cultivation, a fact which—along with the familiarity of tobacco to European observers—no doubt contributed to this uneven representation of this crop (Spinden 1917).

Like the explorers who preceded them, most early anthropologists described human uses of the Northwest coastal environment with a number of a priori assumptions regarding the region, fueled by explorers' accounts such as those of James Cook and William Douglas. Early researchers also arrived with a priori assumptions regarding the nature of cultivation (most now discredited), and drew their conclusions primarily on the basis of brief encounters with indigenous peoples. Very few researchers indeed ventured into the field with indigenous consultants to view traditional resource sites or to participate in harvests, and a remarkable percentage of early Northwest Coast ethnographic research was conducted in hotel rooms, consultants' homes, and other nominally urban settings. In line with the conventions of Boasian anthropology, investigation of traditional resource management was simply not a significant concern among most Northwest Coast ethnographers, who instead focused on the collection of oral texts and detailed information regarding ceremonial, social, and artistic conventions. When resource use and management did appear in the work of anthropologists, it was typically a product of post-hoc analyses of the oral accounts gathered in studies that focused on such incorporeal topics as ceremonialism.

Despite this bias against documentation of resource use and management, some noteworthy early studies did discuss Northwest Coast plant use. The detail of Boas's records of Kwakwa̱ka'wakw plant use, cited in many places throughout this book, is quite remarkable, and was typical of Boas's many "Kwakiutl" publications. Certain authors of the late 19th and early 20th centuries, including naturalist Charles F. Newcombe, geographer George Dawson, and archaeologist Harlan I. Smith (1928), were also interested in documenting plant uses and incorporated this data into their writings. Erna Gunther (1973; orig. 1945) and Francis Densmore (1939), two well-known ethnographers, also contributed substantially to our more recent understanding of Northwest Coast plant use, publishing plant-use data gathered in the course of investigations on other topics. These researchers paid little or no attention to methods of plant gathering or management in these works, and based their accounts largely on interview data to the exclusion of field visits to resource sites. Instead, they simply sought to document the traditional uses of plant materials for foods, materials, and medicines.

Regardless, very little of this potentially relevant work found its way into the broader discussion of Northwest Coast resource use. Quite early in the development of the anthropological literature regarding human subsistence, authors of international stature, with little or no firsthand knowledge of the Northwest Coast, described the region as lacking in plant cultivation. Spinden (1917), for example, equated indigenous New World cultivation solely with the presence of Mesoamerican crops, and accordingly depicted the Northwest Coast as a place without cultivation. Still, of the many scholars who contributed to the image of a nonagricultural Northwest Coast within the academic literature, none contributed as much as the "father of American anthropology," Franz Boas. Boas, ironically, was aware, at least, of the rudiments of the cultivation practices we describe here (Boas 1966, 1934, 1921, 1909).[12] Nonetheless, he chose to advertise these peoples' lack of cultivation and—despite his fierce debunking disposition on many anthropological questions—chose to reiterate rather than to challenge European explorers' claims on this point. While Boas was clearly aware of some forms of plant cultivation on the Northwest Coast, and seemed to accept them as precontact phenomena, he devoted little time to their investigation, and even less to their discussion in print.[13]

Boas's own dismissal of Northwest Coast cultivation seems enigmatic, but was deeply rooted in the intellectual history of his times and the sometimes contradictory aims of his own ethnographic project. Trained in physics and physical geography, Boas sought to employ the most rigorous traditions of European natural science in order to document and to celebrate the cultures of the non-European world. A European-born Jewish intellectual much concerned about the outcomes of anti-Semitism and other forms of Western racism, Boas compiled volumes of data in order to challenge the arguments of environmental determinists, eugenicists, racial determinists, and theorists of crude, unilinear cultural evolution (Herskovitz 1953; Hyatt 1990). Yet, as one of his students, Melville Herskovitz (1953: 112) suggested, "as a Nineteenth Century liberal he rejected, in principle, the colonial system. But as a Nineteenth Century scholar with a European orientation, he tended to think in [European] terms."

As with most other authors of his time, Boas equated "agriculture" with the cultivation of a small range of domesticated crops. Like Spinden, Boas took a strictly diffusionist attitude toward cultivation, and asserted that Mesoamerica was the core from which all New World agriculture had emanated. Mesoamerican seed crops, namely "Indian corn, beans, and squashes," were therefore diagnostic of the presence of true "agriculture" in North America (Boas 1933: 347). With his bias toward the seed-based Mesoamerican agricultural complex, Boas was reluctant to term patterns of plant management based on vegetative transplanting and use of perennial species as true "cultivation."[14] Northwest Coast plant management practices must have presented Boas with challenging exceptions to his own preconceived categories of culture and cultivation. To Boas, the cultivation practices of the Northwest Coast were ambiguous, caught in-between the dichotomous categories of "hunter-gatherer" and "agriculturalist." Although Boas denied that these two states represented necessary stages along an "evolutionary continuum," he did not effectively question the validity of the categories themselves. (Today, these categories are commonly viewed as discredited remnants of evolutionary-stage theories of cultural development—see Smith, this volume.) On the Northwest Coast, Boas saw an absence of Mesoamerican crops or widespread, seed-based agriculture, and a people not primarily dependent upon cultivation for their dietary needs. Not recognizing an intermediate spectrum of cultivation activities between the polar categories "agriculturalist" and "hunter-gatherer," Boas opted to designate Northwest Coast peoples as being in the latter category.

To understand Boas's oversight, it is also important to note that Boas was deeply averse to investigating most forms of human–environmental interaction, and openly discouraged detailed environmental studies among his students, as a result of his own disenchantment with environmentally deterministic literatures of his day. His research on Baffin Island, soon after the completion of his Ph.D. dissertation, had been initiated to demonstrate environmental influences on indigenous cultures of the far north. By the end of this research Boas had rejected the "anthropogeographic" theories that initially drew him to Baffinland; his theoretical position had shifted to a possibilist view of human–environment relations, and his methodological efforts likewise centered on the "mental life" of indigenous peoples. Soon thereafter, Boas arrived on the Northwest Coast, where this new theoretical position was most fully expressed. His vehement early rejection of environmental determinism arguably fostered a reluctance to engage environmental topics, generally, among both Boas and his students, who dominated Northwest Coast studies for many decades (Codere 1959; Speth 1977). Boas's minimal attention to Northwest Coast cultivation practices was additionally complicated by his heavy reliance on the post-hoc analysis of indigenous oral literatures provided by male elites in the study of indigenous subsistence practices. Like his students, who came to dominate the subfield of Northwest Coast studies in the years to come, Boas engaged in very little field observation of these practices, and tended to underplay the practices and testimonies of women or nonelite men, who were the primary plant cultivators (Fiske 1991; Ray 1989).

Boas's distinct theoretical and ideological agendas also actively (if often implicitly) shaped his ethnological and ethnographic writings. Proponents of unilinear, evolutionary theories of human cultural development, including prominent anthropologists such as E. B. Tyler and Lewis Morgan, had posited that agriculture was a universal precondition for large communities, social stratification, and complex ceremonialism, in the progression from savagery to a fundamentally Northern European model of civilization. These models dominated Western social science during Boas's formative years, both manifesting and shaping the agendas of social scientists and colonial authorities alike. Boas actively sought to undermine these deterministic models of cultural development, and his ethnographic project on the Northwest Coast can be viewed, in no small part, as a massive effort to disprove nineteenth-century academic dogma (Lakatos 1970; Popper 1959). In a Northwest Coast without plant cultivation, Boas and his students found one of their most striking and widely publicized exceptions to these deterministic models (M. Harris 1968). In his widely read challenges to environmentally deterministic models of cultural development, Boas employed the Northwest Coast example to prove that "the same environment will influence culture in diverse ways," and that "the most fertile soil will not create agriculture" (Boas 1930: 266). He compared the Northwest Coast to Norway, "where climatic and geographic conditions are similar . . . Evidently contact with the rest of Europe was sufficient to teach the early Norwegians the tilling of the soil. The Northwest Coast of America was not so favored" (Boas 1966: 23). Thus, as Boas claimed, the absence of cultivation on the Northwest Coast could be traced solely to historical rather than environmental causes.

In the late nineteenth and early twentieth centuries, economic determinism was similarly widespread, and posited that societies evolve inexorably from simple to complex modes of production, a process involving certain standard and predictable features at each stage. Agriculture, in this view, was a precondition for the development of social and economic complexity, a point alluded to earlier in this introduction. Boas challenged this dogma as well, depicting the peoples of the Northwest Coast as *exceptional hunter-gatherers.* Unlike most hunter-gatherers around the world, he indicated, the hunter-gatherers of the Northwest Coast possessed so much material wealth that their needs for subsistence scarcely influenced social behavior. Rather, life was "dominated by the desire to obtain social prominence by the display of wealth and by occupying a position of high rank" through lavish exhibitions of accumulated wealth, or public squandering or destruction of surplus wealth (Boas 1928: 154). While many noncultivating peoples experienced severe privation, and "have not produced much that would help towards the enjoyment of life," the hunter-gatherers of the Northwest Coast "enjoy seasons of rest during which they live on stored provisions," and "have developed a complex art and a social and ceremonial life full of interest to themselves" (Boas 1928: 218). Unlike most hunter-gatherers, those of the Northwest Coast lived in socially stratified, sedentary villages, and individuals or clans could own resource sites (Boas 1928: 192, 238–39). Unlike hunter-gatherers elsewhere in the world, who were forced to cope with more precarious food supplies, the hunter-gatherers of the Northwest Coast valued generosity and encouraged pity for the less fortunate (Boas 1928: 224–25). These hunter-gatherers even exhibited degrees of religious specialization, with full-time shamans and a host of "secret societies." There were few generalizations that could be made about "hunter-gatherers" that were not, in some manner, undermined by the example of the noncultivating Northwest Coast peoples. Clearly, in the noncultivating Northwest Coast, Boas had found his most potent evidence in his effort to undo the excesses of nineteenth- and early twentieth-century social science.

Subsequently, the region's presumed distinctiveness as a place of wealthy, noncultivating hunter-gatherers was accepted and promoted by successive generations of scholars, particularly Boas's preeminent students such as Ruth Benedict, Robert Lowie, and Alfred Kroeber. Boas's status within the field of anthropology assured that there would be few effective challenges to this view. After Boas's early pronouncements, and indeed, long after his death, the Northwest Coast would continue to serve as the preeminent textbook example of exceptional hunter-gatherers among some of the world's foremost anthropologists, archaeologists, and others, often based on citations many times removed from Boas' empirical work.[15] As a result, this view of Northwest Coast cultures has proven impervious to subsequent and more inclusive redefinitions of cultivation, used to reevaluate other parts of the non-European world, such as California and Amazonia, where this label may have been less intimately tied to longstanding scholarly agendas. Further, this designation has so influenced contemporary anthropologists and archaeologists that even Northwest Coast specialists commonly initiate their research on human-environmental relations with the assumption that not only were the peoples of this region wholly noncultivating, but further that "Northwest Coast peoples did not rely heavily on plant foods" (Huelsbeck 1988: 166). A handful of Northwest Coast specialists have challenged aspects of this orthodoxy, such as the region's presumed resource superabundance (Suttles 1968) or the assumption that plants were unimportant in the diet (Keely 1980; Norton 1981; Turner 1995). Nonetheless, this critique has been presented only sporadically and within the regional literature, while not gaining the attention of broader, international academic audiences. And arguably, a number of biases—including ethnocentrism and gender bias—have continued to pervade scholarship on Northwest Coast plant management, limiting discussion of these issues in a manner reminiscent of earlier academic discourses. This book, we hope, represents an important step in the critical reconsideration and revision of these long-standing academic biases.

Cultivated Places become Contested Spaces

While the depiction of the Northwest Coast as a region without plant cultivation may have seemed a benign semantic exercise among Western explorers and scholars, this designation was not without its material consequences. First emerging within the journals of explorers, and legitimized through the works of Boas and other anthropologists, the designation would become unexamined orthodoxy among Northwestern policymakers and the public alike during the late nineteenth and early twentieth centuries. Colonial authorities, in turn, openly invoked this "noncultivator" designation as partial justification for the colonial dispossession of tribal lands having enduring patterns of management and tenure, as the nations of North America consolidated political and economic control of these far western hinterlands. Serving as Commissioner on the Joint Committee on Indian Reserves in the nineteenth century, Gilbert Malcom Sproat (1868 [1987]: 8) asserted:

> My own notion is that the particular circumstances which make the deliberate intrusion of a superior people into another country lawful or expedient are connected to some extent with the use which the dispossessed or conquered people have made of the soil, and with their general behavior as a nation. *For instance we might justify our occupation of Vancouver Island by the fact of all the land lying waste without prospect of improvement* . . . Any extreme act, such as a general confiscation of *cultivated* land . . . would be quite unjustifiable." [emphasis added]

Indeed, based on the documentary record of explorers and anthropologists, colonial administrators questioned whether the native peoples of British Columbia required any land whatsoever, save village sites, in order to survive. British Columbia Indian Agent William Halliday, for example, noted that the Indians had never required land to survive prehistorically, and there was no reason why they would need it now (in the early 1900s). Echoing the anthropological literature of the time in his reports on the land question, Halliday suggested that their claims to the land were precarious. The reasons were simple. The documentary record indicated that their traditional diet consisted almost exclusively of fish and the "waters of the coast teem with fish"—moreover, with their lack of experience in caring for crops, "it will take more than one generation to make agriculturalists out of them" (Halliday 1910: 236–38).

Early in the twentieth century, British Columbia established the McKenna-McBride Royal Commission to resolve the Indian land question, by inviting indigenous land claims and making legally binding judgments on their apparent merits. In the process, the Commission created the pattern of Indian reserve lands that persists to this day. Although the potato and other European crops had replaced many indigenous food plants at this time, and some indigenous peoples had been relocated far from plant-resource sites, many such sites were still actively maintained. Accordingly, the Commission received multiple claims on traditional plant resource sites. "Chief Humseet" of the Knight Inlet Kwakwaka'wakw, for example, testified to the Commission that his people wished to maintain several plots of plants that his "forefathers [had] planted" on the central mainland coast of British Columbia (McKenna-McBride Royal Commission 1913–1916: 188). When asked to identify the plants in the plots, he listed a number of root crops, including silverweed and springbank clover. Under Commission guidelines, had he identified these plots as potato patches, this may have represented a valid claim; but as these plants were not cultivars recognized by the Commission the plots alone were not eligible for protection. Despite evidence of long-standing indigenous maintenance and claims on these sites, the Commission designated land claims made for "cultivation" on traditional plant-harvest sites as being for "proposed" or "potential" land uses rather than as land with "existing" uses. Most such claims were summarily denied unless the sites happened to coincide with the location of a village or fishing station that was approved by the Commission. Invoking the Northwest Coast anthropological literature, and echoing the words of Halliday, the final findings of McKenna-McBride (quoted in Galois 1994: 74) explained their decision, suggesting that access to the sea was necessary to the survival of native peoples, but that resource *lands* were not "reasonably required." Accordingly, the Crown authorized only a diffuse pattern of small reserves, encompassing nineteenth-century village sites, and little else (Galois 1994; Tennant 1990; Wagner 1972). At this time, most terrestrial resource sites—including almost all sites where traditional plant-management practices still endured—fell out of Aboriginal control.[16] In turn, this loss of direct control opened traditional plant-resource sites to white reoccupation, and contributed to the demise of many of the plant-management practices that had sustained these people for countless generations.

Arguably, scholars and policymakers who depicted the Northwest Coast as a place without cultivation did so without appreciation for the sophistication of Northwest Coast plant use, specifically, or non-Western systems of cultivation, generally. Certainly, Northwest Coast peoples were cultivating plants in some manner: they seeded or transplanted propagules, intentionally enhanced garden soils, altered local hydrology of garden sites, and weeded out a wide variety of competing plants. Whether an appreciation of traditional plant management would have dramatically changed the direction of colonial policy is unclear; what is clear, however, is that the misrepresentation of Northwest Coast plant use has been a long-standing, pervasive aspect of both colonial and academic discussions of the region, and its effects have been diverse. Clearly, it is time for both a reevaluation of Northwest Coast peoples' reputation as noncultivators, and a reevaluation of policies that may have been predicated on this reputation.

"Keeping It Living": The Cultivation of a Volume

Early in 1996, Douglas Deur and Kent Mathewson (Louisiana State University)— recognizing the abundance of compelling evidence of plant cultivation, as well as its dramatic implications—proposed a symposium for the annual meeting of the American Association for the Advancement of Science (AAAS). When participating in this symposium, they hoped, participants might critically reevaluate Northwest Coast plant management. Their proposal ultimately was approved by the AAAS, and the symposium, "Was the Northwest Coast of North America "Agricultural"?: Aboriginal Plant Use Reconsidered," was held on February 16, 1997, at the Seattle Convention Center. This symposium consisted of six papers, three by Northwest Coast specialists and three by specialists in low-intensity plant cultivation elsewhere in the Americas. The three Northwest Coast papers were by Douglas Deur (then of Louisiana State University), Nancy J. Turner and Sandra Peacock (University of Victoria), and Wayne Suttles (Portland State University). The three comparative papers were by Bruce Smith (Smithsonian Institution), Suzanne Fish (Arizona State University), and William Gartner and William Denevan (University of Wisconsin-Madison).

This book developed out of the discussions surrounding that event. Following the AAAS symposium, Douglas Deur and Nancy Turner solicited essays from some of the region's foremost research specialists in indigenous resource use, identifying forms of plant management they had encountered in the course of their studies. In the writing and editing processes that followed, contributors and editors maintained lively discussions of traditional resource management, and these discussions found their way into the volume in myriad ways. By design, the volume's geographical span is broad, covering almost the full length of the ethnographic Northwest Coast, from the cultures of the northern Oregon coast to the peoples of Southeast Alaska. In addition, by design, the volume brings the insights of multiple disciplines to bear on the issue. It therefore contains the work of ethnobotanists, archaeologists, anthropologists, geographers, ecologists, and First Nations scholars and elders. The evidence that this multidisciplinary group of scholars has uncovered, we feel, is compelling.

Our book begins with a preface by Dr. Richard Atleo (Chief Umeek), a First Nations scholar and leader; here, he considers the differences between European and First Nations ways of knowing the world, and encourages a dialogue between the two. His words encourage all Western scholars to reevaluate the meaning and context of their research and to try to understand First Nations' perspectives on the land and land use issues that they study. In the introductory section that follows, we set the stage for individual case studies by situating the Northwest Coast within a broader context, geographically, ethnographically, and temporally. Bruce Smith provides an overview of cultivation practices found around the world in his chapter, "Low-Level Food Production and the Northwest Coast," and directs particular attention to societies—like those of the Northwest Coast—that seem to sit in the intervening categorical spaces between conventional hunter-gatherers and agriculturalists. Kenneth Ames describes cultural and resource intensification on the Northwest Coast, and proposes alternative models for these processes in his chapter, "Intensification of Food Production on the Northwest Coast and Elsewhere." The two chapters that follow provide a broad introductory overview of patterns of plant use, ownership, and enhancement on the Northwest Coast at the time of European contact. Nancy J. Turner and Sandra Peacock provide an overarching view of the methods used by Northwest Coast peoples in modifying plants and their habitats: "Solving the Perennial Paradox: Ethnobotanical Evidence for Plant Resource Management on the Northwest Coast." Dr. Turner expands on these themes in a chapter co-authored with Robin Smith and James Jones, in which the authors assess traditional ownership and tenure of plant resources and plant-resource sites: "'A Fine Line Between Two Nations': Ownership Patterns for Plant Resources among Northwest Coast Indigenous Peoples."

In the following Part II, a collection of case studies is presented, addressing the modification and intensive use of different plants by different peoples around the Northwest Coast. We begin in the southern portion of the Northwest Coast. Wayne Suttles, in his chapter "Coast Salish Resource Management: Incipient Agriculture?" reviews evidence of Coast Salish camas management that he has compiled during his long and distinguished career as a Northwest Coast specialist. Here, Suttles subjects his ethnographic data to a critical reexamination in an effort to discern whether certain cultivation practices existed prior to European contact. Melissa Darby discusses Chinook use of wapato on the lower Columbia River in her chapter, "The Intensification of Wapato (*Sagittaria latifolia*) on the Lower Columbia River." Darby focuses her attention on the development of efficient harvesting and processing methods as evidence of plant intensification within this uniquely resource-rich part of the Northwest Coast. Next, readers are presented with a review of an ambitious project among the Stó:lō of the lower Fraser River, which has brought together researchers and First Nations elders to document precontact burning, traditionally carried out in mountainous environments to enhance the output of huckleberries (*Vaccinium* spp.). This chapter, "Documenting Precontact Plant Management on the Northwest Coast: An Example of Prescribed Burning in the Central and Upper Fraser Valley, British Columbia," is by Dana Lepofsky, with contributions by Douglas Hallett, Kevin Washbrook (Stó:lō Nation), Sonny McHalsie (Stó:lō Nation), Ken Lertzman, and Rolf Mathewes.

The chapters that follow provide case studies from the northern Northwest Coast. James McDonald, in his chapter "Cultivating in the Northwest: Early Accounts of Tsimshian Horticulture," summarizes his findings on traditional plant management, recorded in the course of his ethnographic research with Tsimshian communities. McDonald notes that even the earliest records on the Tsimshian of the northern British Columbia coast depict these peoples as cultivators, who actively modified patches of berries, edible roots, and fruiting trees, as well as planting tobacco and potatoes. The Tlingit of Southeast Alaska also cultivated potatoes very early in the contact period, a point that is explored by Madonna Moss in her chapter, "Tlingit Horticulture: An Indigenous or Introduced Development?" Dr. Moss suggests that tobacco cultivation and other endemic-plant management methods may have provided the Tlingit with a model for potato cultivation methods, fostering the rapid diffusion of this crop following its introduction during the early fur trade. In the following chapter, "Tending the Garden, Making the Soil: Northwest Coast Estuarine Gardens as Engineered Environments," Douglas Deur asserts that plant communities were effectively engineered to achieve specific, anticipated ends. He emphasizes the creation of estuarine root gardens among the Kwakwaka'wakw, Nuu-chah-nulth, Nuxalk, and others, through the modification of both estuarine plants and their habitats. Together, Deur and Turner conclude with a brief assessment of the implications of these findings, within the academic literature as well as within the ongoing debates over First Nations' sovereignty and land claims.

Taken in total, we contend that the practices documented in these pages constitute a level of resource production not adequately encompassed within the category "hunter-gatherer." Northwest Coast peoples, and perhaps many other societies classified as hunter-gatherers, practiced food production techniques in a variety of forms. Though these practices may not have been "agricultural," in the conventional sense of sowing seeds of annual plants and reaping the harvest of staple grains or other vegetables, they arguably do represent diverse methods of "plant cultivation," as that term is now commonly employed. These methods are aptly summarized in the translation of the Kwak'wala word sometimes used to describe the full range of cultivation methods described here. Shared with both Deur and Turner by Hereditary Chief Kwaksistala Adam Dick, this term is *q'waq'wala7owkw,* or "keeping it living."

Chief Adam Dick (Kwaksistala) (Photo by Lynn Thompson)

By modifying culturally significant plants and their habitats, repeatedly and intentionally, the peoples of the Northwest Coast kept their most important plants living. To prevent this information from being lost to time, we publish this book, to keep the knowledge living too.

We would like to extend our thanks to all of our contributors, as well as to our many fellow researchers who contributed to our knowledge of indigenous resource management but who are not included directly in this volume. We are also grateful to the University of Washington Press, especially Julidta Tarver, Marilyn Trueblood, Sigrid Asmus, and Mary Ribesky, for helping us bring this volume to fruition. Most of all, however, we wish to thank the indigenous consultants, past and present, who have shared so much of their time and knowledge with us and with the researchers who have preceded us. Their help has been invaluable; their patience has been remarkable. *Gila'kasla ninogad. Kleco kleco.* This volume is dedicated to them.[17]

Notes

1. Portions of this introduction were presented in Deur (1999, 2000, 2002a,b). For selected examples of accounts denying the presence of plant cultivation on the Northwest Coast, see Benedict 1934; D. Harris 1977; M. Harris 1968; Kroeber 1939; Linton 1936; Lowie 1920; Netting 1986: 27–40; Sauer 1952, 1936; and Spinden 1917.

2. The term "First Nations" is the preferred term for the indigenous peoples of Canada.

3. On this point, see for example, Harlan 1995, 1975; MacNeish 1992; Cowan and Watson 1992; Ford 1985.

4. The authors recognize that this description of events does not cohere on certain points with the oral traditions of some tribes and First Nations of the Northwest Coast. In an effort to provide all perspectives on the region's traditions of plant management, however, this volume attempts to provide both Aboriginal explanations of these traditions as well as the explanations of archaeologists and other specialists trained in Western conventions of social and natural science. These two groups are in agreement on many points while, on others, more discussion is clearly needed (see Atleo, this volume).

5. While this introduction—with its emphasis on reevaluating the "precontact condition" of Northwest Coast peoples—frequently speaks of indigenous activities in the past tense, many of these practices persist today. Certain traditional resource sites retain tremendous cultural significance among most of the region's First Nations and tribes today. In some indigenous communities, particularly those that are distant from the region's urban centers, people still participate regularly in many of the resource-harvest activities discussed here in primarily historical terms. The use of the past tense should in no way be construed as evidence that the practices being described are no longer extant.

6. This calculation is based on about 30 liters per bushel, with a bushel of potatoes weighing about 50 pounds (Bill Johnstone, personal communication 1999) and therefore about 3,000 camas bulbs per bushel, and four bushels per family (conservatively), yielding 12,000 bulbs.

7. Important, selected reference points within this critique include: papers in Cowan and Watson 1992; Harlan 1995, 1975; MacNeish 1992; Sahlins 1972; and Smith 1995, this volume.

8. One gains a sense of how plant use has been overlooked when recognizing that the levels of salmon consumption proposed in past studies have not been particularly tenable. Past attempts to model settlement or archaeological patterns, using salmon-based settlement models, seldom work. In each case, the modelers have had to conclude that other resources must have also strongly influenced human settlement and seasonal migration (Boyd and Hajda 1987; Hobler 1983). Likewise, purported levels of salmon consumption as documented in stable isotope studies would have proven toxic. As Lazenby and McCormack (1985) have suggested, levels of salmon consumption proposed by some stable isotope studies—with salmon consumption rates in excess of 90 percent of the precontact diet—without an abundance of other foods, including plant foods, would have resulted in hypervitaminosis D and an assortment of other health problems. By this explanation, for 4,000 to 5,000 years, the entire region's population would have experienced incessant vomiting, constipation, heart palpitations, abdominal pains, weakness, mental disturbances and disorientation, to name but a few of the symptoms. This dismal situation is certainly not apparent in the region's oral traditions and ethnographic record. Nor, for that matter, is it evident in the region's paleopathological literature. Other animal species

clearly contributed to the isotopic signatures reported in these studies. This is by no means a suggestion that salmon was unimportant; quite clearly, it was a vital and central resource. However, a discussion of traditional diet that focuses on salmon alone, without reference to other fish or animal and plant foods, is at best incomplete.

9. European explorers consistently overlooked many forms of wetland cultivation and agroforestry throughout the Americas. See Denevan 1992; Doolittle 2000, 1992; and Siemens 1990, 1983.

10. "Potatoes" were also noted by some explorers, but were not viewed as endemic crops (Suttles 1951b). It is clear that a few of the crops referred to by this term were not *Solanum tuberosum* but instead represented certain native plants with edible starchy tubers or root sections; in some cases, these native plants were ambiguously termed "Indian potatoes."

11. Lewis (1973) has made a similar case about the representation of tobacco as a unique cultigen among the indigenous population of northern California, despite the presence of other plants which were subject to similar forms of maintenance.

12. In the posthumously published *Kwakiutl Ethnography*, Boas (1966: 17) grudgingly did ascribe cultivation to the Northwest Coast, suggesting that: "The only trace of agriculture found in this area is a somewhat careless clearing of grounds in which clover and cinquefoil grow and the periodic burning over of berry patches." Empirical support for these claims had been found in his earlier, data-rich ethnographic writings, but was absent from his more widely read ethnological writings.

13. Discussion of certain cultivation methods described in this volume can be found in Boas 1966: 17; 1934; 1921; 1909.

14. On a similar theoretical basis, Steward (1930) suggested that sites of the American Southwest could be cultivated, with irrigation and transplanting of propagules, but that the cultivation of these sites still did not constitute "agriculture." In these Southwestern cases, the presence of "agriculture" is no longer contested (Doolittle 2000, 1992).

15. There have been so many scholarly proclamations regarding the noncultivating Northwest Coast, derivative primarily of Boas's work, in numerous disciplines, that an overview would be prohibitive here. For a sample of influential secondary citations of Boas's work, see, for example, Marvin Harris (1968, 1979: 308–10), Robert Lowie (1920: 128–29, 1937), Ralph Linton (1936: 213), Alfred Kroeber (1939), Marshall Sahlins and Elman Service (1960: 77–80), Carl Sauer (1936: 295, 1952: 55), and Andrew Vayda (1961, 1968).

16. This denial of First Nations' claims on traditional plant sites has persisted into the present; in the era of *Delgamuukw*, the burden of proof is still on the First Peoples to prove, to a skeptical audience, that they have always been careful managers of their plants and their lands.

17. These expressions of thanks are in the languages of the Kwakwa̱ka'wakw and Nuu-chah-nulth, the two First Nations peoples who contributed most directly to the authors' ongoing research on these topics. Thanks could (and should) be extended in many other indigenous languages here, if space permitted. In all chapters, we have used the preferred orthographies of the First Nations/tribal populations being discussed; this represents an effort to manifest the many voices of Northwest Coast peoples and to respect the diverse traditions of our many indigenous consultants.

PART I

Concepts

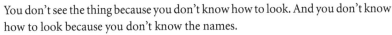

Chapter 2

Low-Level Food Production
and the Northwest Coast

BRUCE D. SMITH

You don't see the thing because you don't know how to look. And you don't know
how to look because you don't know the names.

—Don DeLillo, *Underworld*

The indigenous societies of the Northwest Coast of North America have
drawn the attention of scholars for well over a hundred years. Their remarkable art, rich mythology, complex kinship systems, elaborate Potlatch ceremonies, and salmon-centered economies have all been topics of exhaustive
research and writing back as far as Franz Boas and beyond. In terms of both
the vibrancy of expression of their cultures and the intense scrutiny to which
they have been subjected, Northwest Coast groups often have been placed in
a class by themselves.

They also have been perceived as distinct and different, and somewhat
anomalous in another important respect. Northwest Coast peoples have long
been considered to be hunter-gatherers, with an exclusive reliance on wild
plants and animals for their food. But they have also been acknowledged as
being qualitatively separate from the more typical hunting and gathering societies, which are generally characterized as living in small temporary settlements,
owning few material possessions, and having an uncomplicated social organization. Northwest Coast societies, in contrast, are more closely comparable
to agricultural groups in that they live for a good part of the year in relatively
large, sedentary villages that exhibit substantial labor investment in the construction of houses and other permanent structures, and they are known for
their considerable accumulation of material goods and their complex sociopolitical organization. Clearly they are both more "affluent" and more "complex"
than "typical" hunter-gatherer groups. As a result, societies of the Northwest
Coast have frequently been invoked as emblematic of the economically
affluent and culturally complex way of life that hunter-gatherer groups can

attain when they occupy environments endowed with abundant wild food resources. They have been held up as proof positive that material abundance and cultural complexity are not limited to societies with farming economies (see the discussion in Deur and Turner, in the Introduction to this volume).

Northwest Coast societies have also occupied a high-profile central position within this special category of complex hunting and gathering groups in that most of the other cultures scholars have labeled as "complex" or "affluent" hunter-gatherers are known only from the archaeological record (e.g., Calusa, Hopewell, Jomon—see the discussion below). These archaeologically known cultures can only be partially described, however, and they pale in comparison to the richly documented societies of the Northwest Coast. Most of the archaeological cultures that have been categorized as affluent or complex hunter-gatherers also turn out, based on recent research, to have been growing domesticated annual seed crops, making them poor candidates for hunter-gatherer status, affluent or otherwise (Crawford 1992a,b; Newsom and Scarry in press; Smith 2001, 2002: 205). While these dimly perceived crop-raising cultures from the archaeological record no longer qualify as hunter-gatherers, they also do not fit easily under most definitions of agriculture in that they don't appear to have obtained a significant percentage of their food from domesticates. Thus they fall somewhere between hunting and gathering and agricultural economies.

In a similar manner, this volume reconsiders what could be termed the "economic identity" of Northwest Coast societies, calling into question their frequent classification as "complex" or "affluent" hunter-gatherers. And like the Calusa, Hopewell, and Jomon of the past, Northwest Coast cultures too are found not to fall comfortably under the heading of complex hunter-gatherers, but rather to also belong out in the middle, between hunting and gathering and agriculture in terms of their economic base. As the chapters that follow indicate, a number of different species of perennial starchy root-crops, including Pacific silverweed (*Potentilla anserina* ssp. *pacifica*), rice-root (*Fritillaria camschatcensis*), springbank clover (*Trifolium wormskjoldii*), camas (*Camassia* spp.), Nootka lupine (*Lupinus nootkatensis*), and wapato (*Sagittaria latifolia*), were cared for or tended by Northwest Coast societies. This care took a variety of forms, including weeding, transplanting, soil improvement, and habitat expansion. The labor invested in specific locations also led to the establishment of precisely delineated plots of root crops, and clear claims of family ownership. This level of management goes far beyond the simple harvesting of "wild" plants. However, while terms such as "managed," "cultivated," or "husbanded" might be used to describe the manner in which these food crops benefited from human nurturing and encouragement, in the absence of any actual observed morphological change they would not be considered to be "domesticated" plants under most definitions of the term.

As chapters in this volume demonstrate (see Turner and Peacock, and Deur and Turner's Introduction, this volume), the concepts of domestication, agriculture, and farming that Europeans brought with them to the Pacific Northwest were based on their long-standing traditions of substantial clearing of

natural vegetation and the yearly planting of stored seed-stock of morpho-
logically distinct annual crop plants. As a result, Europeans did not recognize
the widespread tending of perennial root crops by indigenous societies as anything approaching their idea of farming, and easily relegated these activities
to the default category of hunting and gathering. But when these processes
are reconsidered in detail, as the contributors to this volume have done,
Northwest Coast societies and their management of root crops are illuminated
as carrying on something other than simple hunting and gathering or agriculture—placing them in the great unnamed and largely unknown middle
ground separating these two categories. And while there are certainly other
cultures, both past and present-day, that populate this middle ground between
hunting and gathering and agriculture, none are as well known or as well documented as the societies of the Northwest Coast.

So in the process of providing more accurate descriptions and more illu-
minating analyses of the management of perennial root crops by Northwest
Coast societies, this volume also brings into clear focus a number of inter-
twined, very basic, and very challenging questions regarding this middle
ground. In general terms, What labels, what names, for example, might be
applied to societies that are neither hunter-gatherers nor agriculturalists? How
should the dividing lines or borders on either side of this middle ground be
drawn? Are the transitions abrupt or gradual? What role do domesticated plants
and animals play as boundary markers? What terms can be used to refer to
plants and animals that are neither wild nor domesticated, judging from most
definitions? How numerous and how varied are the past and present-day societies of this middle ground? Are they abundant and diverse, or few and anomalous? Are such in-between economies short-lived—a temporary unstable
situation between the steady states of hunting and gathering and agriculture—
or do they represent successful long-term, end-point solutions and strategies?

This chapter addresses many of these more general questions raised by this
volume in an effort to more fully recognize and open up the middle ground
for consideration. It will help place into context the unique role that Northwest
Coast groups play as a window on the other, only dimly visible, category that
comes between a hunting and gathering way of life and one largely dependent on the management and production of domesticated plants and animals.

Does a Middle Ground Exist?

Is there any middle ground between hunting and gathering and agriculture?
Over the years, many researchers have comfortably employed an either–or,
cooked or raw, conceptual dichotomy, and have classed present-day and past
human societies as either hunter-gatherers or agriculturalists, with no intervening options. Hunn and Williams (1982), for example, show a bimodal distribution of relative dependence on agriculture in a sample of 200 societies
drawn from Murdock's *Ethnographic Atlas*, which shows a very low frequency
of societies having a 5–45 percent reliance on agriculture (Figure 2.1). The con-

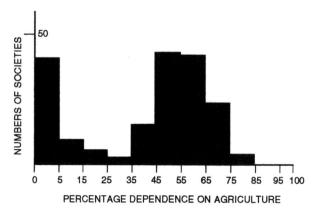

2.1. The relative dependence on agriculture of 200 societies drawn from Murdock's *Ethnographic Atlas* (Murdock 1967). Redrawn from Hunn and Williams, 1982, fig. 3.

clusion drawn from this bimodal distribution is that few viable long-term subsistence solutions exist between hunting and gathering on one side, and agriculture on the other, and that the developmental transition between the two was both radical and rapid (Hunn and Williams 1982: 5–6). Alternatively, the absence of societies in the 5–45 percent range could simply reflect their relatively recent wholesale replacement by rapidly expanding fully agricultural economies. Roy Ellen (1988: 127) argues that hunting and gathering and agriculture are "polarizing categories which tend to obfuscate the analysis of subsistence," and Zvelebil (1995: 80) proposes that "many societies, both ethnographic and prehistoric, filled the hitherto ill-defined gap between hunter-gatherers and farmers through a close management of their resources."

A dualistic perspective, however, continues to appeal to a wide range of scholars, particularly archaeologists interested in explaining, with very limited relevant information, what is viewed as the short developmental pathway leading across from hunter-gatherers to agriculture. Accompanying this central assumption of duality are a number of implicit correlates that have shaped, to a greater or lesser degree, widespread perceptions of the middle ground (or lack thereof) between hunter-gatherers and agriculturists. The boundary between hunter-gatherers and agriculturalists is often viewed as a thin line. This thin border is also often considered to be a one-way membrane. There is no turning back for those crossing over to agriculture. The question of how hunter-gatherer societies transformed themselves into agriculturalists is also considered to be a universal question—for which a single explanation that universally applies to all cases should be sought (e.g., Hayden 1995: 294). Hunter-gatherers and agriculturalists are also considered as being in a steady state, with the transition between them being necessarily rapid. Between these two generally stable and successful solutions or adaptations, there are no intervening solutions, but only societies in transition from one steady state to another. Those past or present-day societies that do not fall into either of these two steady-state categories should therefore be few in number, and they should either represent societies observed during what is a brief process of transformation, or represent a sort of stunted transitional failure that somehow stalled or stumbled in mid-stride between the "solution" states of hunter-gatherers and agriculturalists. Should Northwest Coast societies be viewed as being in a transition phase between hunting and gathering and agriculture (see Suttles, this volume), or as developmental failures of some sort? As Ellen warns in his discussion of the in-between Nuaulu of Seram, it is "dangerous to think of them (or any other similar group) as if they occupied some classificatory (and by implication evolutionary) space . . . transitional between foraging and agriculture" (Ellen 1988: 127).

This basic conceptual dichotomy—hunting and gathering versus agriculture, along with its correlates—constitutes an overarching dualistic epistemology. As a result of this epistemology's considerable popularity, scholars often try to move to one side or the other any anomalous in-between societies, both past and present-day, that neither rely exclusively on wild species nor strongly depend on domesticated species. In this way, they effectively depopulate and shrink down the territory between hunting and gathering and agriculture until it is transformed into a thin boundary line. Such efforts at categorical cleansing can be seen, for example, in the edited volume *Prehistoric Hunter-Gatherers: The Emergence of Cultural Complexity* (Price and Brown 1985). Contributors to the book explore and define the upper end of the range of cultural complexity on the hunter-gatherer side of the thin dichotomous boundary line. Occasionally they can be seen to shift the boundary so as to relocate societies of the middle ground that clearly do not rely exclusively for their sustenance on wild species of plants and animals under the "complex hunter-gatherer" label.

The most notable of these border crossings occurs in the chapter by Barbara Bender, in which archaeological (Middle Woodland, ca. 2150–1750 years ago) Hopewell societies of eastern North America are identified as a paragon of hunter-gatherer cultural complexity. Hopewell societies are confidently placed on the hunter-gatherer side of a dichotomous boundary line, even though they "display social and ideological features usually attributed to farmers" (Bender 1985: 21). Bender does hedge somewhat, however, in the classification of Hopewell as hunter-gatherers by suggesting that "perhaps they are farmers, but in a surreptitious and disappointingly invisible way" (Bender 1985: 21). Even as Bender was presenting the Hopewell as hunter-gatherers, however, other researchers had already clearly established that Hopewell societies were far from surreptitious in their farming efforts and that evidence for their cultivation of maize (*Zea mays*) and at least seven different indigenous seed crops was neither disappointing nor invisible (see for example Asch and Asch 1978, 1985; Ford 1979; Smith 1984, 1985a, 1985b; Stoltman and Baerreis 1983).[1] It is more than a little ironic, then, that in the process of attempting to deconstruct one paired set of dualistic conceptual boxes (hunter-gatherers are naturally simple, agriculturalists are culturally complex), Bender was endorsing and reinforcing another equally reductionist and dichotomous worldview—that societies must be either hunter-gatherer-foragers or farmers.

Hopewell groups are not alone in being shifted across into hunting and gathering. The Jomon societies of Japan, for example, cited as another example of archaeological complex hunter-gatherers by Price and Brown (1985: 10–11), have since been identified as having food-production economies based on domesticated crop plants (Crawford 1992a,b). Present-day Amazon lowland societies having a mixed economy of forest hunting and manioc cultivation are identified as hunter-gatherers (Gould 1985: 433). Clearly of most interest in the context of the present volume, Northwest Coast societies are also characterized as hunter-gatherers (Ames 1985); (see Harlan 1995: 15–16 for continuing characterizations of Jomon and Northwest Coast societies as sophisticated and "luxurious" hunter-gatherers). It is thus not at all difficult to find scattered throughout the literature varied examples of scholars struggling with the problem of how to classify both past and present-day societies that don't comfortably qualify as either hunter-gatherers or full-scale agriculturalists.

On a regional scale, the transition from a hunting and gathering way of life to agricultural economies in northern Europe has similarly long been viewed as a rapid replacement of one steady-state adaptation by another. Peter Bogucki (1995) has recently challenged this hunter-gatherer / agricultural duality, arguing that viewing the transition as a shift between two stable adaptive solutions may capture the beginning and end points of the process but fails to consider what comes in between, as populations that had previously relied exclusively on wild resources initially adopt cultivation and animal husbandry. He suggests that, by viewing this in-between process of change in a linear sense as leading quickly and irreversibly to agriculture, the dualistic perspective "suppresses consideration of the more interesting period in which this process occurred," and "obscures the complex shifts in subsistence behavior that must have been repeated in countless different ways around the world throughout the Holocene" (Bogucki 1995: 105). While it is easy enough to find evidence of the extent to which the dualistic epistemology continues to hinder recognition of the existence of the middle ground, this in-between category has in fact drawn the attention of some scholars.

Defining the Middle Ground

Over the past 30 years, a number of scholars with diverse disciplinary and regional perspectives have attempted to describe different aspects of the middle ground between agriculture and hunting and gathering. The most notable in this regard are Ford (1985b) (Figure 2.2), Harris (1989, 1990, 1996a,b), Higgs (1972, 1975), Hole (1996), Jarman et al. (1982), Peacock (1998), Rindos (1984), Smith (1985a), and Zvelebil (1986a,b, 1993, 1994, 1995, 1996). It is important to recognize that none of these researchers seek out and document in

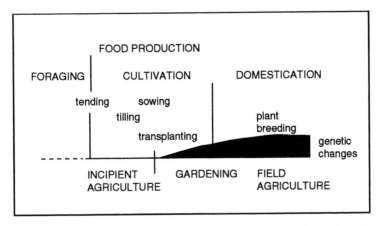

2.2. Ford's stages and methods of plant food production. Redrawn from Ford 1985a, fig. 1.1.

detail any *societies* that occupy this neither–nor territory. Rather they focus on a finer scale of analysis—identifying and defining, in isolation, those types of *activities,* those categories of human-plant and human-animal interaction, that could be considered as characteristic of such middle-ground societies (e.g., Figures 2.2–2.5). These various types of activity are usually placed along a perceived continuum of increasing human intervention or involvement in the life cycle of target species—a continuum that spans the middle ground between hunting and gathering and agriculture. This "behavior pattern" scale of inquiry has several obvious advantages over larger-scale consideration of particular societies. When considered in isolation, such examples of human intervention in the life cycles of plants or animals need not be drawn only from the limited pool of well-documented middle-ground societies, but can also be found in the repertoire of agricultural societies. Such general categories of human patterns of intervention can in turn be used to construct a template or profile, in the abstract, of what societies within the middle ground between hunter-gatherers and agriculturalists would look like—to create "virtual" societies of the middle ground.

This middle ground is not an easy landscape to describe, nor is it a simple matter to reconcile the definitions and vocabularies used by Ford, Harris, Hole, Jarman et al., Rindos, Smith, and Zvelebil, since there is less than complete consistency in the terminology each employs, and even in the meanings they assign to the same term. As a result, it is easy to lose your way, to be " . . . bedeviled by confusion over the meanings attributed to such terms as agriculture, cultivation, domestication, and food production" (Harris 1989: 11):

> The published literature on "agricultural origins" is characterized by a confusing multiplicity of terms for the conceptual categories that define our discourse. There is little agreement about what precisely is meant by such

terms as agriculture, horticulture, cultivation, domestication and husbandry. This semantic confusion militates against clear thinking about the phenomena we investigate. [Harris 1996a: 3]

This is particularly true in regard to the terminology of partition and boundary. It is easy enough to suggest seemingly straightforward, common-sense boundary definitions for both hunter-gatherers—"economies based exclusively on wild plants and animals," and agriculturalists—"strongly dependent on domesticated species as food sources." But when these boundary conditions are considered more closely, a number of more complex and elusive questions come into clearer focus. On one side of the middle ground, for example, along the boundary line for agriculturalists, what exactly is meant by "strongly dependent" on domesticates? Should perhaps a consistent annual caloric budget reliance on domesticates of, say, 40, 50, or 60 percent be the dividing line between nonagriculturalists and agriculturalists, or should some other minimal qualification for agricultural status be employed? Rindos offers little advice, stating only that "the common-sense definition of agricultural subsistence is a dependence upon domesticated plants for a substantial part of the diet" (Rindos 1984: 236). Similarly, Harris (1996b: 446) marks the boundary of agriculture as "denoted when domesticated plants (cultivars) are the main or exclusive components of systems of crop production . . . and when more human labor is invested in cultivation and the maintenance of agricultural facilities . . . " In his consideration of the transition to farming in the circum-Baltic region, Zvelebil places the agricultural border—"the shift to full dependence on agriculture"—at a 50 percent reliance on domesticates. Whether the agricultural boundary is drawn as a specific "percentage contribution of domesticates" line, or defined as a relatively broad clinal zone of transition (e.g., 40–60 percent), it is clear that placement of societies relative to this border will not be a simple task. It is important to recognize that this minimal qualification boundary definition for agriculture centers on the term "domesticate." But what exactly distinguishes domesticated plants and animals from their wild cousins, and from entities and interactions that exist in a not-wild, yet not-domesticated realm between wild and domesticated?

Interestingly, if we turn in the opposite direction and approach the hunter-gatherer boundary line on the other side of the middle ground, the same set of questions again comes into focus, in mirror image. Is the hunter-gatherer boundary natural or artificial? Is the border zone empty or crowded and clinal? Are there boundary markers that apply broadly, in descriptive and developmental terms, to both archaeological and present-day situations? And, of central importance, imbedded in the boundary condition for hunter-gatherers—that they rely exclusively on wild plants and animals—is the question of what, exactly, distinguishes "wild" from "non-wild"? Does "non-wild" equate with "domesticated," or are there categories between wild and domesticated? These elusive and complex mirror-image sets of questions that arise as one attempts to approach and characterize the boundary zones of hunting-gathering and agriculture (and thus define what lies between) highlight the extent to which "domestication" dominates conceptualizations of the middle ground. A closer consideration of the ways in which the terms "domestication" and "domesticate" have been defined provides an important guide to the middle ground. As might be expected, Rindos, Harris, Hole, Ford, and others have each defined "domestication" differently, and situate the concept differently in their interpretation of the intermediate landscape that stretches from hunting and gathering to agriculture. From a dualistic perspective, in which only a thin line separates hunting and gathering from agriculture, and there is no intervening territory, domestication, of course, was situated right on, and in large measure defined, the boundary line between the two. Hunter-gatherers had no domesticates, and any societies with domesticates had agriculture. As scholars began to recognize the existence of a conceptual territory between hunter-gatherers and agriculturalists, however, their use of domestication as a boundary marker changed accordingly.

In his 1989 study, for example, Harris placed domestication squarely on the border of agriculture (Harris 1989, Figure 1.1). In so doing, he retained the conventional partition—that hunter-gatherers have no domesticates and any societies with domesticates are agricultural. At the same time, however, he recognized a broad middle ground of societies that are nonagricultural, yet are not hunter-gatherers, since they do not rely exclusively on wild species of plants and animals. Harris continued to equate domestication with agriculture in a publication that appeared the following year: "the distinction between cultivation and agriculture rests on the presence or absence of domesticated crops . . . therefore, if it can be shown archaeologically that the plant remains recovered at a given site are from domesticated taxa . . . then there is a secure basis for inferring that agriculture was practiced in the vicinity of the site" (Harris 1990: 13) (Figure 2.3). In the figures of his 1996 articles, however, Harris shifts the placement of initial domestication of plants from one side of the middle ground to the other, relocating it from the border of agriculture to the boundary of hunter-gatherer economies (Harris 1996a: Figure 15.1; Harris 1996b: Table 1.1) (Figure 2.4).

Accompanying this relocation of domestication, Harris redefines "agriculture" in a manner that is close to the basic definition offered earlier: "agriculture is denoted when domesticated plants . . . are the main or exclusive components of systems of crop production . . . and when more human labor is invested in cultivation and the maintenance of agricultural facilities (field systems, tools, storage, etc.)" (Harris 1996a: 446). With this graphic relocation of domestication from one border to another, Harris on the one hand keeps the presence of domesticates as a boundary marker for hunter-gatherers (a society is no longer on the hunter-gatherer side of the line if domesticates are present) while at the same time acknowledging that there is a broad and diverse middle ground where societies have some reliance on domesticates, yet are not agriculturalists. In relocating domestication of plants to the hunter-gatherer border, at least in the figures of his 1996 articles, Harris also would

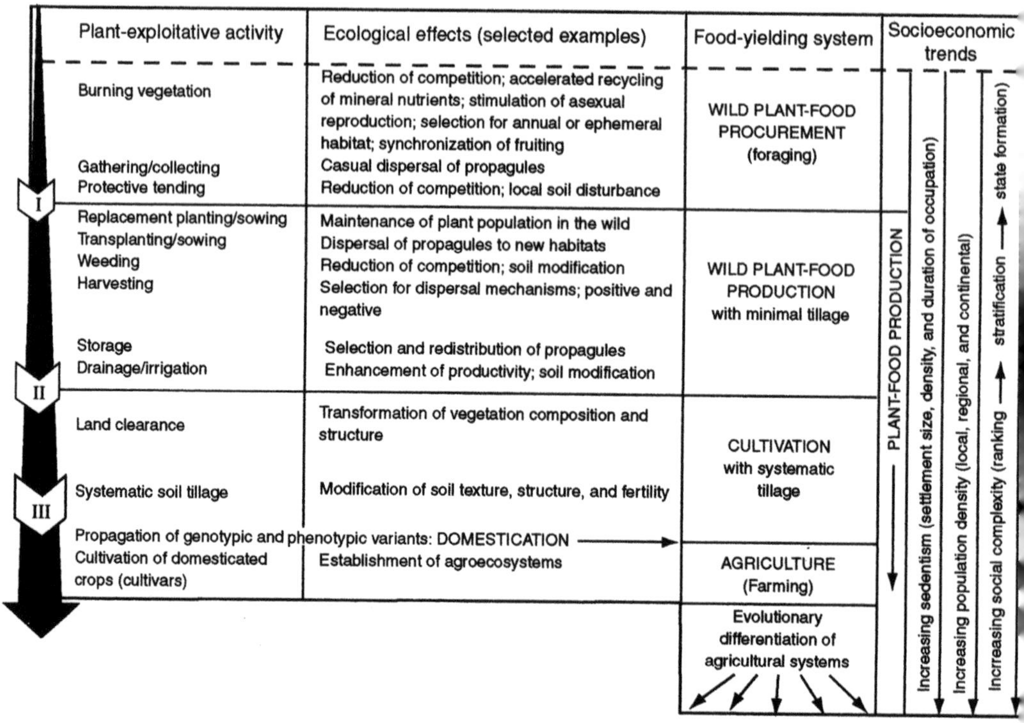

Plant-exploitative activity	Ecological effects (selected examples)	Food-yielding system	Socioeconomic trends
Burning vegetation	Reduction of competition; accelerated recycling of mineral nutrients; stimulation of asexual reproduction; selection for annual or ephemeral habitat; synchronization of fruiting	WILD PLANT-FOOD PROCUREMENT (foraging)	
Gathering/collecting	Casual dispersal of propagules		
Protective tending	Reduction of competition; local soil disturbance		
Replacement planting/sowing Transplanting/sowing Weeding Harvesting	Maintenance of plant population in the wild Dispersal of propagules to new habitats Reduction of competition; soil modification Selection for dispersal mechanisms; positive and negative	WILD PLANT-FOOD PRODUCTION with minimal tillage	
Storage Drainage/irrigation	Selection and redistribution of propagules Enhancement of productivity; soil modification		
Land clearance	Transformation of vegetation composition and structure	CULTIVATION with systematic tillage	
Systematic soil tillage	Modification of soil texture, structure, and fertility		
Propagation of genotypic and phenotypic variants: DOMESTICATION ⟶ Cultivation of domesticated crops (cultivars)	Establishment of agroecosystems	AGRICULTURE (Farming)	
		Evolutionary differentiation of agricultural systems	

Left margin markers: I, II, III (with descending arrow, PLANT-FOOD PRODUCTION)

Socioeconomic trends (right margin, increasing upward): Increasing sedentism (settlement size, density, and duration of occupation) → Increasing population density (local, regional, and continental) → Increasing social complexity (ranking → stratification → state formation)

2.3. Harris's classificatory and evolutionary model of plant-food yielding systems. Redrawn from Harris, 1989, fig. 1.1.

appear to bar from the middle ground any societies that lack domesticated plants. Without such domesticates they would be placed on the hunter-gatherer side of the line. In his discussion of the landscape between hunter-gatherers and agriculturalists in his 1996 articles, however, Harris makes it very clear that there is considerable territory allocated to societies that are not hunter-gatherers yet lack domesticated plants (Harris 1996a: 460). Similarly, in his consideration of animals, Harris clearly places domestication at a distance from the hunter-gatherer border rather than on it (Harris 1996a: Figure 15.2) (Figure 2.5). As a result, even though several of his charts of the middle ground suggest otherwise, it is clear that by 1996 Harris had decided that domestication of plants and animals, as he defined the term, was not situated on the boundary line of either agriculture or hunting-gathering, but rather was located out on the broad landscape between the two.

As shown in Figure 2.2, Ford too places domestication between the boundaries of hunting and gathering and agriculture (Ford 1985b: Figure 1.1), whereas Rindos addresses this issue by defining three different forms of domestication. The first of these, incidental domestication, encompasses the entire human experience: "domestication has been present, to a greater or lesser extent, in all cultures and at all times" (Rindos 1984: 258). As defined by Rindos

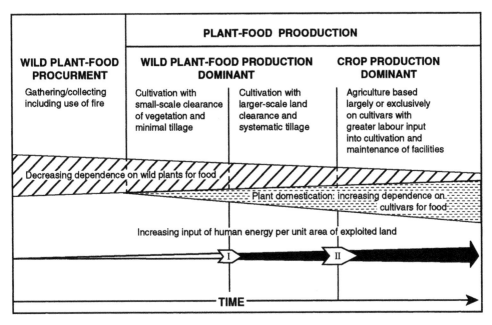

WILD PLANT-FOOD PROCURMENT	PLANT-FOOD PROODUCTION		
	WILD PLANT-FOOD PRODUCTION DOMINANT		**CROP PRODUCTION DOMINANT**
Gathering/collecting including use of fire	Cultivation with small-scale clearance of vegetation and minimal tillage	Cultivation with larger-scale land clearance and systematic tillage	Agriculture based largely or exclusively on cultivars with greater labour input into cultivation and maintenance of facilities

Decreasing dependence on wild plants for food

Plant domestication: increasing dependence on cultivars for food

Increasing input of human energy per unit area of exploited land

I   II

TIME

2.4. Harris's evolutionary classification of systems of plant exploitation. Redrawn from Harris, 1996b, fig. 15.1.

PREDATION	**PROTECTION**	**DOMESTICATION**
Opportunistic and planned procurement of prey	Enhancement of the reproductive potential of selected species through ecosystem manipulation	Maintenance of breeding populations isolated genetically from their wild relatives
Generalized hunting **Specialized hunting** **Fishing**	**Taming** **Protective herding** **Free-range management**	**Livestock raising by settled agriculturalists** **Transhumance** **Nomadic pastoralism**

Decreasing dependence on wild animals for food

Increasing dependence on domesticates for food

TIME

2.5. Harris's evolutionary classification of systems of animal exploitation. Redrawn from Harris 1996b, fig. 15.2.

(1984: 159), "specialized domestication" involves a range of deliberate human behaviors that alter the general ecology and local environment in ways that place certain species at a distinct advantage, while "agricultural domestication" is "very close in concept to *domestication* as it has been used in most of the literature . . . " (Rindos 1984: xv, 164). From his discussion of "specialized" and "agricultural" domestication, it is clear that Rindos places both between the borders of hunting and gathering and agriculture, even though the term "agricultural domestication" could easily be assumed (incorrectly) to be a boundary marker for agriculture.

Zvelebil, like Rindos, attaches a range of modifiers to the term "domestication." Many of these ("full domestication," "genetic domestication," "complete domestication" [Zvelebil 1986b: 174]; "economic domestication" [Zvelebil 1993: 151, 157]; and "full biological domestication" [Zvelebil 1995: 86]) refer to the generally accepted definition of the term (to be discussed below)—"genetic selection which resulted in the establishment of desired characteristics" (Zvelebil 1986b: 174). At the same time, Zvelebil, like Rindos, also defines another general category of human interaction with plant and animal resources that he labels as "cultural domestication" (Zvelebil 1995: 98) and "behavioral domestication" (Zvelebil 1995: 96). This recognition of two categories of domestication, both of which involve close human interaction with and management of plant and animal resources, but only one of which results in easily observed morphological changes, is of critical importance in further describing the middle ground and the societies that occupy it.

What Is Domestication?

Domestication has been viewed and described from a number of different directions over the years. Many of these varied characterizations, including those of Rindos, Ford, Harris, and others, have emphasized two prominent features of domestication: phenotypic (physical-morphological) changes in the domesticated species of plants and animals; and the reliance of these modified species on humans for survival (Figure 2.2). These two features of domestication are also often considered to be linked in a cause-and-effect relationship, with domesticates being, above all, a purely human creation. Human societies cause physical changes in target species of plants and animals such that these new domesticates are no longer viable without continued human protection and care:

> Cultural selection for useful phenotypic characters resulted in new plants dependent upon humans for their existence. Domesticated plants are cultural artifacts. They do not exist naturally in nature; they cannot normally survive without human assistance. [Ford 1985b: 6]

> Domestication meaning that genetic and/or phenotypic selection has led to morphological change and a degree of dependence on human actions for the plant's survival. [Harris 1996a: 446]

Morphological change in the plant is the most important indicator of domestication pressures and serves to tie the survival of the plant to activities of humans (Rindos 1984: 140).

Domestication, then, is essentially a new category of relationship of interaction established between a population of humans and a population of plants or animals. This new relationship is initiated and maintained by humans, with the plant or animal species occupying a dependent role. Central to this relationship is the sustained control and responsibility taken by humans over some aspects of the life cycle of the domesticate. This human intervention often focuses on reproduction, and is often directed toward increasing the reliability and yield of a food resource (or other product—e.g., wool). Humans can deliberately and directly increase the harvest yield of domesticates, for example, by making sure that only the larger individuals (i.e., big bulls, big seeds) participate in creating the next generation. Humans and their life cycle intervention and modification of the surrounding and environment of domesticates also reshape the world of domesticates in many ways. Inadvertently and indirectly, humans redefine what kinds of attributes or characteristics increase or decrease an individual plant or animal's likelihood of producing successful offspring. They change the milieu of selective pressures acting on the "domesticate" population to something quite different from that which shapes populations of the same species in the wild. In so doing, humans change the genetic composition of the domesticate population—altering the relative representation of different attributes in the gene pool, while also changing the rules regarding what constitutes an advantageous versus deleterious mutation among domesticates. Some of the changes in the domesticate gene pool will result in easily observable physical or behavioral changes in the domesticated plants or animals involved, while other genetic changes will be much less evident phenotypically.

The clear, unidirectional, causal chain in "domestication" and the creation of domesticates is: (1) humans establish a new relationship with a target species, one that involves intervention and support, which can cause: (2) changes in the genetic composition of the domesticate population, which can in turn lead to: (3) observable behavioral or physical changes in the domesticates. It is important to keep the obvious directionality and nature of these causal chain linkages clearly in mind when considering domestication. Genetic, behavioral, and physical changes produced in the domesticate populations are the result of, are dependent upon, and are evidence of, domestication. Domestication itself, the independent variable in the causal chain, is the ongoing relationship of intervention, initiated and sustained by humans.

The existence of domestication—this new relationship between humans and favored species—can be identified all along this causal chain. The behavior set itself could be observed firsthand, or physical evidence relating directly to it might be documented in the archaeological record (e.g., llama corrals, irrigation canals). Further along the causal chain, genetic evidence of domestication—specific changes in the genetic code that can be used to distinguish between wild and domesticated populations of a species—are now being identified (e.g., Doebley et al. 1997; Dorweiler et al. 1993; Piperno et al. 2002). When such genetic changes resulting from domestication are manifested by either behavioral changes (e.g., increased docility in livestock) or physical changes (e.g., larger seeds), they too can be observed and documented. It is this broad category of physical, phenotypical changes that has been generally acknowledged as providing the most obvious and the strongest evidence of domestication. Domesticated plants and animals are often visibly distinguishable from their wild relatives—they look different. Because physical changes provide such obvious and compelling evidence of domestication, however, it unfortunately has been commonly concluded that the opposite is also correct: the absence of observed morphological change has been seen as evidence that the plant or animal populations in question are not domesticated. But a relationship of domestication does not necessarily result in easily observable morphological changes. Morphological changes that are evidence of the existence of a relationship of domestication can be so small, for example, as to have escaped recognition up to now. Archaeologists and archaeobotanists are continuing to discover and document a range of new micro-morphological markers of domestication (e.g., Piperno et al. 2000). As Deur and Turner, and Peacock (this volume) point out, it may be that the root crops cultivated by Northwest Coast societies do in fact exhibit morphological changes associated with domestication, but that they have not yet been recognized.

Turner and Peacock (this volume) also provide an excellent discussion of why Northwest Coast root crops might not exhibit physical changes—that substantial differences in fact exist between species in the extent to which they are likely to express a relationship of domestication in terms of phenotypic or physical changes. Recent research by Zeder provides strong supporting evidence for the dangers inherent in relying on physical changes as a marker of domestication (Zeder 1999, Zeder and Hesse 2000). She shows that skeletal changes (e.g., size reduction)—long thought to be good archaeological indicators of initial domestication in livestock species—often do not appear until thousands of years after distinctive shifts in age and sex profiles indicate that the target species were fully domesticated and under direct and sustained human management. The early histories of a number of domesticated animal species, then, provide clear and compelling evidence that they remained physically (i.e., skeletally) indistinguishable from their wild relatives for long after they were initially brought under domestication. A number of Near Eastern livestock species thus provide clear evidence of domestication in the absence of morphological change.

As Turner and Peacock (this volume) point out, morphological change has similarly long been the standard of proof for documenting domestication in plants. This has been due in large measure to the remarkable success achieved over the past three decades in documenting the causes and kinds of morphological changes that reflect domestication in the cereals and other annual seed plants, and the apparent rapidity with which a range of morphological markers appear once deliberate and sustained annual planting of stored seed stock has begun. In seed crops, such control and management of the cycle of reproduction entails the harvesting, storage, and planting of seed stock, which not only releases the managed plants from selective pressures acting on wild populations in regard to seed dispersal, germination, dormancy, and so on, but also puts into place a quite different set of selective pressures. Under the general heading of "the adaptive syndrome of domestication," such managed seed-plant populations automatically respond to these new sets of seedbed and harvesting selective pressures in ways that have been described and explained in some detail (e.g., Harlan et al. 1973; see Smith 1995, 2002 for fuller discussion), resulting in very distinctive morphological changes (such as larger seeds and compacted seed heads), such as ears of corn, that are in turn very good markers of domestication and deliberate planting, and which are visible in the archaeological record.

But perennial plants represent a very different situation. A relationship of domestication in which perennials are the target species, as along the Northwest Coast, would not involve the harvest, storage, and subsequent planting of seed stock. As a result, a set of selective pressures very different from that established for annual seed plants would be established: "the adaptive syndrome of domestication" would take a different form. Any associated morphological markers of the relationship of domestication in perennial domesticates might be much slower to appear, and be much less obvious than is the case in annual seed-propagated plants. For perennial plants, as has been documented for a number of livestock species, it is quite possible to have long-extant relationships of domestication, and "domesticated" plants, in the absence of any morphological change.

Moving back along the causal chain of domestication, the genetic code and DNA sequencing represents an obvious core database from which the domesticated status of individual organisms and populations could be established. Some changes in the genetic code that turn out to be valuable in differentiating between wild and domesticated might not be reflected in any observable behavioral or physical changes. The vexing genome challenge to be faced, of course, is to both identify the full set of genetic changes that comprise the signature of domestication in any particular species, and to resolve the inevitable questions that will arise regarding which particular genetic changes, in what combinations, are the minimum for designating "domestication" of a species.

There is of course no clear and straightforward answer to this question of how to determine exactly when a relationship of domestication has been established. Nor will it be particularly worthwhile or rewarding to approach this question in terms of placing too narrow a focus on seeking some sort of litmus test or definitive demarcation of exactly what constitutes a "relationship of domestication." What is important at this point is to acknowledge that it is these relationships of interaction between human societies and target species that are the beginning of the causal chain, the independent variable, and that are the actual focus of study. These relationships of interaction can be observed firsthand, or by looking at different kinds of data sets at different points along the causal chain, from direct evidence (see for example Deur's discussion of Northwest Coast relict garden or cultivation plots, this volume), to oral and written descriptions (Turner and Peacock, this volume), to the search for subtle genetic and morphological changes associated with domestication (in both chapters).

A number of attributes combine to make Northwest Coast societies, and this volume, so unusual and so significant in terms of efforts to gain a better general understanding of such relationships of interaction, and of societies of the middle ground. First, in contrast to most previous studies (e.g., Figures 2.2–2.5), this volume looks at such relationships of interaction not through an isolated activity-set scale of analysis, but within the fuller and richer context of substantial understanding of the complete cultural systems within which the relationships are sustained. Second, unlike the agricultural societies that have been the source of many of the activity-set resource-management examples cited in previous studies, Northwest Coast societies are clearly situated in the middle ground between hunting and gathering and agriculture. Third, and of great importance, Northwest Coast societies provide an opportunity to consider evidence from all along the causal chain, from interviews with elders who observed management of root crops firsthand, to direct observation of relict garden plots, through analysis of written descriptions of root-crop management, to the still to be undertaken search for genetic and morphological markers of domestication (Turner and Peacock; Deur, this volume). Fourth and finally, Northwest Coast societies actively managed perennial root and berry crops rather than annual seed plants, providing an opportunity to consider relationships of interaction that would not have resulted necessarily in rapid or substantial morphological changes in the target species.

How Large Is the Middle Ground?

In order to fully appreciate the interpretive value of Northwest Coast societies and the importance of this volume in helping illuminate the middle ground between hunting and gathering and agriculture, it is necessary to sketch in some of what is known of this general landscape based on partial studies of other middle-ground societies, particularly some known only from the archaeological record. How expansive, how wide, is this middle ground, for example? Has it been inhabited at any point in time by only a few societies, as they make a rapid transition from hunting and gathering to agriculture? Were Northwest Coast societies interrupted at European contact, in the midst of a rapid developmental shift over to agriculture? Were they on their way to agriculture? Do they provide a window on what is usually a brief transitional state? Or, alternatively, does their limited reliance on managed (domesticated?) root and berry crops (as well as their management of clams and other shellfish, and salmon) represent a rare case-study example that is generally representative of a very large set of societies that flourished over long periods of time on the basis of stable and successful economic adaptations that did not involve substantial reliance on domesticates?

Well-documented archaeological sequences in both eastern North America and Mesoamerica indicate that a transition from hunting and gathering to agriculture is not necessarily a rapid process. While it is usually not possible to accurately gauge the relative contribution that domesticates make to the total caloric budget of societies documented archaeologically, eastern North America provides an interesting exception to this general rule. Stable carbon isotope analysis of human skeletal remains has shown a broad and dramatic shift to maize agriculture and a strong reliance on corn between 1050 and 850 years ago. Although it is not possible to directly translate stable carbon isotope values into specific caloric input figures for maize (e.g., 40, 50, 60 percent), this shift in carbon isotope ratios between A.D. 900–1100 is now widely accepted as marking the initial appearance of maize-centered agricultural economies over a broad latitudinal zone of the East, from southern Ontario to northern Florida. The broad-scale increase in the dietary contribution of corn in eastern North America roughly 1000 years ago, when combined with the equally well-dated evidence for morphological changes associated with domestication of at least four local annual seed plants in the East at ca. 5000– 4000 years ago (sunflower, squash, chenopod, marshelder—Smith 2002) provides one of the best-documented regional records of the long time-span separating hunting and gathering and agriculture. In the midlatitude deciduous forests of the eastern United States, the appearance of morphological markers of plant domestication and the subsequent shift to maize agriculture are separated by a full 4000 years of time. This 4000-year-wide developmental landscape is occupied by human societies relying on a rich diversity of different economic systems—solutions that involved various combinations of wild plant and animal species and domesticated crop plants. It is possible that at least some of these societies may have been situated close to or within the proposed 40–60 percent transition zone of the agricultural border prior to the shift to maize-centered agriculture in the region. A number of Middle Woodland Hopewellian pre-maize farming societies, for example (those classed as complex hunter-gatherers by Bender [1985], and still considered hunter-gatherers by some, i.e., Yerkes [2000]) appear to have substantially increased their reliance on indigenous seed crops by ca. 2300 years ago. Some scholars have proposed that the seed-filled human paleofeces from Mammoth Cave, Kentucky, reflect an even earlier society-wide reliance on local crop plants (as opposed to representing a specialized caver's trail mix). It unfortunately isn't yet possible to convincingly establish the dietary importance of domesticates, or the exact placement of these societies on the vast landscape of the middle ground.

In Mesoamerica, the temporal and developmental distance between the earliest morphological markers of domestication and agriculture is even greater.

While squash (*Cucurbita pepo*) was domesticated by 10,000 years ago in Oaxaca (Smith 1997a), the village-based societies with maize-beans-squash farming economies that have been viewed generally as marking the agricultural border do not appear until about 4500 years ago, a full 5500 years later. Interestingly, a similarly expansive and complex temporal and developmental landscape covering perhaps 2000–3000 years has been recognized as separating the initial appearance of morphological and nonmorphological domestication and the subsequent initial emergence of agricultural economies in the Near East (Hole 1996: 266; Smith 1995).

Here then, is a very basic and very important point to be made regarding the scale of the territory that lies between domestication and agriculture. These three regions—Mesoamerica, the Near East, and eastern North America—are the best documented of the world's recognized primary centers of domestication and subsequent agricultural emergence (Smith 1998). And in each of these three areas, where the temporal-developmental placement of both initial domestication and the subsequent transition to agriculture can be determined with a reasonable degree of accuracy, they are separated by large, still mostly uncharted territories stretching across 2000–5500 years of time. In each area, these vast expanses are occupied by in-between societies that are neither hunter-gatherers nor agriculturalists, even though domesticates contribute to their economies.

Against this broader, if briefly sketched, developmental backdrop, Northwest Coast societies can be seen as not being in transition, not anomalous or rare; they are not "incipient" agriculturists (Smith 2001). Rather they are representative in general of a very large, very diverse category of middle-ground societies which, although not abundant today, occupied much of the earth for thousands of years. These diverse, vibrant, and successful human societies developed long-term solutions for deriving sustenance from a wide range of environments that combined low-level reliance on domesticates with continued use and management of wild species. So Northwest Coast societies, like those of the middle ground in general, should not be viewed simply as reference points on the way to agriculture, as roadside markers of progress, but rather as stable and progressive solutions, as end points and destinations worthy of study in and of themselves. Yen (1989: 66) underscores the errors inherent in viewing such middle-ground societies as "transitional or proto-agricultural." They are not "the backward relics of a single evolutionary line which most accounts seem to suggest." Rather they should be recognized not as being "pristine hunter gatherers, but groups who, in achieving qualitatively distinctive cultural end points, have followed different pathways of subsistence-system development from a common beginning."

Problems with the Terminology of the Middle Ground

Over the years, a number of different terms have been proposed as descriptive labels for the human activity sets considered as characteristic of the middle ground—as falling between hunting and gathering and agriculture. Although many of these have met with various degrees of acceptance, and some continue to be used today, most have problems of one kind or another.

The use of the terms "plant husbandry" and "animal husbandry," for example, has declined considerably since the 1980s. Generally defined as involving the cultivation and production of crops and animals, the term "husbandry" came under attack in 1980 as being androcentric because it: "implies a skewed division of labor in favor of men and arbitrarily narrows the multitude of relationships between people and their biological environment" (Ford 1985b: xii).

Three other terms—cultivation, gardening, and horticulture—that are often used as labels for the landscape between hunting and gathering and agriculture also are each surrounded by a variety of problems. While there are good reasons for applying these labels to the middle ground, they all suffer from a long history of usage and an accompanying multiplicity of meanings that blur and largely defeat their value as descriptive labels that can be understood easily and unequivocally.

Take, for example, the various meanings and applications of different forms of the term "cultivation" (e.g., cultivator or cultivating societies, quasi-cultigen, cultigen, cultivated plants, etc.). Certainly "cultivation" could be accurately and appropriately applied to the management of root crops by Northwest Coast societies. Moving from the general to the specific, "cultivation" can be defined in common usage as: (1) promoting or improving the growth of a plant by labor and attention; (2) preparing and tilling the land in order to raise crops; and (3) working the earth around growing plants to loosen the soil and destroy weeds. At all three of these levels of specificity, "cultivation" involves a lesser degree of intervention in the life cycle of plants than does the term domestication. Thus the terms "cultivated," "cultigen," and "quasi-cultigen" are often used to refer to plants that do not exhibit any morphological markers of domestication, yet because of their abundant representation in archaeobotanical assemblages are suspected to have been generally encouraged by humans in a manner that carries them beyond the realm of simple harvesting of wild plants. This use of the term "cultivation," which follows from common usage and focuses both on the nature and intensity of human-plant interaction and on the territory between wild-plant collecting and domestication, is clearly evident in Ford's (1985b: 4) exploration of the territory between hunting and gathering and agriculture. Under the heading of cultivation, Ford places any and all human undertakings involving plants, short of full control of reproduction and resultant domestication, including tending (weeding, pruning), tilling (soil preparation), transplanting, and sowing of seed or other propagules (at harvest) (Figure 2.2).

In clear contrast to Ford, however, Harris, in his most recent statements (1996a,b) expands the coverage of "cultivation" in several respects. First, he extends cultivation from the border of hunting-gathering ("wild plant food procurement"), past domestication, all the way to agriculture. At the same time he adds another major dimension to the meaning and use of the term "cultivation" when he combines the extent of land clearance and soil preparation with the various forms of human encouragement of plants identified by Ford as cultivating activities. Subdividing the term into "cultivation with small-scale land clearance and minimal soil tillage, and cultivation with larger-scale land clearance" (Harris 1996b: 446) (Figure 2.4), Harris draws the boundary between cultivation and agriculture in terms of "when domesticated plants (cultivars) are the main or exclusive components of systems of crop production . . . " (Harris 1996b: 446). Thus Harris places a greater emphasis on an increasing level of human energy investment and expanding areas under crop (and indirectly, the contribution of cultivated plants to the economy), rather than on the specific nature and degree of human intervention in the life cycle of plants.

To further complicate matters, yet another application of the term cultivation is clearly evident in the "cultivating ecosystem type" as formulated by Stoltman and Baerreis (1983: 257) for eastern North America. Following Bronson (1977: 26) and in a manner similar to Harris et al. (1983: 257) distinguish between agriculture ("reserved for contexts of substantial dependence on [domesticated? –B. Smith] plants by humans"), and cultivation ("only that a useful species has been deliberately caused to reproduce by man"). They differ from Harris, however, in that, for Stoltman and Baerreis, cultivation does not extend all the way from hunting and gathering to agriculture, but only from domestication to agriculture. Although not explicitly stated, it is clear from their discussion that "cultivation" begins with the initial appearance of domesticated plants in eastern North America, and covers the subsequent period of time during which domesticates were present, but agricultural economies had not yet developed.

In summary, while all of these researchers employ the term "cultivation" in their descriptions of the landscape between hunting and gathering and agriculture, they define it in different ways and apply it to different parts of the landscape, substantially undermining its usefulness. Ford, for example, applies "cultivation" to the region between hunting and gathering and domestication, Stoltman and Baerreis to the region from domestication to agriculture, and Harris encompasses the entire in-between territory from hunting and gathering to agriculture. The value of the term "cultivation" is even further reduced when one takes into account its long-established common application to fully agricultural crops and economies. Consider, for example, a standard farm implement called a "cultivator," which is drawn between rows of field crops to turn the soil and uproot weeds. At the same time, while some scholars would argue that, by definition, any plant(s) that are the subject of various forms of human encouragement that could be classed as cultivation should in turn be considered as cultivated plants, it is not at all difficult to find reference to the cultivation of wild plants (e.g., Figures 2.3, 2.4).

Like "cultivation," several other terms in widespread common usage have limited value when used in reference to the middle ground because of their multiple meanings and past applications. The related terms "gardening," "garden," and "horticulture," for example, have been used as general labels to characterize societies that are less than agricultural in terms of both their investment of human labor in land clearance, crop management, and so on, as well as in the limited contribution of domesticated crop plants to their annual caloric budgets. This simple dichotomy between small and large scales of human labor investment and reliance on crop plants (i.e., horticulture versus agriculture, garden versus field, crop diversity versus monoculture) is clearly presented in Stoltman and Baerreis's characterization of the "cultivating" societies of eastern North America that preceded the transition to maize agriculture in the region: "rather than having to depend primarily on wild stands of plants, humans prepared, planted, and cared for garden plots, which can be distinguished from fields by their smaller size and greater botanical diversity . . . it seems unlikely that yields from these gardens were substantial enough to be considered true staples . . . " (Stoltman and Baerreis 1983: 257; see also Ford 1979: 236, 1985b: 6; Crawford 1992b: 17–18).

The use of the terms "horticulture" and "garden" in this way, however, is problematic from several perspectives. In current common usage these words are limited in application to particular categories of plants, referring to the small-scale raising of ornamentals, herbs, flowers, fruits, and vegetables. At the same time, as Harris points out in his discussion of the "definitional difficulties" of "horticulture" and "gardening," they also have quite different meanings in some scholarly circles: "in some of the literature on agricultural systems and their evolution, particularly that which relates to Melanesia and the Pacific islands, the term horticulture or 'gardening' has come to be used as a synonym for agriculture . . . rather than as a means of distinguishing between 'field' and 'garden' cultivation" (Harris 1989: 19). Harris (1989: 19) goes on to point out that when used in reference to "traditional," "door-yard," or "house" gardens, the terms "garden" and "horticulture" carry additional definitional baggage—that these practices bring with them a host of adventive wild and weedy taxa. Rindos, in turn, adds to the definitional confusion by pointing out that, as originally developed, "the concept of horticulture [was meant] to describe an early stage of agriculture in which plants, notably trees, were domesticated by selective preservation of the plants" (Rindos 1984: 101). Given the multitude of overlapping and conflicting definitions and applications assigned to "horticulture" and "garden" over the years, these terms tend to confuse rather than to clarify when they are employed in attempts to characterize in general terms the middle ground between hunting and gathering and agriculture.

Like the terms cultivation, gardening, horticulture, and husbandry, the term "incipient agriculture" has been frequently affixed as a label, particularly in developmental terms, to the region between hunting and gathering and agriculture. Steward coined the phrase "era of incipient cultivation" as part of his effort to identify cultural regularities and establish a developmental sequence for five centers of world civilization. He characterized this era of incipient cultivation as follows: "It must have been very long, passing through several stages, which began when the first cultivation of plant domesticates supplemented hunting and gathering, and ended when plant and animal breeding was able to support permanent communities" (Steward 1949: 10).

In the 1950s and 1960s, MacNeish applied the term "era of incipient agriculture" to the long temporal and developmental span leading up to fully agricultural societies in Mesoamerica (MacNeish 1958, 1967, 1991; Mangelsdorf et al. 1964). As the term is employed by MacNeish and other scholars (e.g., Flannery 1986), incipient agriculture began between 10,000 and 8,000 years ago, with various regions of Mexico witnessing different species of plants (maize, beans, squashes, etc.) being locally brought under domestication. Gradually, over the millennia, these plants contributed ever-increasing percentages to annual caloric budgets until the appearance, about 3500 years ago, of village-based agricultural societies. In the half century that has passed since Steward coined the phrase and MacNeish first applied it to Mexico, landmark archaeological research (e.g., Flannery 1986; MacNeish 1967) has supported and sustained the basic elements of the era of incipient agriculture—that over a long period of time various domesticated crops contributed in a limited way to economies largely based on wild species of plants and animals.

While there is considerable empirical support for the basic outline of the era of incipient agriculture in Mesoamerica, however, there are also a number of strong reasons to question how appropriate it is to use the label "incipient agriculture," in either general or specific terms, for the middle ground between domestication and agriculture. In developmental terms, to be sure, domesticated species are a prerequisite for, a necessary first step toward, agriculture. And some societies in Mexico did transform their low-level reliance on domesticates into fully agricultural economies. So it is not unreasonable to say that agriculture, in some sense, begins with domestication. But these "incipient agricultural" societies were not truly agricultural in terms of reliance on domesticates, and extending the term agriculture so far over into the middle ground, even with a qualifier like "incipient," only serves to compress and obscure the vast and richly variable sociopolitical and economic landscape that stretches from hunting and gathering to agriculture. In some respects, it places domestication back on the boundary of agriculture. Labeling this region as incipient, as "beginning," as a developmental precursor to agriculture, casts it into the pale and partial illumination reflected from full agriculture. The modifier "incipient," I would argue, carries the imbedded implication that domestication to agriculture is a route rather than a region, consisting only of a dim developmental pathway between the steady states of hunting and gathering and agriculture.

But these vast and largely uncharted regions are not just uninhabited territory crossed on the way to an anticipated agricultural destination by evolutionary interstates without exits. Given the considerable temporal and developmental scale of such middle-ground landscapes and their great uncharted diversity, both within and between different world areas, one should not expect to always find the same standard boilerplate route across to agriculture. Rather each region and its particular pathways to agriculture need to be approached and understood within the context of their specific natural and cultural constraints and possibilities (Fritz 1990). Quite diverse developmental pathways in all likelihood existed in different regions. Some of these pathways might qualify for "incipient agriculture" designation, since they ultimately led more or less directly up to and across the boundary with agriculture. Others trace a more leisurely meandering course, and of these, a good number never approach the border zone for agriculture. Societies on these pathways never do develop agricultural economies, but rather sustain successful and appropriate solutions to local environmental settings that involve only a limited use of domesticates. Not all roads lead to agriculture. Harris alludes to this when he characterizes the phrase "incipient agriculture" as "vague and by implication deterministic" (Fritz 1989: 19).

Since all of these terms and labels—husbandry, cultivation, gardening, horticulture, and incipient agriculture are, for one reason or another, confusing and misleading when applied to the middle ground, what terminology can and should be employed? To be appropriate, such terminology should (1) have a history of usage in specific reference to this in-between territory, (2) have few if any other applications either in general usage or in scholarly contexts, (3) be applicable to human use of both plants and animals, and (4) have been used in a clear and consistent manner by a range of researchers. Although it needs some amplification, the single label that I think satisfies these requirements is the term "food production."

Food Production

As discussed by Harris (1989: 13), the term "food production" has a long and relatively consistent usage in archaeology and the study of agricultural origins. Childe coined the term more than 60 years ago as he contrasted "food producing" with the "food gatherers" of earlier times in his accounts of the transition from hunting and gathering to agriculture—"the Neolithic revolution" (Childe 1951: 61, 70–71). Childe defined the boundary between food gathering and food production in terms of humans gaining control over their own food supply (Childe 1951: 59), and while there are certainly varying kinds and degrees of such control, Childe appears, without ever being specific, to equate such control with the initial domestication of plants. The term food production is subsequently adopted by Braidwood (1952, 1960) and employed as a broad general heading in dichotomous distinction with earlier hunting and gathering societies. Binford (1968: 318) and Flannery (1968: 80, 82) also employ the term in this general way, distinguishing between societies with "food procurement" and "food production" economies, without considering exactly where and on what basis to draw the boundary between the two.

This consistent, if generally vague, broad-scale use of the term thus provided no difficulties or constraints for Ford (1985b) when he employed the term food production as an Alpha level dichotomous label to distinguish utilization of wild plants (foraging) from any and all forms of human intervention in the life cycle of plants (food production) (Figure 2.2). Although earlier scholars as far back as Childe had made some attempts at lower-level labeling under the banner of food production (e.g., Childe's "primitive nomadic garden culture cultivation"—1951: 64), Ford offered one of the first systematic and schematic efforts to further partition the broad conceptual category of food production (see also Smith 1985a). In a similar fashion, Harris subsequently developed and refined detailed organizational frameworks that filled in under the general heading of "food production" (Harris 1989, 1990, 1996a,b). But as discussed above, there are problems of epistemology, definition, noninclusion of both plants and animals, and nonstandard usage, with many of these lower-level terms used by Ford and by Harris, as well as those used by Rindos, MacNeish, and other scholars over the years.

A simple three-part framework does exist, however, which on the one hand maintains the established, general-consensus, overarching terminology of dichotomy between food procurement and food production, while also establishing a lower-level tripartite partitioning of "food production." This common sense and minimalist three-part division of "food production" employs labels that admittedly are neither elegant nor easy to acronym, but do have the advantage of being clear, unequivocal, and historically unencumbered. These labels and categories also uniformly encompass human interaction with both plants and animals. In this classification system, food-production societies having a reliance on domesticates that is below the proposed 40–60 percent annual caloric contribution isobar agriculture border zone are simply referred to as low-level food production economies. The presence or absence of observed phenotypic changes in food species, in turn, provides a rough separation between low-level food production economies with morphological domesticates and those in which human activities that qualify as domestication have not yet been shown to have resulted in observable changes in target species.

Such a simplistic either–or division of low-level food production societies into those with morphological domesticates and those with "non-morphological" domesticates (i.e., "behavioral" or "cultural" domesticates, Zvelebil 1995), of course immediately underscores the likely frequent existence, in reality, of middle ground societies that had both: they created and sustained parallel and comparable relationships of intervention with a number of target species, some of which reflected such a relationship of domestication through recognizable morphological changes, while others did not. Such a situation has long been suspected, for example, during the Middle Woodland period in eastern North America, where the seeds of some annuals (e.g., little barley [*Hordeum pusillum*], erect knotweed [*Polygonum erectum*], maygrass [*Phalaris caroliniana*]) frequently are more abundant in archaeobotanical assemblages than those of the better known "morphological" domesticates (i.e., chenopod, marshelder, squash, sunflower), and are believed to have contributed more to the diet. Despite their abundance and apparent economic importance, these possible "nonmorphological" domesticates are invariably relegated to the shadowy status of "quasi-cultigens." In drawing this distinction between groups with morphological versus nonmorphological domesticates, I do not mean to suggest that any focus on sorting middle-ground economic systems into either one or the other of these categories is a very worthwhile goal. Instead, I simply wish to underscore the existence and likely abundance of low-level food production societies based primarily or even exclusively on "behavioral" or "cultural" domesticates.

Low-Level Food Production Economies Without Morphological Domesticates

Because behavioral/cultural/nonmorphological domesticates can be so difficult to identify, particularly in the archaeological record, low-level food production economies without morphological domesticates represent the most elusive and interesting category of the middle ground. It has long presented the difficult and often avoided challenge of boundary definition between food procurement and food production, between wild and domesticated. Fortunately, both Ford (1985b) and Harris (1996b) offer clear and comparable starting points for consideration of how to distinguish between food procurement and food production, both in general terms and in reference to particular landmarks— behavior sets and particular societies that fall on either side of the border. In looking at human patterns of intervention in the life cycle of plants and animals in this border zone, Ford and Harris consider not only the relative level of energy investment. They also consider the casual or inadvertent versus deliberate intent of such actions, and the degree to which they are broadly scattered as opposed to focused and sustained, both on particular target species and on particular parcels of land. Ford, for example, draws the line between food procurement and food production in terms of the "deliberate manipulation of specific floral species by humans for domestic use or consumption . . . " through " . . . activities affecting the biological growth by means of cultural practices" (Ford 1985a: 2). He goes on to further define food production as not only involving deliberate human intervention in the life cycle of target species, but also as having a spatial focus—that such "deliberate actions were undertaken to assist the growth of a plant species in a particular location" (Ford 1985a: 3). In a similar vein, Harris identifies food production, as opposed to food procurement, in terms of a range of different forms of human intervention in the life cycle of plants: "planting, sowing, weeding, harvesting, storing, and even the drainage and/or irrigation of undomesticated crops" (Harris 1996b: 446), and ties these "cultivation" activities to particular cleared plots of land. Ford, too, identifies and discusses a number of categories of human activity which, when deliberate and sustained, can serve to distinguish low-level food production economies from the food procurement economies of hunting and gathering societies. All of these actions, in various ways, involve the beneficial disruption of the life cycle of a plant population in order to ensure easier, more reliable, and more abundant harvests. Ford (1985a: 3–6) arranges these diagnostic activities of low-level food production into four categories: tending, tilling, transplanting, and sowing.

Tending, defined by Ford as the encouragement of plant growth both by direct care of target species and by limiting competition, primarily focuses on weeding—the removal by hand of competing vegetation near useful plants. Tilling, in turn, is defined by Ford as deliberate soil disturbance with a digging stick or hoe to facilitate and encourage the appearance or germination of target species. Tillage could involve deliberate efforts to expand the size of stands of seed plants through soil disturbance around such stands in advance of natural seed dispersal, or the churning of soil and detachment of bulblets and lateral tubers during the harvesting of roots and bulb-bearing species. Transplanting is defined as the movement of a plant, usually perennial herbs, shrubs, or trees, from one locality to replant in another for easier access or other purposes. Transplanting can be quite casual and spatially or temporally scattered, or can involve considerable long-term protection and care, perhaps in designated spaces where a variety of plants from different habitats are brought together (Ford 1985b: 4–5). Sowing too can range from the casual broadcasting of mature seed at the time of harvest to more extended degrees of life-cycle intervention where sowing is done in new locations or even new habitats, perhaps in combination with soil tillage for seedbed preparation and seed storage.

Complementing these considerations of human–plant interactions that are characteristics of low-level food production economies in the absence of domesticates, Harris (1996b: 447–456) and Hole (1996: 264, 276) identify and discuss those forms of human intervention in the life cycle of animals that focus on protection and enhancement to ensure continued or increased availability, and which could occur on the landscape between hunting-gathering and domestication. These include efforts to reduce predator populations and to increase or enhance pasturage, as well as the raising of tame animals and various types of free-range management systems.

It is important to keep in mind, however, that of these general categories of behavior identified by Ford, Harris, and Hole as being representative of lower-level food-production economies, none are restricted to only one side of the border. All should be considered as having casual and scattered tails (in terms of the low-occurrence end of a probability curve) that trail across the border and attenuate in the realm of hunter-gatherer societies. The capturing, taming, and raising of young wild animals, for example, or the casual broadcast sowing of seed or other propagules at the time of harvest, are not the exclusive behavior of food-producing societies, but rather have also been observed in hunter-gatherer contexts. Some of these cross-border behaviors, in fact, such as inadvertent stand-enhancement as a result of the digging of tubers and roots, and the broad-scale burning of vegetation cover to enhance habitat for favored animal and plant species, are encountered frequently on both sides of the border, and thus are considered as characteristic of both food procurement and low-level food-production economies (Harris 1996b). As a result, any effort to determine on which side of the food procurement–food production boundary to place a society involves more than a matter of simply ascertaining the presence or absence of certain forms of life-cycle intervention activities on the part of humans, but rather should include consideration of the intensity, intentionality, species focus, and total range of such activities that are present in a group's economic repertoire. The Kumeyaay Indians of southern California, for example, provide a good test case, and may qualify as having a low-level food-production economy without morphological domesticates, judging from ethnohistoric analyses and interpretations (Shipek 1989). They not only burned over extensive areas to both improve forage for deer, and to remove competing species of plants prior to broadcast sowing of a wild grain grass, but they also had an extensive and far-reaching program of transplantation and tending of a select yet broad assemblage of wild plant species (e.g., oaks, pines, palms, mesquite, agave, yucca, wild grapes, cacti, etc.). Other scholars in recent years have documented similar efforts by human societies to reshape or "domesticate" their environments (to load the term in yet another way) in order to increase usable plant products. In contrast to the Kumeyaay research, however, these studies focus either on hunter-gatherer societies having limited and attenuated sets of life-cycle intervention activities (e.g., Australian aborigines—Yen 1989), or on clearly agricultural societies who also carry out activities that could be considered characteristic of low-level food producing societies without morphological domesticates (e.g., Moran 1993, 1996; Posey 1985). Moran, for example (1996: 538–41), documents the considerable extent to which Amazonian agriculturalists, in the wake of slash-and-burn agriculture, have left behind fallow-cycle vegetation communities that are substantially enriched in forest species of economic value in comparison to their composition prior to clearing.

When such activity-level case studies of life-cycle intervention, potentially diagnostic of low-level food producers without morphological domesticates, are drawn from agricultural societies, it raises the obvious question of whether or not such activities developed as supplemental additions to already established agricultural economies, or if they were preexisting stand-alone practices that survived the agricultural transition. Are, for example, the rock mulching and habitat extension of agave by pre-Hispanic agricultural societies of southern Arizona documented by Fish (1995) a supplemental subsistence extension developed by the region's maize-centered agricultural societies, or a surviving practice of earlier low-level food-producing societies? Castetter and Bell (1951: 177–78) raise this issue in concluding their description of one of the most frequently cited examples of food production involving non-morphologically domesticated plants—the "semicultivation" of several seed-bearing species by Cocopa societies of the lower Colorado (Alvarez de Williams 1983; Castetter and Bell 1951; Kelly 1977). This "semicultivation" of wild grasses by the Cocopa, which ceased with the twentieth-century construction of upstream dams, involved the broadcast sowing, in *decrue* fashion (Smith 1995b:

113; see also Smith 2002: 249–66) of seeds harvested the previous fall, on thin, muddy, nutrient-rich riverbank soils exposed by the receding floodwaters of the Colorado. Plots planted in this manner, which could be 50–100 meters wide and extend up to a mile along the river, and received no further attention prior to harvest, could include any of five or more different identified species. Of these, three were historic-period introductions of Eurasian origin, while two species of panic grass (*Panicum*) were indigenous, and known to have been grown at least as far back as 1541, leading Castetter and Bell to raise the very interesting and tantalizing question: "Did the semicultivation of grasses precede or follow the introduction of maize-tepary [bean]-pumpkin cultivation?" (Castetter and Bell 1951: 177–78).

These various examples of human intervention in the life cycle of non-morphologically domesticated species, often of necessity drawn from agricultural societies, help to underscore several important points. First, it is abundantly clear that the low-level food-production societies without morphological domesticates remain very poorly described. Second, even given the extremely limited information that is available regarding these societies, it is enough to suggest that the boundary between food procurement and food production is not abrupt but clinal, and that it is crowded with a heterogenous assemblage of societies that created food-production relationships along a number of different developmental pathways representing stable if changing economic-subsistence solutions to different cultural and natural environmental settings. Some of these pathways will lead to morphological domestication of target species; others will not. Third, given our low level of current knowledge of this region of the middle ground, rather than focusing too closely on exactly where the boundary lies and on which side to place particular societies, effort instead should be focused on the search for and analysis of additional archaeological and ethnohistorically described low-level food-production societies without morphological domesticates. Fourth and finally, this brief consideration of low-level food-production without morphological domesticates underscores how important it is to attempt to build a larger reference class of full-society-scale case study examples (as opposed to activity level examples) of low-level food production based on behavioral or cultural domesticates.

The importance of building a richer reference class of low-level food-production societies having behavioral domesticates for the purpose of comparative analysis can be underscored by considering one final example: the Owens Valley Paiute of eastern California. Interestingly, even given the considerable environmental and cultural differences, the Owens Valley Paiute share a number of similarities with the Northwest Coast societies that are the focus of this volume.

Based on Steward's informant interviews of the 1920s and 1930s, the Owens Valley Paiute, prior to the massive disruption of their way of life in 1863, were similar in many respects to neighboring Great Basin Shoshoni hunter-gatherer groups (Bettinger 1977; Liljeblad and Fowler 1986; Steward 1930, 1933, 1938). Their economy, for example, was largely structured around the procurement of a wide range of wild species of plants and animals, with pine nuts and rabbits playing a prominent role. The Owens Valley Paiute, however:

> were distinctive for their band ownership of hunting and seed territories . . .
> the population was comparatively dense, stable, and settled in unusually
> permanent villages. The country was fertile, so that subsistence activities
> could be carried on according to a comparatively fixed routine within a small
> territory. Each territory was only large enough to embrace all the natural
> resources habitually exploited and included both game and vegetal foods
> [Steward 1938: 255–56].

Among the environmental features of the Owens Valley that set it apart from other areas of the Great Basin, and which allowed the development of such relatively stable and fixed camping sites and seasonal rounds, were the extensive tracts of swampy, low-lying meadows near the Owens River. The vegetation of these waterlogged river-valley resource zones included a number of bulbous hydrophytic food plant species, including nutgrass (*Cyperus*) and spikerush (*Eleocharis*), which had long been an important component of Owens Valley subsistence economies (Bettinger 1977). Major permanent camping sites were often located near these floodplain meadows. Owens Valley groups both enriched and expanded these water meadow resource zones through irrigation, thus increasing the size and the reliability of each year's wild root crop harvest. This enhancement and expansion of the natural habitat of the water-meadow root crops was carried out on a large scale in several locations. The construction and subsequent removal of temporary diversion dams each year along some of the tributary creeks of the Owens River called for the labor of all the men of the communities, under the direction of a communal irrigator. Feeder ditches up to four miles long carried nutrient-rich early summer mountain runoff from dams to the river valley plots, the largest of which were two to four square miles in size (Steward 1933: 247). No efforts were made at planting, tilling, or tending either the wild root crops of these water meadows, or the adjacent downstream stands of wild seed plants (including sunflower [*Helianthus* sp.], chenopod [*Chenopodium* sp.], and lovegrass [*Eragrostis oxylepis*]), which also benefited from the irrigation efforts.

Even a brief comparison of the Owens Valley Paiute and Northwest Coast societies illuminates a number of intriguing parallels in terms of their respective patterns of low-level food production involving nonmorphologically domesticated crop plants. In both the Owens Valley and along the Northwest Coast, for example, food production includes perennial species—hydrophytic tuberous/root crops (see Darby, Deur, this volume). In addition, rather than being monocrop in nature, up to perhaps a half dozen food plant species are involved. And in both areas these targeted suites of food-crop species include components of natural wet-soil communities, with food-production efforts directed toward the deliberate and sustained enrichment and expansion of the habitat zones of these plant communities. In both situations, too, such human labor investments in habitat expansion and improvement, resulting in increased harvest yields and reliability, involve getting nutrient-rich water to the target species. There is also in both regions an investment of labor, sometimes communal, sometimes family, on particular, often demarcated, parcels of land associated with food production. And, not surprisingly, there are associated parallel shifts in, and strengthening of, concepts of community and individual ownership of such land parcels and the yearly harvests that they yield. In this regard, Steward (1938: 106), in his discussion of seed sowing by Great Basin Shoshoni groups, states that: "ownership of sowed plots accords with the Shoshoni principle that there are property rights only in things in which work has been done." Perhaps such changing concepts of ownership may turn out to be a good boundary marker of the transition zone between food procurement and low-level food production.

The Owens Valley Paiute are also quite relevant to this volume's reconsideration of Northwest Coast societies in a more basic way. They highlight the considerable problems such low-level food-production societies still present in terms of categorization and characterization, and the almost complete lack of an appropriate reference class. Liljeblad and Fowler (1986: 418), for example, when faced with an either–or situation, place the Owens Valley Paiute on the hunter-gatherer food procurement side of the border, and characterize their low-level food production focused on nondomesticates as " part of a perfected gathering complex, a spontaneous extension and prolongation of an observable natural process . . . well within the scope of the hunter-gatherer adaptations and utterly unlike any other horticultural development in the region . . . " Steward faced a similar dilemma in developmental terms, and resolved it by placing the Owens Valley Paiute in the brief interval of transition between the two steady states of hunting-gathering and agriculture, caught "on the verge of horticulture" (Steward 1933: 248, 250).

This volume should go a long way toward freeing scholars from such difficult conceptual dilemmas. Northwest Coast societies, and in particular their management of root crops, provide clear and detailed full-society case study examples of low-level food production involving behavioral/cultural/nonmorphological domesticates. The studies here represent a major conceptual atlas of the middle ground, and provide an essential guidebook for anyone interested in considering the rich and long history of societies that are neither hunter-gatherers nor agriculturalists.

Note

1. These crops include erect knotweed (*Polygonum erectum*), maygrass (*Phalaris caroliniana*), little barley (*Hordeum pusillum*), sunflower (*Helianthus annuus*), squash (*Cucurbita pepo*), chenopod (*Chenopodium berlandieri*), and marshelder (*Iva annua*)—the last four listed being domesticated.

Chapter 3

Intensification of Food Production
on the Northwest Coast and Elsewhere

KENNETH M. AMES

The techniques of plant food production described in the chapters that fol-low are very likely the result of an evolutionary process known as intensification, or, in other words, producing more food. The causes and pos-sible effects of increasing food production are central research question to many disciplines, Northwest Coast anthropology and archaeology among them. This chapter examines intensification on the Northwest Coast and among com-plex hunter-gatherers. I first very briefly outline several issues in the archae-ology of the Northwest Coast and of complex hunter-gatherers to which intensification of food production is directly relevant. In the sections that fol-low, I place the concept of intensification in a broader perspective, both in terms of theory and application. From this general consideration of intensi-fication, I discuss the intensification of plant food production on the Northwest Coast using models either developed by evolutionary ecologists or based on their work. One set of models is quite general in its application, and a second set of two models specifically focuses on the intensification of root harvest-ing. The archaeological record for intensification is then examined against the predictions of the general models. Following that discussion, I review the evi-dence for the intensification of root food production on the Intermontane Plateau of south central British Columbia and Interior Washington and Idaho. I also review the evidence for plant use during the Locarno Beach and Marpole phases of the Gulf of Georgia region of the southern Northwest Coast.

Issues in Northwest Coast Social and Economic History

It has only been within the last thirty years that the subsistence economies and ecology of Northwest Coast peoples have been of central concern to anthropologists and archaeologists working on the coast. Prior to that time, the coast was assumed to be an exceptionally rich and productive place, one that permitted, but did not cause, the development of Northwest Coast societies. Wayne Suttles (e.g., 1962, 1968) and others (Vayda 1961) placed the interplay among the coast's dynamic environments and coastal peoples' complex social organization at the center of anthropological inquiry in the early 1960s. Knut Fladmark (1975) put salmon production at the center of archaeological investigations of the evolution of Northwest Coast societies in 1975.

Nineteenth-century Northwest Coast societies are world-famous for, among other things, their extraordinary art. Other attributes of these societies include complex forms of sedentism (people living in one place year-round); large towns and villages; some degree of occupational specialization among at least some groups; and ranking of their members. These are all traits that anthropologists and other social theorists long assumed were associated exclusively with agriculture; that, indeed, in order to develop, they require the high levels of food production that only agriculture can produce. These traits can be placed together under the term "social complexity." Since Northwest Coast peoples were not farmers, as classically defined, then how did they evolve social complexity? The question that Suttles tried to answer remains: What are the relationships between subsistence production, ecology, and social complexity on the Northwest Coast (and, by extension, among other, so-called "complex hunter-gatherers" in the world)?

The issues arising from these questions have been extensively reviewed elsewhere, and the reader is referred to Ames (1994) for a thumbnail sketch of them. However, to introduce both this chapter and those that follow, I will quite briefly outline some of the research questions immediately relevant here (for more complete discussions, and bibliographies, see Ames [1994], Ames and Maschner [1999], and Matson and Coupland [1995]).

SOCIAL COMPLEXITY: INTENSIFICATION AND FOOD STORAGE

For most of the past century, researchers around the world have assumed a direct link between the evolution of complex societies (such as those on the Northwest Coast, as well as modern society, the Roman Empire, etc.) and increased levels of food production in the form of farming. It was long held that hunter-gatherers simply could not produce enough food to create the surpluses required to enable the development of social complexity. In a significant intellectual shift, it is now recognized that intensification of food production can lead to social complexity, regardless of economy.

All the debates about intensification on the Northwest Coast, until this volume, focus on the role and relative importance of salmon. Many researchers regard increasing salmon production as the crucial economic change driving almost all other social and economic changes in the long history of the coast's peoples. Others, myself included, regard salmon as only one among many resources whose intensification has been important in causing, or affecting, social and economic change. In other words, we argue that salmon intensification is not the key economic change on the coast. I have termed this the debate over "secondary resources" (Ames 1994). This continuing debate, to which this book is a contribution, is essentially over our basic understanding of Northwest Coast economies.

Northwest Coast subsistence economies of the nineteenth century were heavily dependent on food storage, and the development of this heavy dependence was also probably central to many other social and economic changes on the coast. As with salmon production, key questions are when did Northwest Coast societies become heavily dependent on food storage, and why.

Food storage must be understood to mean: (1) the acquisition of the food to be stored; (2) the methods and tools used to process it for storage, and then (3) the methods and facilities used to store it. It also involves the social and economic relationships that storage creates: Who controls the stores, who looks after them, who has rights to them? It is also important (but far beyond the scope of this chapter) to distinguish between food storage to sustain a household through the winter, and the production of a surplus (Ames 1985; Miller 1997; and Sahlins 1972). Food storage also requires storable foods.

SEDENTISM

A useful, simple definition of sedentism is that the members of a community live in a single settlement or place for over one year, or they may occupy it regularly (i.e., seasonally) over several years or generations (see Ames 1991a). Sedentism, like social complexity, has long been seen as a major development in human history worldwide, making its presence on the Northwest Coast of considerable interest to archaeologists. It, too, has long been regarded as a cause of social complexity. Not all sedentary peoples are socially complex, but most socially complex peoples are sedentary to one degree or another. Successful sedentism on the coast would require food storage, since most food resources are highly seasonal and very little is available for harvest during the winter and early spring.

SUBSISTENCE HISTORY

Archaeologists working on the Northwest Coast have long debated the relative roles of aquatic versus terrestrial resources in the evolution of Northwest Coast subsistence economies. They have focused exclusively on animal resources, dismissing plant resources as being of little consequence (see Lepofsky in press for a full discussion of these issues). As will be shown in the chapters that follow, this has been a significant error.

Over the past 10,000 years of known Northwest Coast prehistory, the peoples of the coast increased their production of food resources. What we now wish to know is what resources, how they did it, and why. We also wish to know: What effects did those changes in production have on how their societies were organized, and vice versa? And we wish to know, What effect did the coast's ecology have on the way their subsistence economies changed (and vice versa: How did changing subsistence economies affect the coast's ecology)? We have some answers to these questions, but much remains to be learned.

Answers to these questions are of concern not only to archaeologists and anthropologists working on the coast, and to the people living on the coast, but to the development of our understanding of the evolution of complexity and of hunting and gathering worldwide. The coast's peoples are the world's best-known examples of complex hunter-gatherers, and the more we understand of their history, the better we are able to use that understanding to elucidate similar developments elsewhere in the world over the past 30,000 years or so.

Intensification Among Complex Hunter-Gatherers: Issues and Problems

Despite an apparent theoretical focus on intensification of food production among hunter-gatherers over the past two decades, we still know little about it. This chapter, in part, is an attempt to engage potentially significant questions of production and intensification in the archaeology of complex hunter-gatherer societies. Binford has recently defined intensification as "any tactical or strategic practices that increase the production of food per unit area. Production can be increased by investing more labor in food procurement activities or by shifting exploitation to species occurring in greater concentration in space" (Binford 2001: 188). His definition is close to how intensification is generally understood by archaeologists. As useful (and precise) as it is, it has some conceptual limits. A more general definition would be that intensification is the processes by which one or more elements of production (e.g., labor, land, technology, skill, knowledge, organization) are increased relative to other elements in order to maintain or increase food production (or the production of some other commodity). This definition, unlike Binford's, recognizes intensification through increased efficiency. It also stresses that intensification is always relative to some measure (see below).

INTENSIFICATION AND AGRICULTURE

Conceptually, researchers have tended to blur and link the concepts of hunter-gatherer intensification and the origins of agriculture. One basic reason for this is that over the past half-century most of the research producing useful insights into increased food production by foragers has been research on the origins of farming, not on the study of highly productive hunter-gatherer economies as phenomena in their own right—phenomena that may or may not lead to full-time agriculture.

There is a yet more fundamental underlying cause of conceptual problems. Hunter-gatherers and farmers are viewed as the polar extremes of a single continuum (see Smith, this volume, for a fuller but different treatment of this issue). At some point along this continuum, hunter-gatherers pass over a threshold and are farmers. We sometimes recognize gradations in the continuum: we may speak of gardening, horticulture, a domesticated environment, and then finally agriculture. This continuum has certain qualities that limit our ability to see our subject matter. It works in only in one direction, from not-a-farmer to fully-a-farmer. It does not go the other way, from not-a-hunter-gatherer to fully-a-hunter-gatherer, for example. We discuss proto-farming, but not proto-foraging. This is reminiscent of the linguistic continuum in the English language between wet and dry. I can be wet, sodden, soggy, soaked, damp, moist, or dry. I can say "I'm a bit damp," and the sentence does not ring odd to English-speaking ears. But the continuum does not go the other way. How do I contrast in English between exceedingly dry and only slightly dry? I can say, "It is very dry," as opposed to dry, but if I say, "I am a bit dry," the sentence is odd, even though grammatical, and does not really make sense. However, the continuum between wet and dry actually does go both ways. The unidirectionality of the forager-farmer continuum of course reflects deep-seated notions of progress, in which farming succeeds hunting and gathering as a natural corollary of complexity inevitably arising from simplicity.

Beyond that, this conceptualization leads to a focus on the boundary (Smith, this volume) rather than on understanding variation in subsistence patterns and what that variation means in the evolution of economies. Rather than looking at that variation, we focus on locating where along the continuum a particular ancient economy lies. This has long been a crucial first step in interpreting ancient cultures. If they were hunter-gatherers, we expect one kind of economy, population size, and social and political organization. If they were farmers, we expect something very different. While these expectations hold generally (at the broad level of agricultural empires versus small, mobile hunter-gatherer bands), they often collapse at the level of particular societies, such as those of the Northwest Coast.

As a result of these expectations, the research issues become: What constitutes the threshold, and where along the continuum do we put it? How many corncobs do we need to declare agriculture? How many rice grains in pots equals farming? Does any evidence for cultigens, no matter how sparse, mean that these people were not hunter-gatherers? The debate over the economy of the Jomon tradition of Japan offers an excellent example of the difficulties. There is evidence for cultigens, but it is usually quite sparse before about 3000 years B.P. Crawford (1992a, 1992b) suggests that gardening played a minor role in Jomon economies in some places in Japan. Barnes (1993) treats them as complex hunter-gatherers, with some domesticates; Imamura (1996) dismisses what he sees as very thin evidence for Jomon farming, while Smith (this volume) classes them as low-level food producers with domesticates. Our understanding of Jomon is in part dependent on how we classify it, where we place it on the continuum. In the meantime, we have learned very little about the actual dynamics of Jomon subsistence and have added little to our knowledge of the variability in human subsistence arrangements.

There are at least three solutions to this general issue. We could apply Yoffee's rule for establishing whether a particular ancient polity is a state or not: If we have to debate it, it is not a state (Yoffee 1993). The revision here would be that if we have to debate whether an ancient group were farmers, they were not. This, however, merely puts the threshold far over at the farmer end of the continuum, and requires quite clear-cut evidence for agriculture. On the other hand, it places a whole range of subsistence economies into a single category—not-farmers—and continues to obscure important variability elsewhere along the continuum.

The second approach is to reconceive the continuum itself. It should minimally be seen as a field, with at least three points: pure farmers, pure foragers, and something else that is neither. It might be even more useful to see the evolution of subsistence economies as a bush with many branches. We actually know little about these economies, which Bruce Smith calls low-level food producing societies in the previous chapter. I suspect that in terms of production they may have more in common with some forms of farming, like swiddening, than they do with generalized foraging; but they may be distinctive economies quite unlike either foraging or farming, as ethnographically known, in their fundamentals, despite superficial similarities.

The third solution (these last two are not mutually exclusive) is to focus more directly on production itself, as difficult as this can be for archaeology. We often know a great deal about food production but very little about how food production was organized into subsistence systems, with all that that implies. We do not know because our focus has been on these particular societies as transitional to agriculture, or on food production as simply the means by which enough food is produced to permit, or allow for, the evolution of social inequality. In either option two or three, we should abandon (*pace* Smith) the whole notion of a continuum, and think in terms of "bushy variation."

INTENSIFICATION AND HUNTER-GATHERER SOCIAL COMPLEXITY

There have been divergent views on the relationships between intensification of food production and the development of social complexity, and even about what "complexity" means. Arnold (1996) has recently restricted the definition of complex hunter-gatherers to those displaying permanent forms of inequality. Other scholars (e.g., Hayden 1996) define the class more broadly, as minimally including higher population levels and/or densities than are common among generalized hunter-gatherers, and a productive subsistence economy. Maschner and I (Ames and Maschner 1999) propose a longer list, which includes, in addition to the foregoing, some degree of sedentism, production of large amounts of processed and stored foods, a broad diet breadth coupled with an emphasis on a few resources, a complex material culture, and, in addition to social inequality, some degree of social differentiation, which is probably accompanied by some occupational specialization. A final, crucial attribute is participation in large-scale social and economic interaction networks. However, at base, a productive subsistence economy is required to sustain most of these attributes. Recent theory building recognizes this necessity to some extent, but decouples production from the development of social inequality. The recent work of Hayden and Arnold can serve as examples.

Hayden (1996) stresses that the environment, or the economic base, must be sufficiently productive and stable so that no one in the society is threatened by starvation. He makes no strong connection between production (how food is acquired from this rich environment) and social complexity/inequality. Arnold (e.g., 2001) argues that social inequality arose on the Channel Islands of southern California in part as a response to environmental stress in the local ecosystem, and through efforts to gain access to food production on the California mainland through the production and exchange of craft items, such as shell beads. Thus, while production is a factor in her thinking, it is not food production. However, food production and its intensification are implicitly central, although unaddressed, issues. In Arnold's model, for example, people on the mainland either had to intensify their own food production to meet the needs of the Channel Islanders, or their subsistence economies produced a sufficient surplus to meet those needs without intensifying. Either way, food production is crucial.

Until rather recently, intensification of food production was treated as a central process, along with population growth and social and economic exchange, among others, in the development of complex hunter-gatherers (e.g., papers in Price and Brown 1985). It is now clear that while relatively high levels of food production are enabling conditions for the development of social and economic complexity among hunter-gatherers, they do not lead ineluctably to it. At this level, I am in agreement with Hayden, Arnold, Maschner (e.g., 1992), and many others. With this recognition, researchers searching for the causes of the evolution of social complexity shifted their attention away from food production. Arnold (1996) has emphasized how emerging elites reinforce and strengthen their positions by gaining control of labor and of the production of crucial craft items. Hayden (1996) shows how ambitious individuals can build up relationships of inequality through debt, social obligations, exchanges of prestige goods, and so on. With this new focus on the development of inequality among complex hunter-gathers, there has been correspondingly little direct attention paid to subsistence and economic production among these people.

Consequently, there has been declining interest in production and intensification as issues and as significant concepts in hunter-gatherer archaeology over the past fifteen years. Bender (1978) made intensification an important idea for archaeologists, but its theoretical use-life seems to have been quite short. To my knowledge, the last extensive discussions of hunter-gatherer (as opposed to agricultural) intensification were in Price and Brown's volume on complex hunter-gatherers (Price and Brown 1985). As Gould observes in his summary commentary on the book, "(t)he idea of intensification pervades this collection . . . " (Gould 1985: 429). The word "intensification" appears 22 times in the index, for example. Intensification of production was seen as a significant process in the evolution of social complexity among hunter-gatherers, leading, among other directions, to permanent social inequality.

In contrast, the term "intensification" does not appear at all in the index of a more recent volume co-edited by Price (Price and Feinman 1996) on the development of inequality. There are only seven entries for it in a recently published edited volume on hunter-gatherers (Burch and Ellanna 1994), and only one paper in that volume that discusses intensification to any extent (Yesner 1994). The term does not appear in the index to Kelly's encyclopedic book on hunter-gatherers, though it does appear here and there in the text (Kelly 1995). A recent key-word search using the CARL Uncover electronic periodicals database found the words "intensification" and "intensifications" used in 415 journal titles over the last decade or so. Of these, 5 dealt with archaeological topics, while 20 treated current issues in agricultural intensification and were published in journals both anthropologists or archaeologists might read (e.g., *Current Anthropology, Human Ecology*). The other 390 titles were papers in chemistry journals. Arnold (1996) does not discuss intensification, except in passing, in her review article on complex hunter-gatherers, and she does not treat it as a significant issue. It does not figure at all in Hayden's recent theory-building on the causes of complexity among hunter-gatherers. Most recently, however, Binford has given it a central explanatory role in accounting for global variation in hunter-gatherer economies (Binford 2001).

Despite (or perhaps in response to) this lack of interest, problems in complex hunter-gatherer production remain unresolved. How did they finance complexity? In other words, How were resources extracted from the environment and converted to wealth? How did they support the large populations that are one of the criteria for being classified as complex hunter-gatherers? How did they produce what they produced and why? How was increased production—intensification—accomplished? And what does "increased production" actually mean?

Intensification and Production

This section discusses the concept of intensification, tying it to the more basic concept of production. As part of this discussion, it also reviews other definitions of intensification as necessary. It makes the argument that the essential archaeological indicators of intensification are technological, economic, and social. Most recent work on intensification, at least on the Northwest Coast, focuses on faunal remains, on the debris produced by the harvesting and processing of fish and mammals. This is essential evidence, but it must be viewed within the context of a system of production, and tied to technological, social, and economic evidence for intensification. In contrast to the emphasis on food remains, Zvelebil (1986b) insists that intensification can only be measured through evidence of increased labor. He argues that shifting or changing frequencies of faunal (and floral) remains may reflect adjustments to resource availability, but not to intensification as an economic strategy. Long-term environmental changes may cause changes in subsistence economies as people adjust to changes in available resources. This is not intensification in Zvelebil's thinking, unless there is evidence for increasing labor effort. He does not provide any means of measuring increased labor, however. We will discuss how that can be done below.

Archaeologists often use the term "intensification" to mean increasing the amount of something being produced. To intensify salmon production is to increase the amount of salmon being harvested (and/or processed for storage); to intensify pottery production is to increase the amount of pottery being made. However, the term carries broader implications. One of these is the issue of measurement: increasing the amount relative to what? The answer often given is "relative to an earlier period." An alternative answer is that some resource, salmon for example, so dominates the faunal assemblages that the resource must have been being intensified. In theory, however, intensification needs to be measured against some standard more precise than simply saying that there is more of something than there used to be, or that there is a whole lot of it.

Jochim (1976) suggests time, space, or labor as these standards for measuring intensification. The amount of pottery made increases per unit time, per unit space, or per unit labor. Zvelebil (1986b) strongly asserts that intensification is an increase in the amount of labor invested in production. In his thinking, the only appropriate measures of intensification are those that track increases in labor (amount or effort). Using Jochim's measures, evidence for salmon intensification might be more salmon caught per fishing season, more salmon caught per linear mile of stream bank, or more salmon caught per person. For Zvelebil, evidence of salmon intensification would be evidence of increasing effort or work to produce more salmon. For archaeologists, of course, these measures can be very difficult to operationalize. We will also return to that issue below.

Boserup (1965), whose work sparked current approaches to the intensification of food production, defined it in terms of units of land. In preindustrial agricultural economies, she thought farmers would intensify food production by increasing the area farmed, and by increasing the number of crops harvested annually from a given piece of land (Boserup 1965: 43). She saw, as a major implication of her definition, that while both of these approaches would probably (*but not necessarily*) result in higher workloads, they would also cause lower overall efficiency. People would work harder, producing more food overall, but less food per hour—agricultural production, then, would increase more slowly than would work effort. Broughton (1997: 646) put this succinctly, "*resource intensification* [is] classically defined as a process by which the total productivity or yield per areal unit of land is increased at the expense of declines in overall caloric return rates or *foraging efficiency* [emphasis his]." Thus there are two aspects to the classic definition of intensification: increased production and (following Boserup) lower overall efficiency. This clearly is implicit in Zvelebil's definition.

Implicit to Jochim's measures, in contrast, is the possibility of intensification through increasing productivity. Bender (1978) first drew the distinction between increasing production and increasing productivity in the archaeological literature. Increased productivity is simply increasing production by increasing efficiency. In her thinking, it might or might not lead to increased production, and people could invest the time and labor saved in social activities, such as participation in exchange networks. Gould (1985) sees the distinction as processually insignificant, arguing that increases in efficiency will almost inevitably lead to increased production in the long run, regardless of the original reasons for increasing efficiency. The point here is to stress that in contrast to the "classic" definition of intensification, production can be intensified both by increasing labor and by increasing efficiency.

There are other ways in which increasing production is a broader concept than just "making more." Increasing salmon production is not the same thing as just harvesting more salmon. According to Halperin (1994: 41), "All production processes require assembling and allocating resources." Production includes the raw materials used, the tools, techniques, technological processes, work effort, and labor organization required to extract other raw materials from the environment and turn them into something humans use (Ellen 1982, Elster 1986). It also includes the allocation and distribution through the society of both raw materials and the end products of production.

Production also creates value. Things produced by humans have value within the social and cultural context in which they are produced. Their production gives them their value (or enhances the value of the raw material) and the value of that class of object affects production (Ellen 1982). An object's value may reflect its rarity, or the effort or skill required in its production or acquisition, and so on. Nonindustrial diamonds are extremely valuable in part because they are relatively rare, but also because of the incredible skill and knowledge required to cut them to make them suitable for jewelry. They are thus markers of wealth and status, which in turn fuels their value and their continued production. The values an object has in a culture need not be restricted to any single sphere of life. A food resource may be revered for its links to a spirit world, convey social prestige, and be eaten with gusto in large amounts. A Mercedes-Benz station wagon can still convey a message of wealth while it hauls babies, groceries, and fertilizer.

Production can also be conceptualized as a technological process, as a chain of steps or tasks. In the case of pottery, for example, the chain includes a range of activities from the acquisition of the clay at the clay pits to the use of the pot. Thinking of it in this manner is important, since initial increases in levels of production are likely to involve overcoming bottlenecks along the production chain. Increasing the salmon harvest for storage, for example, would be pointless if the fish rot before they can be transported. Solutions to bottleneck problems can be technological (reorganize the old technology, invent new tools), or organizational. For example, Donald (1983, 1997) argues that the sexual division of labor on the Northwest Coast was a potential bottleneck in the processing of foods for storage (men harvested foods, women processed them), one that was resolved by slavery. At some points in the chain, production may be increased by a straightforward increase in production (working longer, harder, or adding more workers), while at others production may be increased through an increase in productivity. While technological innovations can increase productivity (the potter's wheel vs. coiling), specialization can also increase productivity (specialized potters are likely to be more adept at making coiled pots than are nonspecialists).

How tasks themselves are accomplished can also be changed. Wilk and Rathje (1982) draw a useful distinction between linear tasks and simultaneous tasks. Linear tasks are those in which the steps of the tasks can be accomplished in a sequence, while simultaneous tasks are those in which the steps are done by a group at the same time. They further distinguish between simple simultaneous tasks and complex ones. In simple simultaneous tasks, everyone performs the same step in the task at the same time. In contrast, complex simultaneous tasks require that different steps in the process be done at the same time "all work at the same time, but do different parts of the job" (Wilk and Rathje 1982: 622). Intensification can proceed if linear tasks are performed as simultaneous tasks. This increases the amount of labor devoted to the task, and can increase productivity. Instead of one person coiling pots, have ten do it. A simple simultaneous task can be a complex one, with specialists. No change in technology may be required.

Intensification, then, is more than more food, and has wider implications. In a sense, though, the issues can be boiled down to two questions: "How is more food produced?" and "What happens to it then?" A third question, of course, is "Why?"

Measuring and Modeling Intensification

The definition of intensification outlined above allows two broad lines of evidence about intensification: evidence of increased labor input (either more labor, or greater labor efficiency, either of which may involve changes in technology and/or changes in social organization), or evidence of changes in what is produced. In this second instance, demonstrating intensification of food production often becomes a problem in zooarchaeology and paleobotany. There is also the issue, noted above, of distinguishing between intensification and subsistence adjustments to short- and medium-term environmental changes. (Did they take more herring because they needed more food, or because of declines in the availability of sea mammals, or increased numbers of herring?)

One can also look for artifacts that may indicate increasing emphasis on a particular class of resource, or greater efficiencies at some crucial task (e.g.,

Mitchell's [1971] suggestion that a certain kind of spaul tool might represent a fish knife, speeding up the filleting of fish prior to drying them), or the capacity to harvest more of a resource. Moss et al. (1990) dated fish weirs in Southeast Alaska, and found that they dated to ca. 3500 B.P. and later. Moss and others (e.g., Ames 1994) suggested that the presence of these weirs indicated the presence of the technological capacity to catch large amounts of fish by that time, and therefore served as evidence of the intensification of fish production. Eldridge and Acheson (1992) replied, arguing they had evidence of an older fish weir along the lower Fraser River, so that intensification of fish production may have been earlier on the Fraser than in Southeast Alaska. This raises issues in the history of Northwest Coast technology, which is rather poorly understood. Croes (e.g., 1991, 1995; Croes and Hackenberger 1988) makes what are among the few general statements explicitly linking changes in technology with economic changes along the coast. Often, local technological changes are treated as the result of shifting ethnic boundaries (e.g., Mitchell 1988). We will return to these issues in more detail below.

All of these approaches are employed in a rather ad hoc fashion. With a single, notable exception (Croes and Hackenberger 1988), there has been no effort to develop a comprehensive model of what intensification of production on the coast might have looked like. Rather, there has been an implicit and powerful assumption that intensification of food production will be indicated by ancient subsistence economies becoming increasingly like those of the Early Modern period (ca. from about 1775 to 1875; see Ames and Maschner [1999], Suttles [1990]). However, sufficient evidence has accumulated to indicate that this assumption is not tenable. It is clear that Early Modern economies may have differed in important ways from those of the centuries just before contact (e.g., Acheson 1991; Hanson 1991). It is equally clear, as the chapters in the rest of this volume make evident, that we do not yet have as good an understanding of Early Modern economies as we have thought.

One powerful alternative approach to these questions is modeling. Modeling can be a rigorous means to develop predictions about the way economic changes, such as intensification, may appear in the archaeological record. Given the nature of the archaeological record, multiple lines of evidence are almost always required to address and answer archaeological research questions. Archaeologists therefore need several parallel answers to a question such as "What will the intensification of plant resources look like in this place and at this time?" There are two ways to answer this question and to develop predictions: either examine modern cases—analogies—that parallel the ancient one (e.g., find modern, complex hunter-gatherers who are intensifying production of plant foods and compare the modern cases with the archaeological case), or develop models. There are no obvious analogies in the modern world for the processes of interest here,[1] so it is necessary, even essential, to construct models. Both the utility of models and the predictions made from them rely on a number of factors, including the basic assumptions underlying the model itself, as well as the rigor with which predictions are derived from the model.

Three models of intensification are presented in the rest of this chapter. None are original to this chapter; they are taken from the literature. The first is a general model of intensification and domestication developed by Winterhalder and Goland (1997). The second and third are more narrowly focused models, developed by Thoms (1989) and Peacock (1998), of the causes and effects of the intensification of root harvesting, including of camas (*Camassia quamash*) and balsamroot (*Balsamorhiza sagittata*), on the Intermontane Plateau. The first model (with three outcomes) permits an evaluation of the evidence for intensification along the entire coast, while the Thoms and Peacock models are the most detailed models and tests for the intensification of a plant food ever developed for western North America. The modeling approach used by Winterhalder and Goland, and that of Thoms, are based on evolutionary ecology. The first models are very simple and general, though they can be made quite explicit as the camas model shows. The approach differs from that used by Croes and Hackenberger (1988) in their simulation of economic changes at the Hoko River archaeological complex. As I argue elsewhere (Ames 1998), this last model, while extremely powerful and thought provoking, is too specific and too complex to permit its use to model economic changes along the entire coast. In contrast, the approach developed in evolutionary ecology permits the formulation of predictions that have widespread applicability but that can also be tailored to fit very specific circumstances.

Peacock's (1988) model focuses on root intensification on the Canadian plateau, which includes the part of the Intermontane Plateau along the Washington State–British Columbia border and extending north into central British Columbia. She explicitly rejects the assumptions about optimization that underlie evolutionary ecology. Her work and predictions offer an interesting comparison to those developed here.

Both the Winterhalder and Goland and the Thoms models are based on the diet breadth model, a very general model focusing on how foragers select resources to exploit, along with a consideration of the effects of risk on these choices and on the outcomes of these choices. The reader is referred to E. A. Smith (1991: 1–64) for a thorough discussion of evolutionary ecology. Bettinger (1991) analyzes the approach and outlines several of the models applied by evolutionary ecologists to human foraging behavior. Broughton (1994a,b; 1997) presents particularly powerful applications of these approaches to the faunal records of northern California.

In the diet breadth model, foragers rank resources to exploit according to the decreasing net efficiency in acquiring the resources. Resources vary according to: "(1) their abundance; (2) amount of energy produced per item; (3) amount of energy needed to acquire the energy from each [resource]; and (4) the amount of time needed to acquire that energy once the item is selected" (Bettinger 1991: 84). The net efficiency of a resource is the total overall energy costs of those activities less the energy gained by harvesting the particular resource. According to Winterhalder and Goland, "the summary rule is this: add the next item if its pursuit and handling efficiency is greater than the overall efficiency of the diet without it, and, conversely, stop expanding the diet and ignore the [next] item for which the return . . . is less than the average return . . . of higher ranked items" (Winterhalder and Goland 1997: 128). Top-ranked resources are those for which net efficiency is the highest of the available resources. Top-ranked resources are always in the diet, irrespective of the resource's density, and low-ranked resources are never in the diet, irrespective of their density (Winterhalder and Goland 1997). In other words, top-ranked resources are always pursued, whether they are common or rare; low-ranked resources are never pursued, whether they are common or rare.

Some of the foregoing may seem to contradict aspects of the discussion of intensification—they do. One of the goals in constructing models of this kind is to keep their basic assumptions as simple as possible. These models assume that people make their decisions about resources on the basis of net energy return. While this may not seem realistic, models that attempt to be realistic and include such concepts as "value" become increasingly complex. The approach in evolutionary ecology would not be to ignore value, for example, but to construct a separate, but still simple model that uses value as its currency, not net energy return, and to compare the results. For the purposes of the discussion here, it is best to continue with the models based on net energy return.

Items will be added or dropped from the diet for two kinds of reasons, one reflecting the entire diet, and a second having to do with the particular resource. If the overall net efficiency of acquiring all items currently in the diet declines, resources will be added to the diet (their net return is no longer less than the overall efficiency); if overall efficiency of acquiring items increases, then low-ranked resources will be dropped from the diet. On the other hand, if the net efficiency of acquiring a single resource goes above the overall net efficiency, it will be added to the diet. If it goes below the overall efficiency, it will be dropped.

"Risk" refers to variability in the outcome of resource harvesting (E. A. Smith 1991: 53; see also Hayden 1981). In this sense of the term, the returns of an activity vary over some period. One goes hunting once a week, for example. Sometimes the hunter finds nothing; other times the hunter kills several animals, and yet other times, only one. Calculated over a year, the hunter averages two animals per trip. However, there is considerable variation (or not) from trip to trip. That variation between trips is one element of what is meant here by risk. Over some period, harvested resources have a mean return and a variance in return that can be measured, for example, by the standard deviation in return. This is not the same meaning the word "risk" has in everyday speech, where one might be concerned about one's risk in flying in an airplane (How likely is the plane to crash?), or in investing money in the stock market (How likely is it that the stock market will crash and the person will lose all their money?).

Bamforth and Bleed (1997) have argued that the concept of risk should be expanded to include risk as the cost of failure. This is closer to the everyday sense of the term than the notion of variability in outcome. It is important to consider the cost of failure. Planes may not crash very often, but the costs of failure are high for the participants. Salmon runs on a particular stream may not vary much year to year (have low risk in that sense), but as human populations grow, the cost of failure, the risk, from even a rare failure, increases. Implicit here is that the costs of failure will change with changing economic and social conditions, while the variability in the resource does not. The cost of failure will be considered here, but not included in the definition of risk, in order to maintain definitional clarity. "Cost of failure" will be termed cost of failure rather than risk.

As noted above, risk can have other meanings, including danger. Northwest Coast whaling was risky in that it was dangerous, and if Jewitt's accounts of whaling among the Nuu-chah-nulth of western Vancouver Island are any indication, even when everyone returned home, they often failed to strike a whale (Jewitt 1807 [1987]). Danger may be one of the costs of failure, but we are primarily concerned with other costs, such as starvation. Considerations of risk (and variation in risk) have led to the following generalizations: When faced with lower average returns than needed, foragers should be risk seeking (because a single major success or windfall may cover their needs); when faced with higher returns than required, foragers should be risk avoiding (to avoid a shortfall) (E. A. Smith 1991; Winterhalder and Goland 1997).

Foragers can reduce, or buffer, the effects of variance in relatively few ways (e.g., Winterhalder and Goland 1997). It has long been recognized that one major way is through food sharing among the members of a social group. Foragers are also often highly mobile, shifting residences according to the availability of food. They develop social ties to allow them access to the territories of other groups (Kelly 1991). And they can also improve methods and technologies of food storage. On the Northwest Coast food storage was central to the economy for at least 3500 years and is a crucial factor in considering the effects of intensification on the coast.

Storage by hunter-gatherers has one characteristic that makes it, as a form of delayed consumption (Schalk 1981; Woodburn 1980), structurally similar to farming. Many modern foragers, particularly those in the tropics, consume food immediately or shortly after it is harvested. Immediate food consumption tracks closely with the availability of food resources, and the cycles of search, pursuit, and processing are short. People can respond rapidly to failure. With storage, the production interval lengthens to the months that pass between the time the food is acquired and when it is consumed. If resources fail, or stored foods are insufficient, months may pass before sufficient food is available again, or even before it is realized there is not enough food to pass the winter. This can increase the costs of failure associated with storage, particularly given the higher population densities usually associated with storage economies. As an economy becomes more dependent on storage, it becomes more vulnerable to failure or even periodic low returns on the stored resource. With this background, we can now turn to discussing specific models (Table 3.1).

THE WINTERHALDER-GOLAND MODEL

There are three outcomes in the Winterhalder-Goland model.

TABLE 3.1. MODEL OF PLANT INTENSIFICATION

Rank of resource

	Low	High

RISK →

1. Intensified resource dominates diet; diet breadth broad; human population increases in density: highly ranked resources depleted.
RISK HIGH

2. Intensified resource dominates diet; diet breadth narrow; human population increases in density; some highly ranked resources unexploited.
RISK HIGH, but lower than #1

3. Intensified resource small part of diet; diet breadth broad, no significant increase in human population density, minimal additional depletion of highly ranked resources.
RISKS RELATIVELY LOW

Density of resource

(from Winterhalder and Goland 1997)

Outcome 1. In the first outcome within this model, the intensified resource comes to dominate the diet, though diet breadth remains broad. Human populations increase in density, and highly ranked resources, since they are always exploited, are depleted as populations grow. Risk is high (there is a great deal of variation in returns). The cost of failure may increase as populations grow. The intensified resource was abundant and dense, and may have originally ranked low. It would have been added to the diet only if overall subsistence efficiency declined, or if there was an increase in the efficiency at which this particular resource was harvested.

At first glance, the first outcome seems closest to the generally accepted picture of Northwest Coast subsistence economies, with salmon dominating the diet, but also with a broad range of dietary resources. The available ethnographic and archaeological data seem to indicate that a very wide array of resources was harvested on the coast for several millennia. Additionally, the same basic suite of mammals and fish was exploited throughout the coast's known 11,000-year prehistory. We will return to that point below. There is no equivalent information for plant use, because archaeologists on the coast have not seriously attempted to retrieve evidence for plant use until very recently (see Lepofsky in press).

However, in this outcome the dominant resource is a low-ranked, very abundant, high-density resource. It is low ranked because of processing costs. We would expect at first to see salmon appear as one resource among many, and then gradually (or rapidly) come to dominate faunal collections. Schalk originally argued (1977) that salmon was at first low ranked and was added to the diet mix when population growth made it worthwhile and possible to harvest large numbers of salmon and to store them. He also suggested this happened because of innovations in storage, which made storing large amounts of fish possible. Archaeological evidence to evaluate this outcome is skimpily and widely scattered along the coast and through time. There is apparent evidence for high levels of salmon harvesting at an early date (ca. 11,000–8000 B.P.) at the Five Mile Rapids in the Columbia River Gorge (Cressman et al. 1960). However, Five Mile Rapids is located at the upstream end of what was the best place to catch salmon in the entirety of western North America, given the available technology. Therefore, finding evidence of salmon fishing there seems at best to indicate that people were taking advantage of an excellent place to catch fish, rather than pointing to anything about regional subsistence patterns. Finally, the faunal record for Namu on the central British Columbia coast indicates relatively early, heavy reliance on salmon (Cannon 1991). However, this does not seem to be the case everywhere. Data from Glenrose Cannery, on the Fraser River near Vancouver, suggests a very broad diet, without heavy reliance on any particular resource (Matson 1976). Evidence for storage is discussed below.

This outcome also predicts continued exploitation of a wide range of resources, a prediction that seems to be met, particularly when the faunal remains from a number of sites in an area are examined (e.g., Calvert 1970; Hansen 1991). Even at sites such as Namu, where salmon comprise the overwhelming majority of faunal remains, the rest of the assemblages are diverse.

The issues of risk and cost of failure arise. According to this outcome, increased reliance on the intensified resource leads to population growth, in turn exposing that growing population to increased cost of failure, and to unpredictable variation that may have been insignificant to a smaller population but that becomes significant with higher numbers of people. From this, we can also predict that we will see both risk-seeking and risk-avoiding strategies, depending on local return levels. If risk increases with increasing population, then we can predict increasing risk-seeking strategies, strategies where the possibility of failure is high but in which the occasional returns are also quite high. Whaling along the west coast of Vancouver Island, Washington State, and at least parts of Haida Gwaii may be exactly this: a high-risk, high-return strategy in areas with low average returns (relative to what is needed).

This outcome also predicts resource depletion. Croes and Hackenberger (1988) review evidence for depletion of intertidal resources at Hoko River during the Late Pacific Period, and shellfish are generally not considered a high-ranking resource (cf Moss 1993). However, the outcome predicts depletion of high-ranked resources. Little work on this topic has been done on the coast. Butler (2000) found evidence for depletion of high-ranked resources on the Lower Columbia River. Broughton (1994a,b; 1997) examined the issue for the San Francisco Bay area. He demonstrates archaeological evidence of widespread reductions in the numbers of high-ranked resources, including sturgeon, salmon, elk, deer, and large pinnipeds, such as sea lions, with resulting shifts to lower-ranked resources, and argues that these changes reflect resource depletion beginning about 2500 B.P. His inclusion of salmon among these resources is interesting, given the widespread opinion among workers on the coast that Native fishing technology could not have depleted major salmon runs to any degree. Large pinnipeds, such as seals and sea lions, are also animals Northwest Coast archaeologists assume would be hard to deplete. However, there has been little discussion or analysis of the effects of Native subsistence practices on Northwest Coast animal resources. Hewes (1947, 1973) has suggested that native fishing techniques significantly lowered salmon productivity. He argued that the very rich runs observed by early Euro-American travelers were actually the result of a "rest period" produced by the catastrophic decline of human populations in the area in the Early Modern period. Schalk (1987) maintains, contra Hewes (1947, 1973), that native fishing methods either had no negative effect, or actually enhanced salmon productivity. At present, no data exist with which to test either of these ideas. However, one of several possible reasons for intensification of plant resources could be depletion of high-ranked animal resources because of population growth in limited geographic areas (i.e., there might be an inverse relationship between depletion of animal resources and intensification of plant resources).

In any case, Outcome 1 has two testable implications for the initial cause of resource intensification: increased foraging efficiency for a resource (such as postulated by Schalk, or listed in Table 3.5), or an overall decline in foraging efficiency, which could have a variety of causes, including increasing human populations.

Outcome 2. In the second outcome within the Winterhalder-Goland model, the intensified resource was always both dense and highly ranked. In this case, the outcome of intensification is a narrow diet breadth (few resources being harvested), dominated by the intensified resource—that is, a specialized economy. Human population densities are high, and, because of the narrow diet breadth, some formerly high-ranked resources are no longer exploited. Risk is high, but not as high as in Outcome 1.

The second scenario meets the expectations of some researchers for salmon intensification (e.g., Carlson 1996a). In this outcome, the resource was originally dense and highly ranked. In Fladmark's (1975) model, for example, intensification of salmon harvesting began on the coast as soon as salmon became dense as a result of sea level stabilization. Carlson argues that what he terms intensive exploitation of salmon begins wherever and whenever the fish are available in large numbers. Consequently (in the general model), populations grow, and the economy becomes increasingly specialized, focusing on this resource, and some high-ranked resources are dropped from the diet. I am aware of almost no archaeological data from the coast that approximate the complete scenario. Diets seem always to have been quite wide everywhere, even where salmon dominate faunal remains. Nor I am aware of any demonstrated, significant cases where previously highly ranked resources were completely dropped from the resource mix. The point here is that while this outcome fits many archaeologists' assumptions about the role of salmon in regional economies through much of Northwest Coast prehistory, that prehistory, as it is presently known, does not fit the predictions of this outcome (Ames and Maschner 1999).

There is emerging evidence suggesting extreme fluctuations of salmon runs through time (Finney et al. 2002). If correct, then this would suggest that salmon were not the reliable resource generally assumed.

Outcome 3. Finally, in Outcome 3 within the Winterhalder-Goland model, the intensified resource remains a small part of the overall diet; diet breadth is broad, populations remain stable, and there is minimal risk. In this option, the resource has low abundance, and may be either low ranked or high ranked. In other words, it may be a very desirable resource but, because of its low abundance in time or space (or other qualities), its intensification has little overall impact.

This scenario is central to this volume. Does it describe the overall role of plant resources on the Northwest Coast? Were plants resources that were exploited when possible, but that had little overall impact on the subsistence economy and on labor organization? If the answer is yes, then the contents of the present volume are interesting, but perhaps of little importance to our understanding of cultural evolution on the Northwest Coast. If the answer is no, then perhaps Outcomes 1 and 2 are applicable to plant intensification along the coast. None of these models would apply to the entire coast, but to particular places at particular times, perhaps.

The rest of this volume discusses methods introduced to reduce the risks associated with plant resources (reduce the variability over time in plant harvests) and to increase the density of desired plants. In western Oregon, for example, regular burning of the floor of the Willamette Valley would probably reduce variation in acorn production (reducing risk) while increasing overall production of acorns by oak trees (e.g., Shipek 1989). Maintaining berry patches by burning would have a similar effect on berry harvests (Lepofsky et al., this volume). This outcome may describe the role of plant foods in early Northwest economies before significant intensification of food production. To explore this possibility, we turn first to the general record of intensification on the Northwest Coast, and then look specifically at plants.

Intensification on the Northwest Coast

I do not intend to review the entire record of intensification, especially the faunal record, for the coast. That has recently been done elsewhere by several authors (Ames 1994, 1998; Ames and Maschner 1999; Cannon 1998; Carlson 1998; Coupland 1998; Erlandson et al. 1997; Matson 1992; Matson and Coupland 1995; Moss 1998). Rather, I wish to make a few points.

The available record clearly suggests that the same animal resources were exploited throughout the last 10,000 years, the current length of the archaeological record for the coast. Nothing in the record suggests that significant new animal resources were added to the diet at any point during that span. This does not preclude shifts in emphasis, either as a response to local and regional environmental changes or to efforts to intensify production. However, there is presently no good evidence for significant expansions of diet breadth (in the number of resource species exploited) during the past 10,000 years. The evidence suggests the regional diet was always broad. Plants may represent the only category of resources on the coast that could be added to the diet as part of intensification. The implication of this for animal resources (including fish) is that significant intensification of animal harvesting on the coast could only have occurred by increasing production of resources already in the diet.[2]

There are a number of possible ways to accomplish this. Broughton (1994a,b; 1997), as well as other workers, stresses that as intensification proceeds among hunter-gatherers, there should be shifts from large-bodied, highly ranked animal resources to smaller-bodied prey. Hayden (1981) argues for shifts from k- to r-selected[3] species, which will also be smaller bodied. If these predictions hold for the coast, smaller-bodied prey species should become increasingly important in the diet. We might expect to see, for example, increased harvesting of herring. Following Boserup (1965), a second approach would be either to more intensively harvest particular resource habitats (with the implication of decreasing prey size) or to exploit new resource habitats, but for the same resource species. Kew (1992) suggests this latter strategy was a significant form of intensification on the coast. These options all involve lowering overall foraging efficiency.[4] A fourth set of strategies would be to raise foraging efficiency or to raise the rank of a particular resource. The increased collection of mollusks along many portions of the coast that occurred ca. 5500 years ago (and which led to the widespread formation of large shell middens) reflects increased harvesting of very small-bodied organisms. It also probably indicates increased harvesting per unit area.

Alternative strategies involve increasing foraging efficiency. Winterhalder and Goland (1997) suggest a number of such strategies, most of which seem to have been employed on the Northwest Coast (Tables 3.2 and 3.3).

The Northwest Coast strategies include ones that are not usually thought of in terms of resource intensification. Perhaps the most important of these is the construction and use of large canoes. While it seems extremely likely that people along the coast always had canoes, what is crucial here is the appearance of the large freight canoes that could carry substantial loads of raw and processed foods over open water, and which would also permit the transportation of entire villages to new locations where resources were available. Current evidence for their development is quite indirect, and may ultimately have to be derived from evidence of the capacity to transport large volumes of material, or very heavy objects, long distances over water. Other indirect evidence includes the evolution of large, mature stands of cedar by ca. 4500 B.P. (Hebda and Mathewes 1984; Hebda and Whitlock 1997), providing raw materials for large canoes, and the presence in the record of the tools and skills to make them. The available record suggests these tools and skills were widespread on the coast not long after 4500 B.P. Evidence for the northern British Columbia coast points to exploitation of small but possibly productive offshore habitats as well as transportation of large volumes of fish by 2000 B.P., if not much earlier (Ames 1998).

Woodworking skills were also very crucial in the development of the wooden storage box, the Northwest Coast's functional equivalent of pottery. Boxes were used for storage of a wide range of processed foods, including oils, as well as for cooking by boiling with fire-heated rocks. Watertight baskets could also be used for these purposes, but boxes have many advantages. They have flat bottoms, and are easily stacked and packed. I have elsewhere discussed the role of Northwest Coast houses in resource intensification (Ames 1994; Ames and Maschner 1999.). The large wooden houses of the coast were, for all intents and purposes, large food processing and storage facilities, in which animals were butchered and, along with the plants, were processed and stored.

While these strategies and others listed in the table may have increased foraging efficiency, they also required increased labor. A 20 m long canoe capable of passage from Haida Gwaii to the mainland required the investment of considerable labor and skill, and such vessels may appear directly in the archaeological record in the form of increased labor (more woodworking tools, specialized woodworking tools, and tools such as celts, which are themselves costly in labor and time).

INTENSIFICATION OF PLANT PRODUCTION: THE THOMS AND PEACOCK MODELS OF ROOT FOOD INTENSIFICATION

Until quite recently, there has been very little interest among archaeologists in plant use on the Northwest Coast, despite long-term interest in the topic by anthropologists and ethnobotanists working in the region. There has been greater interest in the role of plant use, particularly of camas and other roots,

General processes	Possible Northwest Coast Strategies
1. Increase in the density of high-ranked dietary items, increasing the forager's encounter rate, through habitat improvement, game population cycles, release from over-exploitation, etc.	Gardens, patch creation, and maintenance by burning, coppicing, weeding, use of traps and weirs, to concentrate available prey
2. Reduced search costs perhaps due to decreased energy expenditure in movement	Use of boats, positioning strategies, periodic movement of villages, collapsible houses
3. Changes in resource distribution (resources become more spatially aggregated)	Gardens, patch creation, and maintenance by burning, transplanting, traps, weirs, exploitation of new patches
4. Increase of pursuit and handling efficiencies of items in diet.	See Table 3.3

(modified from Winterhalder and Goland 1997)

among researchers on the Intermontane Plateau east of the Coast and Cascade ranges, and in the Willamette Valley of western Oregon. The most detailed model and thorough test for the intensification of plant food production anywhere in western North America was conducted by Thoms (1989). He was particularly concerned with geophytes (plants whose nutritious organs are below the ground), such as roots, bulbs, and corms, and especially with members of the lily family (Liliaceae) of plants. His model had the same theoretical basis as that of Winterhalder and Goland and is briefly reviewed here. Peacock (1988) is also interested in the intensification of geophytes in her model but takes a somewhat different approach than does Thoms. Both models, as well as the archaeological record for geophyte intensification, provide an excellent touchstone for discussing intensification of plant production on the coast. The evidence for root exploitation also indicates what can be expected for long-term trends in plant exploitation on the coast.

Thoms's Model. Thoms developed a general model for geophyte exploitation (Table 3.4) and a more specific model for camas (Table 3.5). These can provide a basis for modeling plant intensification on the coast, particularly since Thoms was interested in the relationship between intensification of plant foods

TABLE 3.3. POSSIBLE CAUSES OF INCREASED RANK OF FOOD ITEMS
(e.g., increased efficiency in the pursuit and handling of that item)

General processes	Possible Northwest Coast strategies
1. Improved transportation in pursuit	Canoes, including specialized canoes, and related equipment
2. Improved technology of harvest	Ground-slate lance heads, toggling harpoon heads, floats for dead sea mammals, increased diversity in nets, expanded use, and improvement of basketry in traps and storage, berry combs
3. Increased capacity for transporting produce.	Large seaworthy freight canoes, boxes, baskets
4. Improved methods of food processing.	
a. More efficient tools for cutting, cracking, grinding, etc.	a. Filleting knives, mullers, and pounders
b. Better fuels (e.g., hotter firewood species)	b. Use of hot-burning fuel woods
c. Improved technology in heat transfer in cooking	c. Wooden boxes for stone boiling, changes in hearth design (?), pit cooking
5. More effective storage methods (e.g., those that reduce storage loss) or storage facilities that are more efficiently constructed	Waterproof wooden boxes in a range of sizes, basketry, large houses with interior fires and storage racks in rafters, subfloor storage and exterior caches, distinctive outdoor processing techniques, such as berry-drying trenches
6. Morphological changes to the resource increasing its profitability	Long-term effects of human selection and maintenance of particular desirable habitats by burning and culling? Short-term effects might include increased size of preferred part of the plant (e.g., corms, nuts) but (see # 1 in Table 3.2)

(modified from Winterhalder and Goland 1997)

TABLE 3.4. THOMS'S GENERAL MODEL
OF GEOPHYTE INTENSIFICATION

The general model of geophyte intensification:

  A. Conditions:

    1. Population circumscription,

    2. Availability of an intensifiable geophyte(s) that are:

      a. Accessible, calorie-rich, capable of sustaining systematic harvests, and;

      b. The raw materials necessary for exploiting the plant, including fuels for processing,

    3. Be part of existing subsistence strategies,

    4. Technology available for efficiently exploiting resource, but these technologies must exist long before they are used for intensification.

  B. Causes: Intensification is ultimately the result of an imbalance between the need for low-cost foods and the nutritional needs of the population caused by population growth, environmental fluctuation, or both.

  C. Consequences:

    1. Increasing sedentism in localities near root grounds;

    2. Increasing use of root grounds,

    3. Increasing frequencies of geophyte-related artifacts and feature assemblages, including pestles, mortars, digging sticks, and earth ovens, and

    4. Increasing storage facilities.

(Thoms 1989: 121-22)

and intensification of salmon. Thoms's models rest on the same basic assumptions about foraging behavior as does Wintehalder and Goland's model. He also assumes that camas will not be a high-ranking resource (#1, Table 3.5). A recent test of his models' applicability to wapato in the Wapato Valley area of the Lower Columbia River suggests this assumption may not be met (Darby 1996 and this volume). His models also require that the resource, in this case camas, be a part of the diet mix before intensification, again, as do Winterhalder and Goland's models. Both the resource and the appropriate equipment are already in use. This precludes explaining intensification as the result of a fortuitous invention, as is often assumed.

TABLE 3.5. CAMAS INTENSIFICATION MODEL

1. Groups relying on camas as a staple should be those lacking adequate supplies of higher-ranked and intensifiable resources. Other things being equal, the intensity of camas intensification should vary inversely with the availability of anadromous fish.

2. There should be a positive correlation between intensity of camas exploitation and the size of productive camas grounds in a group's territories.

3. The degree to which groups rely on camas should vary inversely with transportation costs.

4. There should be a positive correlation between the use of camas and bulk processing, as measured by the use of large earth ovens and storage facilities.

5. There should be a positive correlation between the intensity of management techniques at camas grounds and population density. Such techniques can include ownership, weeding, watering, seeding, and transplanting of root grounds.

(adapted from Thoms 1989: 184)

The basis for the latter point is important because, ultimately, explaining events, such as the intensification of some resource as the consequence of an invention, explains nothing. One still needs to explain why the invention was accepted and used, why people adopted the innovation, and what problem it solved for them.

Thoms's models also stress the importance of the size of resource habitats or patches (#2, Table 3.5), and the crucial role of transportation, arguing, for camas, that the degree to which groups rely on camas should vary inversely with transportation costs. In other words, he argues that people's reliance on camas will be directly dependent (1) on the size of the available camas stands (people are more likely to use big stands), and (2) on the distance to these stands. People will rely most heavily on stands that are close, but will be most willing to travel to distant stands that are large. These points have implications for some of the practices described elsewhere in this volume.

The gardens on Vancouver Island studied by Deur (this volume) are located very close to villages, where transportation costs will be slight. Many of the prairies and other burned areas discussed by, among others, Turner and Peacock (this volume) are sometimes also close to villages. This kind of intensification and location then seems predictable. However, berrying grounds that were maintained by burning were sometimes considerable distances from villages and towns over difficult terrain (Lepofsky et al. this volume). Berries are often highly desired foods, providing variation and sweetness, especially to a winter diet of dried fish and oil. Gatherings at berry grounds are also important socially and ritually. The berries themselves are also small, especially after they are dried, are transported in burden baskets, and sometimes have to be packed over some distances, often out of the mountains. However, the berry patches may also attract other, high-ranked resources such as deer and elk (wapiti). Thus berries themselves may be highly ranked, while in addition it may be that the patches that produce them are highly ranked and worth the costs of moving back and forth to them.

Finally, Thoms links population density with management practices, such as weeding and burning. I would generalize that and say, following Boserup, that these practices should become increasingly common as overall foraging efficiency declines, for whatever reason, population growth being a likely one.

Peacock's Model. Peacock sees the intensification of plant production as a means to reduce the effects of seasonal variation in plant productivity, particularly in temperate zones. In the terms used here, intensification is a consequence of efforts to reduce risk. She focuses on climate changes leading to increased seasonal and annual variability as the primary cause of intensification of plant food production, but other factors, including population growth, could increase the effects of risk by raising the costs of failure. She argues that there are three classes of strategies to reduce risk: cognitive, technological, and social, and focuses on technological strategies in her work. Three main technological strategies include active management of plants and landscapes, changes in food processing to increase the food energy available from foods collected, and, thirdly, increasing the amount of food stored to support winter populations.

Peacock describes a wide variety of management practices, which include ownership, weeding, watering, seeding, and transplanting of root grounds, as also specified by Thoms, as well as burning. She also recognizes management at the level of the individual plant or species of plant, the plant community, and the landscape. In terms of food processing, she, like Thoms, is particularly interested in the effects of cooking of food as a means of intensification that results from extracting more food energy and nutritional value from foods. In this regard, she sees the expanding use of earth ovens for pit cooking as a major step in intensification, as well as an indicator of it. In the terms defined at the beginning of this chapter, Peacock's emphasis is on increasing efficiency of production, rather than on increased labor.

The two models just discussed have a great deal in common, though they differ in details, and in some implications. In both, plants and roots originally rank low in the diet. This is explicit in Thoms's model and implicit in Peacock's. Intensification is a result of increased risk. In Peacock's formulation, risk increases as a consequence of environmental changes causing greater seasonal and annual variation. For Thoms, causes can be climatic change, and/or demographic change. For the latter, he specifies circumscription (Carniero 1970) and population growth. These can each have the effect of increasing either (or both) the risk and the costs of failure. Circumscription can increase risk by limiting or reducing the territory available to a group for harvesting resources, and therefore increasing the variability in resource availability. Rowley-Conway and Zvelebil (1989) show that smaller resource patches are more variable than large ones. Population growth can obviously increase costs of failure simply by producing more mouths to feed. Increased human density, either through population growth or circumscription, can cause environmental degradation through overharvesting of resources, again increasing both risk and failure costs. Peacock treats population growth as a corollary of intensification, but not as a cause.

Thoms requires the presence of both an intensifiable resource, which Peacock assumes, and the means to intensify that resource. While this may not seem significant in regard to intensifying plant production, it certainly may be with regard to other resources, for which intensification was not technologically feasible. Thoms also specifies that the technology for intensification already be available. This may be one of the crucial differences between Peacock and Thoms. Peacock does not make this requirement, or does not stress it. The effect of that absence in Peacock's formulation is to eliminate the possibility of fortuitous invention as a case of intensification.

Peacock identifies three different possible routes for plant intensification: management, increased food energy, and storage (Ford 1985a,b), all of which Thoms treats together. Peacock's approach allows a more nuanced testing of her model, since all three need not be expected. Indeed, the discovery, for example, that increased burning is not accompanied by an increased number of earth ovens (or visa versa) will indicate something of the direction and form intensification took.

Both predict that intensification of the harvesting of roots should produce increased numbers of earth ovens. Thoms also predicts, among other things, increased sedentism, while Peacock sees increasing sedentism as a possible corollary of intensification.

THE EVIDENCE FOR INTENSIFICATION OF GEOPHYTES

Currently, four archaeological data sets exist for geophyte production from Cascadia[5] that can be used to evaluate the various models presented in this chapter, though none are specifically from the coastal zone. These are: the Willamette Valley of western Oregon (Burtchard 1988; Connolly et al. 1997; references in Lepofsky in press); the Calispell Valley of northeastern Washington State (Thoms 1989; Thoms and Burtchard 1986); the Upper Hat Creek valley of south central British Columbia (Pokotylo and Froese 1983); and Komkanetkwa (Peacock 1998), also in south central British Columbia, near Kamloops. Peacock (1998) and Lepofsky and Peacock (in press) review the evidence from these localities (except the Willamette Valley) as well as several other localities in British Columbia in greater detail than I can here. These four data sets include usable numbers of excavated and radiometrically dated earth ovens.

The Willamette Valley evidence suggests that geophytes were harvested and processed there by 11,000 years ago (Connolly et al. 1997). Caution is necessary in interpreting these data since they come from only one part of the valley. Roots may have been sporadically exploited until about 5500 B.P., when their production seems to have expanded, continuing at relatively high levels for another 1500 years. Geophyte production fell to almost nil between 4000 and 3000 B.P. There may be another episode of expansion at ca. 3000 B.P., followed by another period of low production between 2500 B.P. and 1500 B.P. Geophyte production was again increased after that date, though Connolly et al.'s data suggest at least one brief period in those 1500 years when root production may again have ceased.

The Calispell Valley record shows regular but low-level root use after ca. 6500 years ago until ca. 4000 B.P., when production was dramatically intensified. Camas production remained high until about 2500 B.P., when it dropped; though it did not cease. Production increased again after 2000 B.P., but never reaching the levels it had previously. Thoms demonstrates that the fluctuations do not correspond to environmental changes (Thoms 1989).

The dated ovens in the Upper Hat Creek valley span the period between 2300 B.P. and contact. However, the temporal distribution of radiocarbon dates suggests that roots were most intensively processed there ca. 2300 B.P.–1900 B.P. (Pokotylo and Froese 1983). Peacock reports ten dates from Komkanetkwa that indicate the site was used for camas processing between ca. 2400 B.P. and the twentieth century. However, the most intensive period of use was between ca. 1800 and 800 years B.P. She concludes that edible root production began on the Canadian Plateau ca. 3300 B.P. While it is difficult to draw regional trends from these data, some general conclusions are possible.

Most important is that geophytes were a part of the diet in the Willamette Valley by 11,000 years ago. This suggests that edible roots were present in Cascadia where the conditions were right for them by the end of the Pleistocene. It is not too great a leap from this to think that they were exploited when and where available throughout the region, as needed, by the end of the Pleistocene. There is, in addition, other weak evidence for root exploitation elsewhere on the Plateau during the earliest Archaic, or Windust period (Ames 1988). Diet breadth during that period was probably quite broad. A range of small (rabbits) to large (bison and wapiti, or elk) mammals was taken, though the larger mammals were certainly preferred when available. There is also indirect evidence for nets and fishing. There are no data for vegetal foods for this time, beyond what is discussed here. Despite the diet's apparent breadth, human population densities were very low (see Ames 1988 for a review of all of this evidence).

The evidence also seems to be showing that levels of production of camas and other root resources fluctuated at each locality through time, though they were exploited more or less everywhere during the last 2000 years or so. The available record is a record of the use of particular root-digging grounds, not of root resources at a regional level. In fact, it is important to stress how profoundly local in scale these data are, and that they may be but dimly reflecting regional-level processes. I have suggested elsewhere (Ames 1991b) that intensification is a regional-level process that will look very different from place to place. Because edible root production, in the case of Thoms's work, was a part of much broader subsistence economies, then levels of geophyte production might be expected to vary as a consequence of changing subsistence strategies, even if those strategies were always aimed at increasing overall production. This would imply that roots were not a highly ranked resource, and were likely to be dropped from the diet periodically. This conclusion does fit the available archaeological data but not the ethnographic data (e.g., Peacock 1998). A second alternative is that geophytes were always highly ranked, at least within the last 5000 years.

In the latter alternative, geophyte production was always an important part of the resource mix. If this was the case, how do we account for these local fluctuations in production? These fluctuations could be pointing to shifting production locales. One reason to shift from one habitat to another would be local declines in productivity—a particular locale becomes less productive, even though edible roots continue to be available elsewhere. Changes in habitat could be the very local consequence of broad climatic shifts, of the kind known for the Holocene. However, these changes do not readily fit the known climatic sequence for the Plateau (e.g., Chatters 1995). Another alternative is that declines in production were a consequence of human exploitation of edible-root habitats—that is, it is possible to overexploit them, perhaps over time-spans making the overexploitation invisible to humans. At the moment this is speculation on my part and it contradicts what we know about the positive effects of management on geophyte productivity. I raise the possibility because resource depletion is an issue arising from the application of the Winterhalder and Goland models.

One of the interesting difficulties in this evidence is reconciling the camas record for the Willamette Valley with the paleoenvironmental record for environmental management there—i.e., burning. This evidence may indicate that extensive environmental management of the Willamette Valley may have begun as early as the middle Holocene (Boyd 1986), although Whitlock and Knox (2002) dispute anthropic burning in the Willamette Valley. The valley was regularly and deliberately burned during the late Holocene, until Euro-American settlement, producing a distinctive oak-savannah that was highly productive in acorns, deer, elk, and other resources. If burning of the valley did begin ca. 3900 B.P. (and that date is preliminary and even controversial), the initiation of burning may have followed, or been associated with, the decline in camas production observed by Connolly and his associates. Linkages among rising human populations, overproduction of camas and depletion of camas meadows, and valley floor burning are intriguing, but the dates for the beginning of burning are not yet firm. If the events are linked, burning could also simply have changed the foraging efficiency for other resources.

Considering all of this, it is almost impossible to use these data, taken as they are in isolation, to evaluate Thoms's or Peacock's models, let alone those of Winterhalder and Goland. That is also far beyond the scope of this chapter. However, a few summary comments can be made. Following the logic of these models, intensification occurred because of either: (1) an overall decrease in foraging efficiency, or (2) increases in risk and/or the costs of failure associated with food getting. Increasing risk would be an increase in the probability of failure in food getting, while increasing costs of failure might include an unacceptable increase in the consequences of failure, even if the probability of failure (risk) remained low. For Thoms, the cause of intensification is population growth, which could cause either result (1) or (2), while Peacock suggests a decline in regional temperatures at about 4500 B.P. to be the cause for the increase in risk (see also Chatters 1995). However, this date does not seem to fit well with the dates for camas intensification reviewed here.

Chatters' (1995) population curve for the southern Plateau, based on radiocarbon dates (cf Ames 1991a, 2000), can be used here. The curve (which extends only to ca. 7700 B.P.) suggests slow growth until ca. 5300 B.P., when populations rise rapidly, only to crash at 4500 B.P. (as a consequence of the fall in temperatures). Numbers fluctuate after that, with peaks around ca. 3300 B.P. and 2600 B.P., and with rapid, continual growth after 2000 B.P. The fit between his curve and the camas data is not very good, or the connections are not yet obvious. For example, the peak in numbers ca. 5300 to 4500 B.P. does occur when camas intensification first takes place in the Willamette Valley, but before the increase in the Calispell Valley. If Thoms's model is correct, then we should expect to see intensification of salmon across the Plateau at that time, except in salmon-poor areas, such as the Willamette and Calispell valleys, where geophyte intensification should occur. Intensification does occur in the Willamette Valley, but not the Calispell Valley. There is no evidence for the intensification of salmon on the Plateau at that time.

In Winterhalder and Goland's first scenario, geophyte intensification would lead to their coming to dominate the diet, although diet breadth remains broad; human population densities would increase, and other high-ranked resources depleted. In the second scenario, diet width should narrow as edible root production was intensified, and some other high-ranked resources would be dropped from the diet. In their third scenario, roots, despite intensification, would remain a small part of the diet, and we would anticipate there being no corresponding population increases or resource depletion. Based on the available archaeological evidence for the Willamette Valley and the Columbia Plateau, neither their second nor their third scenario seems applicable to intensification of edible roots. The first scenario is supported to some degree by the available ethnographic evidence for camas use (e.g., Peacock 1998; Thoms 1989). The implications of this scenario are that risk (in both senses) was high, and that there may have been problems of resource depletion. These implications cannot be pursued further here.

The Gulf of Georgia region of the Northwest Coast, broadly defined, includes the waters, islands, and mainland coasts of a triangle with points at Victoria, British Columbia, on southern Vancouver Island, Vancouver, British Columbia, and Seattle, Washington. Much of it falls within the area inhabited by the Central and Southern Coast Salish. This region currently possesses the best-documented archaeological sequence for the entire Northwest Coast (e.g., Ames and Maschner 1999; Matson and Coupland 1995). The last three phases of this sequence are the Locarno Beach phase (ca. 3250 B.P.—1500 B.P.), Marpole phase (2500 B.P.—1500 B.P.) and Gulf of Georgia (1500 B.P. to contact). As originally defined (Mitchell 1971, 1990) these three phases differed markedly in their subsistence-related artifacts.

Locarno Beach assemblages include chipped-stone stemmed points, microblade cores and blades, microflakes produced by bipolar technology, large ground-slate points and blades, and net-sinkers. The bone and antler industry includes unilaterally and bilaterally barbed antler points, and composite and single-piece toggling harpoon heads. Most distinctively, hand stones (mullers) and grinding slabs are present, albeit probably in small numbers (it is hard to determine). These tools were probably used for grinding acorns although other plant foods and other purposes are certainly possible. Such tools occur to the south, along the Washington and Oregon coasts, and in sites along the Lower Columbia River, but are rare or completely absent in sites on most of the coast to the north (Ames and Maschner 1999).

The Marpole phase is commonly thought of by researchers in the Gulf of Georgia as the period when what is regarded as classic Northwest Coast culture appeared in the local area (e.g., Matson and Coupland 1995). Both flaked-stone and microblade technologies persist. The variety of toggling harpoons is replaced by large, unilaterally barbed harpoon heads. The grinding slabs and hand stones were still present, however. In the subsequent Gulf of Georgia phase, ground-slate points and blades became smaller and triangular in form. The large, unilaterally barbed harpoon heads were replaced by large composite harpoon valves and by bone bipoints and bone points in general. The grinding slabs and mullers disappear.

As noted above, the Locarno Beach grinding slabs and mullers are similar to those recovered in sites along the Lower Columbia River and in California, where they imply processing and use of acorns. In the Lower Columbia River region, both acorns and hazelnuts were exploited, and acorns at least appear to have been relatively high-ranked resources (Boyd and Hajda 1987). It is plausible that acorns were exploited during Locarno Beach and Marpole times in some areas of the Gulf of Georgia. The other artifacts indicate subsistence shifts. The Locarno Beach tackle—small toggling harpoons and barbed antler points—suggests the exploitation of a wide array of marine habitats and a range of fish and sea mammals, perhaps with a focus on small-bodied ones. The net weights imply the use of gill nets for fishing. The appearance of the large harpoon heads during Marpole points to a shift to larger-bodied sea mammals and fish (see Lyman 1991 for the basis of this reasoning). Some authors see evidence in Marpole for a possible specialization in salmon fishing (e.g. Burley 1980; Matson and Coupland 1995; Mitchell 1971). If this is so, and the evidence is not at all strong, then one could speculate that nuts (or whatever plant foods were ground on those slabs) were sufficiently highly ranked that they continued to be exploited although diet width decreased. One could speculate further that Marpole had a subsistence base rather different from any historically known group on the Northwest Coast, and perhaps somewhat similar to that found in California, with its famous native economy based on salmon, deer, and acorns.

The Gulf of Georgia period appears to have been marked by a wide diet breadth, as indicated by the artifacts and associated faunal remains (Hanson 1991). However, the grinding slabs drop from the repertoire. This could reflect environmental change and a reduction in the range of the appropriate oaks (or other plant foods), or it could reflect shifts in foraging efficiencies for different resources in the diet. The former seems the more likely, but the changes in artifacts do not exclude the latter possibility. It is also interesting to recall that there is evidence of depletion of shellfish beds at the Hoko River Rockshelter during this time period (Croes and Hackenberger 1988). In any case, I offer these speculations to raise the possibility that our assumptions about Northwest Coast subsistence through time may be blinding us to alternatives.

Summary and Conclusions

I have argued elsewhere (Ames 1998; Ames and Maschner 1999) that Northwest economies could best be understood at two levels: a regional level and a local level. Some resources, such as salmon, can best be thought of as regional resources (resources that are essential to maintain high regional population levels), while others are entirely local (resources essential to maintaining high local population levels). In some places, regional and local resources are the same, but in other places not. I have also suggested a similar regional–local dynamic with regard to intensification as a process: production is increased across a region, but the particular mix of intensified resources, or even means of intensification, is entirely local. Thus we may see evidence of increased salmon production near the mouth of a river, but of herring in the next bay north. Or we might see a regional intensification of salmon production (increased levels of salmon production wherever possible), but intensification of other, so-called secondary resources everywhere else.

I have suggested here that most, if not all, the major animal resources exploited in the nineteenth century have been used on the coast for all of the last 10,000 years. This observation is not new with me. Intensification on the coast could not then proceed by adding new animal resources to the diet, but by increasing the efficiency of exploiting some, or by adding entirely new categories of resources. This has implications for the patterns of regional and local intensification discussed above. If, for example, a resource such as salmon was being exploited at maximum levels given available technology, then it could not be intensified. Intensification of overall food production would require increased harvests of other food resources. The technological and organizational problems in this situation might be how to intensify harvests of other resources without a parallel decline in salmon harvests. Some of the ways in which this bind might be solved have been discussed in this chapter, and include shifts in task organization.

There is one further implication of this that I wish to explore here. In a situation where salmon production is at maximum levels and salmon are the dominant resource, we would expect to see salmon dominate faunal remains. If intensification proceeds as described above (that is, of so-called secondary resources), then salmon might continue to dominate the faunal assemblages, especially if plant resources are those intensified. In other words, intensification might not show up in the faunal assemblages.

I have argued a number of points in this chapter. I end by stressing these: (1) finding evidence for intensification requires looking on both sides of the equation stated as *intensification = more food*. Archaeologists often look only for evidence for more food remains, or an increasing diversity of food remains. However, evidence for increased effort, or increasing efficiency of effort, is just as important. (2) Intensification includes an array of strategies, some of which, such as labor reorganization, may be difficult to see in the archaeological record; and (3) general models are useful for developing testable predictions of processes such as intensification. Evolutionary ecology is a powerful source for such models. Finally, I have implied several times in the foregoing that plant foods may have offered one of the few avenues on the coast for people to intensify food production, by providing the only class of "new" resources to be added to the diet, and ones that could be directly manipulated to increase production.

Notes

I would like to thank Doug Deur and Nancy Turner for the invitation to contribute to this volume. I have enjoyed working on and thinking about these issues very much.

I also want to thank Bruce Winterhalder, Dana Lepofsky, and an anonymous reviewer for their comments. They have materially improved this chapter. Any errors are mine.

1. Peasant economies very often provide useful analogies to Northwest Coast economies; however, such analogies require careful theoretical construction, which is far beyond the scope of the present chapter.

2. In a recent paper, Carlson (1998) argues a number of points: (1) that salmon were available in the coast's rivers at the end of the Pleistocene, and that (2) they were intensively exploited at least in some places prior to 6000 B.P. While his first point requires further empirical support, his second raises issues germane to this chapter. *Intensive* exploitation does not necessarily imply that it is the result of the process of intensification. In fact, intensive exploitation of salmon is predicted by Winterhalder and Goland's outcome 2, as discussed in the text. However, the following question arises: If salmon were already intensively exploited by the middle Holocene, then how was their production to be intensified? If they were harvesting salmon at the levels of Early Modern economies on the coast by the middle Holocene, then salmon could not have been intensified to meet the increased needs of what was clearly a growing population and the evolving social systems. Only two options seem to be available, if we accept Carlson's arguments: increase the efficiency of salmon harvesting and processing, or intensify the production of something else.

3. "K"-selected species are those that produce relatively few young, and invest heavily in them (e.g., humans). They often reproduce slowly. "R"-selected species produce enormous numbers of young, investing little in any individual offspring, but reproducing in great numbers.

4. High-ranking habitats would be those that were the least costly to exploit, given the available technology and labor. Any new resource habitats would necessarily then be those that were more costly to exploit, for whatever reason. Any additional habitats would be increasingly expensive to exploit since these would be the only ones left to add to the subsistence economy.

5. The term "Cascadia" refers to a region that includes both the Northwest Coast and the Columbia Plateau (Ames 1991b; Ames and Maschner 1999).

Chapter 4

Solving the Perennial Paradox

Ethnobotanical Evidence for Plant Resource Management

on the Northwest Coast

NANCY J. TURNER AND SANDRA PEACOCK

It was all important. That *texwsus* [springbank clover], and the *tliksam* [silver-weed], and the *q'weniy'* [Nootka lupine], and the . . . *xukwem* [northern rice-root]. See, when they go down the flats, they use little pegs. "This is my area." You got your own pegs, in the flats. And then you continue on that, digging the soft ground . . . so it will grow better every year. Well, I guess, fertilizing, cultivating, I guess that's . . . the word for it. Every family had pegs, owned their little plots in the flats.
—Kwaxistala, Chief Adam Dick, Kwakwa̱ka̱'wakw, from Kingcome Inlet, 1996

People–Plant Interactions on the Northwest Coast

The indigenous cultures of the Northwest Coast are traditionally viewed as being one of the original affluent societies (Sahlins 1968)—hunter-gatherers, who, by exploiting the abundant "natural" resources of the Pacific Northwest, attained a high level of social complexity. Paradoxically, this complexity, it is often said, was achieved in the absence of plant food production and domestication. Consequently, as Ames and Maschner (1999: 13–14) observe,

> These societies confound ideas about the development of social complexity during human history and many of the traits expressed on the Northwest Coast are exactly those traits widely viewed as the basis for the development of civilization. It has always been assumed by historians, anthropologists, archaeologists, and others that farming is necessary for these traits to evolve. The non-farming hunter-gatherer societies of the Northwest Coast possessed all those traits, and did so for at least the last 2,500 years if not longer.

Efforts to explain this apparent anomaly (cultural complexity sans agriculture) have focused on the rich marine environments and the importance of anadromous fish to the economies of the region, suggesting that these plentiful resources served as the cornerstones of complexity. In contrast, few have explored the extensive ethnobotanical literature to evaluate the contributions of plant resources to Northwest Coast subsistence and to assess whether the dichotomy between foraging and farming is a useful distinction for the area (Deur and Turner, Introduction, this volume).

The purpose of this chapter, then, is to provide a broad overview of the nature of people–plant interactions on the Northwest Coast, as a prelude to the case-specific chapters that follow. In doing so, we hope to set the context for the following discussion of plant management traditions. We also hope to dispel the myth that social complexity arose here in the absence of food production by demonstrating that the "hunter-gatherers" of the region were not simple "affluent foragers," but active managers who have cultivated, sustained, overseen, and promoted culturally valued plant resources. The chapter proceeds as follows. We begin with a brief summary of the landscapes and peoples of the Northwest Coast, topics treated in greater detail in the Introduction to this volume, and, in terms of the peoples, in Chapter 5. Then we establish the theoretical basis for our research, drawn from the work of Peacock (1998), by outlining two models of people–plant interactions we have found particularly instructive. With this framework, we turn to the specifics of plant resource use on the Northwest Coast, identifying a continuum of management activities and discussing how these affected the composition and productivity of plant communities. Here, we draw from Turner's extensive ethnobotanical experience in the Pacific Northwest (Kuhnlein and Turner 1987, 1991; Turner 1995, 1997a–c, 1998, 1999, 2003a,b; Turner and Efrat 1982; Turner and Kuhnlein 1982, 1983; Turner et al. 1983, 1990) as well as from other recent recollections and accounts of elders from communities throughout the study area (see Beckwith 2004; Bouchard and Kennedy 1990; Compton 1993a,b; Davis et al. 1995; Deur 1999, 2000, 2002a,b) and corroborative materials from the Canadian Plateau (Peacock and Turner 2000; Turner et al. 2000). We also build upon the work of Suttles (1951a,b; see also this volume) who, several decades ago, drew the connection between precontact plant management practices of Coast Salish peoples and the ease with which they adopted potato cultivation. In our concluding remarks, we summarize the evidence presented and discuss the implications of these findings for our understandings of people–plant interactions on the Northwest Coast.

Plant Management Principles and Practices

The Northwest Coast of North America is well defined both geographically and culturally (Schoonmaker et al. 1997; Suttles 1990). Spatially, the Northwest Coast is identified as the strip of coastline extending from northern California to Yakutat Bay, Alaska, bounded on the inland side by mountain ranges and characterized by a relatively mild climate, temperate rainforest, and rich marine life (see Deur and Turner, Introduction, this volume).

Within each of the four major ecological or "biogeoclimatic" zones of the Northwest Coast as described in the Introduction, are many specific habitats such as river estuaries, peat bogs, marshes, meadows, forests, parkland, and floodplains (Figure 4.1). These are occupied by specialized communities of plants and animals, many of them culturally important to Northwest Coast peoples. As noted previously, too, the various successional stages of the forest communities resulting from disturbance, as well as the climax communities, are important in providing a wide diversity of culturally important species. It is important to keep this geographical and ecological diversity in mind in discussions of traditional plant-resource management, because, although we can generalize management strategies to some extent, specific practices will vary according to environmental as well as cultural factors. It is also essential to acknowledge the cultural diversity of the region.

Despite sharing many cultural features, the peoples of the Northwest Coast are linguistically and culturally diverse. In British Columbia, at least fourteen to sixteen or more languages,[1] classed within five major language families, were spoken along the coast. It is important to recognize that the peoples of this coast were, and are, distinctive, and that there is a danger in too much generalization about peoples' cultures and lifestyles. We acknowledge that information known to one individual or community should not be applied to others without qualifications.

In this chapter we present admittedly relatively sporadic accounts of plant management. It is fair to assume that the practices we have documented are more widespread, but obviously more research is needed to confirm such an assumption and to extend the range of our knowledge and understanding about relationships between people and plant communities in the Northwest Coast region.

Over 300 plant species were utilized traditionally by Northwest Coast peoples as food, sources of material, medicines, and for spiritual purposes (Table 4.1). Our assessment of the traditional plant-management strategies associated with these resources indicates that they vary with species and geographic region, but range along a continuum from foraging activities with minimum or incidental impacts, to cultivation practices such as selective harvesting, tilling, weeding, pruning, and landscape burning, to intensive gardening—leading, apparently, to complete domestication, as in the case of the Northwest Coast tobacco (evidently a variety of *Nicotiana quadrivalvis*; Turner and Taylor 1972).

This range of activities is listed in Table 4.2, which represents our interpretation of models of human–plant interaction advanced by Ford (1985b) and Harris (1989). Like these models, it represents a continuum of increasing human effort and influence on the "natural" landscape. Unlike the others, it includes, as well as food plants, those species used for materials and medicines. We maintain that the same principles and practices guided the harvesting of all types of plant products (see M. K. Anderson [1996a] and

4.1. Kingcome River estuary (Photo by N. J. Turner)

Nicholas [1999] for discussions on the influences of harvesting plant materials).

We turn now to a discussion of specific examples of management practices relevant to each of these categories, remembering, of course, that these categories represent arbitrary divisions across a continuum.

FORAGING STRATEGIES

Foraging activities, according to Ford (1985b), have minimal impacts on plant resources, although they may have incidental or unintentional effects. Included in this category are plants not harvested on a regular basis, or in large quantities. Foraging as such in the Northwest Coast area may have occurred with incidental picking of lesser-used fruits, famine foods, flavorings, materials for makeshift baskets, containers, and cordage, and some medicinal plants, as well as with opportunistic harvesting in a variety of circumstances. However, even in these cases, there were protocols for harvesting and use that would have resulted in non-random selection, which would be expected to produce some directed influence on the plant resources and their habitats.

CULTIVATION STRATEGIES

As depicted in Table 4.2, the range of activities included under cultivation varies from low-intensity techniques, such as tending and tilling, to those techniques requiring greater human input—such as transplanting, landscape burning, and the maintenance of garden plots. The activities had both incidental and intentional impacts on plant resources.

The majority of these plant-management strategies were directed at particular species, and therefore, these were the fundamental unit of resource management on the Northwest Coast. Populations of plants with cultural utility were encouraged through a number of strategies associated with aboriginal harvesting practices. These strategies were based both on biological and cultural considerations, and were employed to ensure the continued productivity of key resources. They include: the selective harvesting of plants based on well-defined criteria, a number of extractive techniques that increased rather than decreased overall population levels, and the creation and maintenance of "garden" habitats for certain key species.

HARVEST CRITERIA

The harvesting of plant resources was and is selective, being neither random nor all-encompassing. Selective harvesting was widely practiced in plant gathering, whether it was for greens, roots, berries, plant materials, or medicines. The criteria used to select plants for harvest varied considerably between species, and depended upon the type of plant resource and its intended use. Cultural preferences, the physiology of the plant, and environmental factors all influ-

(text continues on page 112)

TABLE 4.1. SUMMARY OF CULTURALLY IMPORTANT PLANTS OF THE
NORTHWEST COAST WITH ASSOCIATED HARVESTING AND
MANAGEMENT PRACTICES

Use category	Examples	Notes on management
FOOD: Berries (Total ± 50 species)	*Fragaria* spp. (wild strawberry); *Gaultheria shallon* (salal); *Ribes bracteosum* (gray currant); *Rubus spectabilis* (salmonberry); *Pyrus fusca* (Pacific crabapple); *Shepherdia canadensis* (soapberry); *Vaccinium membranaceum* (black huckleberry); *Vaccinium ovalifolium* (blueberry); *Vaccinium parvifolium* (red huckleberry); *Vaccinium oxycoccos* (bog cranberry); *Viburnum edule* (highbush cranberry)	All except wild strawberry are woody perennials; fruit picking generally non-impacting on plants; diversification and use of alternative species in poor crop years was practiced; seasonal rounds; some enhanced by periodic burning; some (e.g., gray currant, salmonberry, blueberry, red huckleberry, Pacific crabapple) were pruned periodically; salal pruned when stems picked; highbush cranberry and others known to have been transplanted; some berry patches fertilized; productive berry patches often owned by individuals or families
FOOD: Root vegetables (Total ± 25 species)	*Camassia* spp. (edible blue camas); *Conioselinum pacificum* ("wild carrot"); *Dryopteris expansa* (spiny wood fern); *Fritillaria camschatcensis* (rice-root); *Lupinus nootkatensis* (Nootka lupine); *Potentilla anserina* ssp. *pacifica* (Pacific silverweed); *Sagittaria latifolia* (wapato); *Trifolium wormskjoldii* (springbank clover)	All are herbaceous perennials with bulbs, corms, rhizomes, tubers, or taproots; all selectively harvested by season, age, size, life-cycle stage; part of seasonal round harvesting cycles; some (e.g., camas) enhanced by periodic landscape burning; propagules often incidentally or intentionally replanted; some weeded during harvest; specific sites harvested in annual or several-year cycles; patches often owned by individuals or families

Use category	Examples	Notes on management
FOOD: Green vegetables (Total ± 20 species)	*Epilobium angustifolium* (fireweed); *Equisetum telmateia* (giant horsetail); *Heracleum lanatum* (cowparsnip); *Rubus* spp. (thimbleberry, salmonberry, shoots)	All are herbaceous perennials or woody perennials (*Rubus* spp.); all selectively harvested as shoots or leaves by season, age, size, life-cycle stage; part of seasonal round harvesting cycles; some enhanced by picking
FOOD: Tree inner bark (Total ± 7 species	*Alnus rubra* (red alder); *Populus balsamifera* ssp. *trichocarpa* (cottonwood); *Tsuga heterophylla* (western hemlock); *Abies amabilis* (silver fir); *Picea sitchensis* (Sitka spruce)	All tree species; harvested by partial bark and cambium removal, but not girdling; selectively harvested by season and in some cases tidal phase; part of seasonal round harvesting cycles
FOOD: Marine algae (Total ± 6 species)	*Egregia menziesii* (boa kelp: capturing herring spawn); *Porphyra abbottiae* (red laver, nori); *Macrocystis integrifolia* (giant kelp: capturing herring spawn)	Selectively harvested by abundance, taste, season, location; fronds plugged, or plants pulled from rocks, allowing regeneration from remaining base of fronds
FOOD: Casual foods, flavorings and sweeteners, emergency foods, beverage plants (Total ± 50 species)	*Blechnum spicant* (deer fern; fronds as hunger supressant); *Tsuga heterophylla* (western hemlock; branch tips as hunger supressant; boughs for capturing herring spawn); *Ledum groenlandicum* (Labrador tea: beverage); *Picea sitchensis* (Sitka spruce: chewing gum)	Variously herbaceous or woody perennials; leaves, branches, gum, shoots, or other parts selectively harvested by season or at times of need; seldom harvested intensively in any locality
SMOKING: Tobaccos and tobacco flavorings (Total ± 2 species)	*Arctostaphylos uva-ursi* (kinnikinnick: leaves smoked by some groups); *Nicotiana* sp. (Haida tobacco; leaves chewed)	*Nicotiana* an annual, formerly grown in gardens from seed by Haida, Tlingit, Tsimshian, and others; kinnikinnick is perennial, leaves selectively harvested

TABLE 4.1 *(continued)*

Use category	Examples	Notes on management
MATERIALS: Pit-cooking, matting (Total ± 10 species)	*Alnus rubra* (red alder); *Gaultheria shallon* (salal branches); *Polystichum munitum, Pteridium aquilinum* (ferns); grass species	All herbaceous or woody perennials; materials harvested selectively from living plants as plucking or "pruning"
MATERIALS: Woods for construction and manufacture (Total: ± 25 species)	*Acer macrophyllum* (broadleaf maple); *Alnus rubra* (red alder); *Chamaecyparis nootkatensis* (yellow-cedar); *Picea sitchensis* (Sitka spruce); *Pseudotsuga menziesii* (Douglas-fir); *Salix* spp. (willow); *Taxus brevifolia* (Pacific yew); *Thuja plicata* (western red-cedar, intensively used)	All trees or shrubs; some harvested or coppiced as branches or stems from living plants; cedar planks cut from standing trees; trees selectively cut (e.g., cedar for dugout canoes; houseposts, etc.); stumps as habitat for berry bushes
MATERIALS: Woods and others for fuel, tinder (Total: ± 25 species)	*Alnus rubra* (red alder, for smoking fish); *Chamaecyparis nootkatensis* (yellow-cedar); *Juniperus scopulorum* (Rocky Mountain juniper, for smudging); *Populus balsamifera* (cottonwood); *Pseudotsuga menziesii* (Douglas-fir: wood, bark as fuel); *Thuja plicata* (western red-cedar, bark and wood)	Most are trees or woody perennials; materials often harvested from downed/dead trees, driftwood, or selectively as branches from living trees; burns sometimes used to create firewood
MATERIALS: Bark sheets for roofing, or other purposes (Total ± 5 species)	*Prunus emarginata* (bitter cherry, bark peeled off in strips as binding material); *Thuja plicata* (western red-cedar, bark sheets for roofing)	All trees; outer bark only removed, without killing tree; others removed in large sheets, which would kill trees if all bark taken; bark harvested selectively by size, season, tree characteristics
MATERIALS: Stem, leaf, root fibers/fibrous tissues (Total: ± 15 species)	*Carex obnupta* (basket sedge: leaves); *Chamaecyparis nootkatensis* (yellow-cedar: inner bark); *Nereocystis luetkeana* (bull kelp: stipes for fishingline);	All are woody or herbaceous perennials; materials cut (stems, leaves) or pulled in strips (barks) from living plants by pruning; herbaceous

Use category	Examples	Notes on management
MATERIALS: Stem, leaf, root fibers etc. *(continued)*	*Picea sitchensis* (Sitka spruce: roots); *Salix* spp. (willow: stems, roots); *Schoenoplectus acutus* (tule: stems); *Thuja plicata* (western red-cedar: inner bark, roots, withes); *Typha latifolia* (cattail: leaves); *Urtica dioica* (stinging nettle: stems); *Xerophyllum tenax* (bear-grass: leaves)	materials harvested selectively by size, season, plant growth form, habitat; prime populations of kelp, nettles, cat-tails, etc., were recognized and selectively harvested
MATERIALS: Dyes, stains (Total: ± 8 species)	*Alnus* spp. (alders: bark); *Echinodontium tinctorium* (Indian paint fungus); *Mahonia aquifolium* (Oregon-grape: inner bark)	Various types of materials; harvested selectively and sporadically as required
MATERIALS: Adhesives, caulking, waterproofing agents (Total: ± 10 species)	*Pinus contorta* (lodgepole pine: pitch); *Populus balsamifera* (cottonwood bud resin: adhesive); *Picea sitchensis* (Sitka spruce: pitch); *Pseudotsuga menziesii* (Douglas-fir: pitch)	Harvested selectively from living trees
MATERIALS: Scents, cleansing agents, spiritual protection, miscellaneous (Total: ± 50 species)	*Abies amabilis, A. grandis* (silver fir, grand fir: boughs as scent); *Alectoria sarmentosa* (old man's beard lichen: wiping fish); *Aquilegia formosa* (red columbine: roots as good-luck charm); *Asarum caudatum* (wild ginger: leaves, rhizomes as scent); *Cirsium brevistylum* (thistle: roots as good-luck charm); *Equisetum telmateia* (scouring rush: abrasive); *Lysichiton americanum* (skunk-cabbage: leaves for drying/laying food on); *Polystichum munitum* (sword fern: fronds to line eulachon ripening pits); *Salix scouleriana* and other spp. (willow: leafy branches for	Variously herbaceous or woody perennials; most materials selectively harvested from living plants; usually harvested in limited quantities by season, life-cycle stage

TABLE 4.1 *(continued)*

Use category	Examples	Notes on management
MATERIALS: Scents, etc. *(continued)*	cleaning, separating fish; withes as fish stringers); *Sphagnum* spp. (sphagnum moss: infant diapering)	
MEDICINES: Whole plants or leafy branches (Total: ± 100 plant preparations)[1]	*Achillea millefolium* (yarrow: leaves, roots for colds, poultices); *Blechnum spicant* (deer fern: fronds as poultice for cuts); *Maianthemum dilatatum* (wild lily-of-the-valley: leaves as burn medicine); *Plantago major* (broad-leaved plantain: leaves as poultice for cuts)	Variously herbaceous or woody perennials; most materials selectively harvested from living plants; usually harvested by season, life-cycle stage
MEDICINES: Bark tissues (Total: ± 50–60)	*Abies grandis* (grand fir: coughs, tuberculosis, many ailments); *Alnus rubra* (red alder: coughs, tuberculosis, skin wash, many ailments); *Arbutus menziesii* (arbutus: bark for colds); *Oplopanax horridus* (devil's club: diabetes, stomach problems, arthritis); *Pyrus fusca* (Pacific crabapple: stomach problems, other ailments); *Rhamnus purshiana* (cascara: laxative); *Tsuga heterophylla* (western hemlock: internal bleeding, other ailments)	Barks usually removed from whole twigs or as portions from trunk; twigs sometimes rooted; some transplanting; trees not girdled; harvested selectively from a number of individual plants; usually sunrise side of tree; cultural constraints against girdling or overharvesting
MEDICINES: Pitch, resin, latex (Total ± 20)	*Picea sitchensis* (Sitka spruce: pitch as salve); *Pinus contorta* (lodgepole pine: salve, colds); *Populus balsamifera* ssp. *trichocarpa* (cottonwood: bud resin as salve); *Pseudotsuga menziesii* (Douglas-fir: salve, colds)	Pitch removed from injured or insect-damaged trees, or from bark blisters or buds; sometimes permanent "medicine" trees maintained over many decades

Use category	Examples	Notes on management
MEDICINES: "Roots" (Total: ± 50)	*Rumex occidentalis* (western dock: cuts); *Nuphar polysepalum* (yellow pond-lily: heart medicine); *Urtica dioica* (stinging nettle: arthritis); *Veratrum viride* (false hellebore: arthritis and other ailments: TOXIC)	Virtually all are herbaceous perennials; roots selectively harvested by size, life-cycle stage; fragments left behind often capable of regeneration
MEDICINES: Leaves and/or shoots (Total: ± 30)	*Lysichiton americanum* (skunk-cabbage: leaves for burns); *Maianthemum dilatatum* (wild lily-of-the-valley; leaves as poultice for cuts, burns); *Urtica dioica* (stinging nettle: leafy shoots for arthritis)	Herbaceous or woody perennials; leaves/shoots harvested selectively from living plants, which can then regenerate from rhizomes
MEDICINES: Flowers, fruits (Total: ± 15)	*Holodiscus discolor* (oceanspray: fruits for diarrhea); *Shepherdia canadensis* (soapberries: indigestion, ulcers); *Symphoricarpos albus* (waxberry: eye medicine, warts)	Herbaceous or woody perennials; flowers/fruits harvested from living plants, by season, life-cycle stage
MEDICINES: Miscellaneous, or unspecified, including fungi (Total: ± 20)	*Fomitopsis officinalis* (bracket fungus: fungus tissue used as purgative)	Gathered sporadically or incidentally

[1] Note that these numbers are only approximations, based on the summaries calculated for Nlaka'pmx (Thompson) herbal medicines. They are based not on species per se, but on the numbers of particular medicinal preparations used in treating a specific illness or condition (see Turner et al. 1990: 43-54). Although each cultural group has its own traditional medicines, the Nlaka'pmx medicines seem generally representative of the species, plant parts, and applications used in other areas of British Columbia, including the Northwest Coast.

TABLE 4.2. ECOLOGICAL EFFECTS OF INDIGENOUS HORTICULTURAL METHODS ON SPECIES POPULATIONS

Horticultural Method	Ecological Effects
Selective harvesting and replanting	Reduces intraspecies competition; promotes intentional dispersal of propagules
Digging and tilling	Incidental dispersal of propagules; creates local soil disturbance; recycles nutrients, aerates soil; increases moisture-holding ability; possible reduction of allelopaths
Tending and weeding	Reduces interspecies competition; promotes soil modification
Sowing and transplanting	Replenishes population; promotes dispersion of propagules to new habitats
Pruning and coppicing	Removes dead material reducing plant vigor; stimulates vegetative reproduction and eventually flowering and fruiting
Landscape burning	Reduces competition; accelerates recycling of mineral nutrients; blackened ground encourages spring growth; promotes selection for annual or ephemeral habit, synchronization of fruiting; maintains successional stages; creates openings

(from Peacock 1998 and based on Anderson 1993a; Ford 1985; Harris 1989)

enced the selection process (e.g., Peacock and Turner 2000). However, in general, the most important criteria were the yearly growth cycle, reproductive status (e.g., flowering versus nonflowering), and maturity and size.

The yearly growth cycles of culturally important species were well known and carefully monitored, since the desired qualities of a particular resource varied throughout its development, either seasonally (spring versus summer) or yearly as the plant matured. On a seasonal basis, variations in growth cycles

4.2. Cow-parsnip at harvesting stage (Photo by R. D. Turner)

meant certain resources could only be harvested during a short period of time at any given location even though the plant itself might be present throughout the year. The green shoots of cow-parsnip (*Heracleum lanatum*), for example, are harvested in early spring, before the plant flowers (Figure 4.2). After that, the stalks become unpalatable and undesirable (Kuhnlein and Turner 1987). On a yearly basis, variations in growth often meant a particular plant was left to mature for several years prior to harvesting.

The reproductive status of an individual plant, which is linked to growth cycles, was also an important criterion for selection. For example, a number of important root vegetables, such as camas (*Camassia leichtlinii* and *C. qua-*

mash), were frequently harvested only after the plant had started to die back and to go to seed (Arvid Charlie, personal communication to NT 1999; Elsie Claxton, personal communication to NT 1996; Christopher Paul, personal communication to M. Babcock 1967). Edible seaweed (*Porphyra abbottiae*) was harvested at its young, prereproductive life stage; the reproductive plants are considered too old, tough and "rotten" (Turner and Clifton 2002; Turner 2003a). Many medicinal roots were collected after flowering, at which point the roots were considered more potent. Similarly, the leaves of vegetative, non-fruiting cattail (*Typha latifolia*) and basket sedge (*Carex obnupta*) plants were selected for harvest over the fruiting ones. The reasons given are that the vegetative leaves are longer and of better texture (R. Y. Smith 1997; Turner and Efrat 1982), but the result is that the fruiting portions are left to produce seed and propagate. There were also cultural prohibitions against harvesting certain plants at certain reproductive stages.

Plants were also selectively harvested based on size preferences. Some would leave the largest plants to go to seed. For example, in the harvesting of root vegetables, only those of a certain size are taken; smaller individuals, be they bulbs, tubers or rhizomes, are left in place or sometimes replanted, to continue to grow for the next season. For example, Babcock (1967) reports about size selection of camas bulbs as discussed with Christopher Paul of Tsartlip:

> When gathering, the Indians would not collect the immature bulbs; Christopher Paul's mother's mother told him that these small, soft bulbs were "not worth cooking," and Mr. Paul has overheard his own parents discussing the matter. Mr. Paul thinks that these bulbs may have been gathered in another year, when they were mature, by the same family . . . The mature bulbs sometimes get as big as about 2½ inches [>6 cm] in diameter. After the bulbs become too old, however, they aren't any good to eat, either, Mr. Paul's maternal grandmother said. [p. 5]

Leaving the youngest and the very oldest bulbs in the ground would be an advantageous strategy for sustainable harvesting, because the younger bulbs would continue to grow for later harvest, and the oldest bulbs would be among the heaviest seed producers (Suttles, this volume). From our observations, there is a general correlation between the size of bulbs and the number of flowers and capsules it produces (see also Beckwith 2004). For root vegetables such as springbank clover (*Trifolium wormskjoldii*) rhizomes and northern rice-root (*Fritillaria camschatcensis*), leaving fragments behind to propagate and grow into new plants was a common practice (Deur, this volume).

Habitat preference was another criterion used in selective harvesting. Often, plants growing in a specific location were preferred to their counterparts in other regions. Certain places were known for their high-quality products: a stand of western yew trees (*Taxus brevifolia*) along the Cheewaht River was used intensively by Ditidaht implement makers (Turner, Thomas et al. 1983); an island in Clayoquot Sound was called "spruce roots island" after the spruce roots (*Picea sitchensis*) for basketry gathered there (Bouchard and Kennedy 1990). The western red-cedar (*Thuja plicata*) trees at Kanim Lake and a place called "wahiitlmitis" in Clayoquot Sound, and at another location along Tsusiat Lake in Ditidaht territory, were specifically sought by canoe makers because of the high quality of their wood; trees growing away from the ocean were preferred to those along the shore, whose wood would crack due to its exposure to salt (Bouchard and Kennedy 1990; R. Y. Smith 1997; Turner et al. 1983;). Similarly, basket-weavers have noted that red-cedar inner bark was preferred and considered of a finer grade when taken from inland trees than from those closer to saltwater (Compton 1993a; R. Y. Smith 1997). Medicinal plants, such as licorice fern (*Polypodium glycyrrhiza*, an epiphyte), were also preferentially harvested from particular places or from particular substrate species, such as Pacific crabapple (*Pyrus fusca*) or western hemlock (*Tsuga heterophylla*) trees (Compton 1993a; Turner and Efrat 1982). Some medicinal plants were said to be more potent if harvested from remote locations and upland sites. Berries, too, were selectively harvested from favorite places. In Clayoquot Sound, for example, a place called *"tl'uulhapi"* (Tonquin Beach) was a preferred place for Nuu-chah-nulth people to pick salal berries (*Gaultheria shallon*) because the berries are large and sweet here, according to Mary Hayes (Bouchard and Kennedy 1990: 513; R. Y. Smith 1997).

EXTRACTIVE TECHNOLOGIES

The specific tools and techniques used to harvest plant resources varied according to the species and the intended use of the plant. However, the net result of harvesting was to create an anthropogenic disturbance within a selected plant community. This was accomplished through digging, tilling, weeding, pruning, coppicing, and in some instances the selective burning of individual plants.

Digging or Tilling. Digging was the most common harvesting technique for roots and was used to collect a wide variety of edible and medicinal underground parts, as well as tree roots for basket materials. The digging stick (Figure 4.3), made of a hard wood such as western yew, Pacific crabapple, or oceanspray, was the implement of choice for prying up root vegetables such as camas bulbs, silverweed roots, clover rhizomes, and rice-root (Compton 1993a; Turner 1997a, 1998; Turner et al. 1983). Specialized tools, such as the yew wood *k'ellákw*, used by the Kwakwaka'wakw for root digging on estuarine tidal flats, allowed easy penetration of the soil and efficient extraction of the root product, with minimal damage to "root" segments (Deur, this volume; Adam Dick, personal communication to NT and DD, 1998). Generally, clumps of roots— together with any turf—were dug around, then overturned, the appropriate sized edible portions removed, and the other parts returned and covered up again (Turner et al. 1983). Digging activities tilled the ground, breaking the turf or sod, loosening and aerating the soil.

4.3. Yew wood digging stick (Haida; carved by Giitsxaa) (Photo by N. J. Turner)

Replanting. Harvesting activities often had impacts on the redistribution of plant propagules. Propagules could be separated from the harvested portion and replanted, or seeds from mature plants could be knocked unintentionally or sprinkled intentionally into the loose soil. For example, as noted previously, many root-vegetables were harvested after flowering, when their seeds would be ripe. The digging and associated harvesting practices would be expected to assist in the dissemination of seeds. The question is whether this dissemination was incidental or intentional (see Suttles, this volume).

There is also evidence of intentional planting for other species. Chief Adam Dick and Daisy Sewid-Smith discussed the practice of removing the small propagule at the base of each rice-root[2] (*xukwem*) bulb and replanting it at the time the bulbs were harvested. Adam Dick recalled doing this when he was a young boy of about nine or ten years:

> You know, it's on the bottom, called the *gagemp* [literally "grandfather," the underportion of the main bulb] . . . we peeled it off and, throw it back there [to be replanted—Daisy Sewid-Smith].

Fragments and smaller individuals of silverweed (*Potentilla anserina* ssp. *pacifica*) roots, clover rhizomes, and "wild carrot" (*Conioselinum pacificum*) roots were also replanted at the time of digging, according to several sources (Adam Dick, Alice Paul, Daisy Sewid-Smith, Margaret Siwallace, Felicity Walkus, personal communications to NT, 1984, 1996; Bouchard and Kennedy 1990; Compton 1993a,b; Deur 2000, this volume; Edwards 1979; Turner and Efrat 1982). It is notable that similar practices were followed in the digging and replanting of propagules of interior root vegetables like chocolate lily (*Fritillaria lanceolata*) and, especially, yellow glacier lily (*Erythronium grandiflorum*), according to elders like Secwepemc plant specialist Mary Thomas (Loewen 1998; Peacock and Turner 2000; Turner et al. 2000).

Weeding, Clearing, and Fertilizing. Harvest locales were often weeded during harvesting to remove competing or unwanted species. Weeding was reportedly undertaken on occasion by camas-bulb diggers of Vancouver Island (Arvid Charlie, personal communication to NT, 1999; Babcock 1967; Suttles 1951a; Turner and Bell 1971), as it was by harvesters of yellow glacier lily and other root vegetables of the Interior Plateau of British Columbia (Peacock and Turner 2000; Turner et al. 2000). Weeds were also cut from around Pacific crabapple trees owned by Haisla people (see Compton 1993a; McDonald, this volume). Sometimes beds for root vegetables were cleared of loose rocks and other debris to make digging easier. In the Kwakwaka'wakw silverweed and clover plots of the tidal flats, weeding was commonplace (Deur, this volume).

In addition, plant populations were occasionally fertilized to enhance productivity. For example, seaweed was placed on camas patches and potato gardens as fertilizer to enhance production by some people, at least recently (McDonald, this volume; Moss, this volume; Suttles, unpublished notes with

Mary George and Agnes George 1952; Turner 1973; Turner and Efrat 1982). The traditional fertilizing of tobacco plots on the Northwest Coast with rotting wood has been well documented (Turner and Taylor 1972). Leslie M. Johnson (personal communication to NT and DD, 1999) encountered oral accounts of fertilizing nettles (*Urtica dioica*) with waste products from eulachon processing. A significant account of traditional tending of "berry gardens" was provided by Heiltsuk cultural specialists Cyril Carpenter and Pauline Waterfall (personal communication to NT, 2002). They were each told by their grandmothers, Bessie Brown and Beatrice Brown (who were sisters-in-law), about traditional berry gardens, in which people selected sites that were especially productive for (native) blueberries, huckleberries, and other berry species, and carefully maintained and enhanced these patches. One example of such a berry garden was pointed out to Cyril Carpenter when he was about ten years old, by Bessie Brown. It was a wide, bushy ledge beside a waterfall, at Rosco Inlet in Heiltsuk territory. This was above the peoples' houses. Bessie Brown described how the berry gardens were located beside waterfalls, above the communities, because there was a continuous mist there for a good part of the year. This kept the berry bushes moist, even in the summertime. The gardens were also generally situated in locations where they were protected against the prevailing extreme wind conditions. On these ledges, too, they were exposed to the sunlight at certain hours of the day, and this was important for the ripening berries. Long ago, Bessie Brown said, the hunters and fishers took all the remains from cleaning and dressing their salmon, as well as deer, mink, otter, wolf, and mountain goat. They dug large holes in the ground around the berry bushes and buried these remains, covering them over with soil. When herring eggs were harvested, the waste trimmings were washed and dug into these garden plots. This is what made the berries grow so well. Cyril Carpenter recalled that the berries from these bushes were healthier, bigger, more productive, and tastier than untended berries; you could harvest them from the branches in handfuls. These were salal (Figure 4.4), blueberries, and huckleberries. People also scattered ashes from the fireplaces of the long houses around the berry bushes, as well as adding clamshells. In addition, people used to transplant whole berry plants to these productive sites, and when they did they used crushed clamshells in the holes.

The Heiltsuk used similar practices in their home gardens as well, following introduction of potatoes and other domesticated plants. Cyril Carpenter remembered seeing his neighbor's father packing sacks and sacks of salmon bones and guts, digging holes, and burying them beside his plum trees and other fruit trees, a practice that was believed to be traditional. They also used to pack seaweed up to the garden. This fertilizing was done in the fall, just before winter set in, when they were processing the fall runs of salmon. As a young boy, Cyril recalled hammering many sacks of clamshells into small pieces, to spread on the garden. More recently, Cyril fertilized a domesticated cherry tree beside their house; he buried the remains of red snapper, ling cod, and salmon, as well as ash from the fire beside this tree, and it had grown rap-

4.4. Salal berries (Photo by R. D. Turner)

idly, in contrast to another cherry tree nearby that was just left alone and is still much smaller and less productive. Pauline's family spill the salmon blood and other liquids at the base of various plants in their parents' garden. Those plants which receive this seasonally grow tall and strong with rich green foliage.

Other documentation of soil enrichment or fertilizing has been elusive, except in the context of burning, where the ash from the firing of camas beds was said to provide nutrients to the soil and allow the bulbs to grow better the next year. This was also suggested for interior peoples' burning of root-digging and berry grounds (Turner 1999).

It is notable that weeding and clearing of land is often correlated with proprietorship over a harvesting locality. This was commonplace, for example, in the case of camas harvest areas (Suttles, this volume). Straits Salish camas management practices were described to Marguerite Babcock (1967: 5) by Christopher Paul:

> The way that the family group . . . would establish claim to a plot of land [for camas harvesting] would be by clearing it. Once a family cleared a plot, it would "just naturally" become their plot to use, explained Christopher Paul. This clearing was done in the fall or spring before the gathering season, Mr. Paul thinks; in those seasons the soil was soft from the heavy rains, but not muddy (or frozen) as in the winter . . . The plot from which the bulbs were to be gathered would be cleared of stones, weeds, and brush, but not of trees. The stones would be piled up in a portion of the plot where there were no camas plants growing, and the brush would be piled to one side, left to rot or to be burned . . . ; this brush was actually uprooted, not just cut down. . . . The purpose of the clearing, said Mr. Paul, was to make the camas easy to clear [sic: dig?] when the camas was gathered intensively. The piles of stones on the plots are the remains or "markers" of the plots . . . however Mr. Paul doesn't know how or if the Indians set about marking off their plots other than clearing them. . . . He thinks that these plots may have been cleared every year before their use, in order to facilitate the gathering of the camas bulbs. . . .

Pruning or Coppicing. Pruning or coppicing (cutting shrubs down to their base to promote new growth) was another form of harvesting practiced on the shoots and stems of herbaceous and woody perennials used as food and for materials.

Green shoots of fireweed (*Epilobium angustifolium*), salmonberry (*Rubus spectabilis*), thimbleberry (*Rubus parviflorus*), cow-parsnip, and horsetail (*Equisetum telmateia*), all harvested for food in the early spring, were broken off from their rootstocks, and this stimulated the production of more shoots, as several elders have pointed out (Kuhnlein and Turner 1987; R. Y. Smith 1997; Turner 1995; Turner et al. 1983). Sometimes two or more "crops" could be taken from one patch during a single year, and then the plants were allowed to mature for a subsequent harvest, in a manner similar to the harvesting and cultivation of asparagus.

A range of berries were also "pruned" to enhance productivity. Daisy Sewid-Smith and Chief Adam Dick recalled that red huckleberries (*Vaccinium parvifolium*), salmonberries (Figure 4.5), and stink currants (*Ribes bracteosum*) were routinely "pruned" after the annual harvest by breaking the branches off.

> As soon as they clean that tree out [i.e., pick all the berries], we *tl'exw7id* [break off], we breaks them so . . . [the berries would grow plentifully later]. See, a lot of people think we never touched the wild . . . berries. But we did. We cultivated it. We pruned it . . . Especially that *gwadems* [red huckleberries], when they finished picking the *gwadems*, you know, they pruned them.

4.5. Salmonberries, different color forms (Photo by R. D. Turner)

They break the tops off. Salmonberries too. So, when the *qw'alhem* [salmon-berries], it's done, after you pick it, *tl'exwiiy* ["breaking the tops off"] they called that. My grandma tell me that if you let it grow this high [two meters or so], then it doesn't produce much berries. You know. But when you keep it down and, she says, the water, it's hard going up there, I guess, when it's too tall [Chief Adam Dick, personal communication to NT].

This pruning practice for red huckleberries was quite widely applied (Arvid Charlie, personal communication to NT, 1999; Percy and Dolores Louie, personal communication to NT 1998; R. Y. Smith 1997). Sometimes berries, such as soapberries (*Shepherdia canadensis*), were harvested by breaking off branches from the bush and shaking or picking off the berries (Elsie Claxton, personal communication to NT 1997; Compton 1993a). This was a form of pruning, which some people maintained made the berries more productive in the following years (Peacock and Turner 2000). Salal berries, too, were picked by their whole stems, a pruning process that some felt helped the plants to produce multiple stems of berries the following year (Gloria Frank, personal communication to NT, 1996; R. Y. Smith 1997). Crabapple trees were also pruned, according to Haisla elder Bea Wilson, who noted, "Old people would cut the branches from the top—this would ensure that there were lots next year— just cut the ones that had lots of fruits, not all of them" (Davis et al. 1995: 29). Pruning crabapple trees was also practiced in the territory of the neighboring Gitga'at (Coast Tsimshian) First Nation (Helen Clifton, personal communication to NT, 2002; McDonald, this volume).

4.6. Devil's club (Photo by R. D. Turner)

Devil's club (*Oplopanax horridum*) (Figure 4.6) and some of the other medicinal plants whose branches were harvested from the main plant could be considered to have been pruned as well (Captain Gold, personal communication to NT 1997; Marilyn Walker, personal communication to NT 1997, from her work with Tlingit people in Alaska). Not only might this stimulate growth in the plants, but one suggestion is that it may also stimulate the production of pharmacological compounds. Lantz (2001) has investigated the ecological characteristics of devil's club and demonstrated how traditional selective harvesting and other practices like replanting sections of the stem by harvesters can maintain the populations of this highly valued shrub.

Plants like basket sedge, bear-grass (*Xerophyllum tenax*), cattail, tule (*Schoenoplectus acutus*) and stinging nettle (*Urtica dioica*), sought for their leaves or stems as weaving and cordage materials, were cut in the late summer when the plants were at full maturity. Since these plants are perennials and their root systems were not impacted in harvesting, they would continue to grow the following year, and some people say they are even more prolific with cutting or pruning:

> Before I heard they didn't allow it [picking basket sedge in Pacific Rim Park Reserve], but I told him [the warden] that it's always better to get it cut, and prune it, it'll come out nice because the way it was, the grass it was brown from years back, dry [because people hadn't been gathering it] . . . it's like pruning it [Lena Jumbo, personal communication to Juliet Craig and Robin Smith, 1996; R. Y. Smith 1997: 149].

It used to be known that if you pick them, a better crop would always be coming out. Just like my Aunty Lena was talking about the grass [basket sedge]. You have to pick them in order to have a better crop the following year . . . [Arlene Paul Jumbo, personal communication to Juliet Craig and Robin Smith, 1996; R. Y. Smith 1997: 149].

The bark, wood, and roots of a wide range of trees were also important to Northwest Coast peoples and in a sense these materials were "pruned" from the living trees in a manner that would ensure the survival of tree. For example, in harvesting cedar bark, it was customary to remove bark from only one-third or so of the circumference of the tree. In this way, the tree would continue to grow. Boas (1921: 616–17) defines this practice clearly for the Kwakwaka'wakw:

Even when the young cedar-tree is quite smooth, they do not take all of the cedar-bark, for the people of the olden times said that if they should peel off all the cedar-bark . . . the young cedar would die, and then another cedar-tree near by would curse the bark-peeler so that he would also die. Therefore, the bark-peelers never take all of the bark off a young tree [see also Schlick 1994].

The thousands of "culturally modified trees" still standing in the Pacific coastal temperate rainforests with peeled bark or removed planks from past decades are a testimony to the success of this partial harvesting as a strategy for "keeping it living" (see British Columbia Ministry of Forests 1997; Garrick 1998) (Figure 4.7). Bark of bitter cherry (*Prunus emarginata*), which was used for wrapping implements, was stripped away from the inner bark, leaving the growing cambium layer intact and protected. Spruce roots and cedar roots for baskets were dug selectively as well, since it was recognized that taking too many roots from an individual tree would be harmful to it (Nelson 1983; Turner 1996a).

Bark removed totally, either to access inner bark and cambium for food or for use as medicine (e.g., cascara, *Rhamnus purshiana*), was generally taken in small patches or strips, so that the tree was not girdled and would be able to heal after awhile. One Nuxalk elder, Edward Tallio, is quoted by Edwards (1980: 10), discussing cascara bark harvesting: "If you just peel the tree as it stands the whole tree dies. The roots must die too I guess. Nothing comes up. Therefore we have to chop the tree down in order to have young ones come out again, or they just peel one side . . . ." Many trees were used for their bark medicine and, almost always, the bark was cut in a narrow vertical strip from the sunrise or river side of the tree (see Turner and Hebda, 1990). This was to allow the tree to continue to grow, and the sunrise side was said to heal over most quickly (Tom Sampson, personal communication to NT, 1987; Daisy Sewid-Smith, personal communication to NT, 1994).

Of course, trees needed for the wood of their trunks, or those whose bark

4.7. Haida weaver Florence Davidson peeling red-cedar bark (1971)

could not be removed in part, were cut or girdled, but this was evidently done selectively, and, by at least one account, people sometimes purposefully used windfalls and dead or dying trees to avoid killing living ones.

There are many other examples of particular parts, individuals, or populations of plants being pruned off, or selected in some other way, leaving the main portion of the perennial plant intact to continue to grow and reproduce. These practices are documented in many sources (Boas 1921; Compton 1993a,b; Scientific Panel for Sustainable Forest Practices in Clayoquot Sound 1995; R. Y. Smith 1997; Stewart 1984; Turner 1997b; Turner and Efrat 1982; Turner et al. 2000). Also, there is some ethnographic evidence of the pruning back of certain shrubs or trees to eliminate competition with culturally preferred species (McDonald, this volume). Deur (personal communication to NT, 2000) notes that there is evidence of similar pruning practices around the perimeter of small estuarine root grounds, where adjacent trees and shrubs would have shaded or overgrown patches of edible roots.

Transplanting. Transplanting, the removing of a growing plant, seedling, cutting, or other propagule from one place and replanting it in another, was also practiced on occasion, at least within the memories of contemporary elders. Adam Dick noted that highbush cranberry (*Viburnum edule*) could be transplanted and that patches that he had transplanted as a young man multiplied so that they became a prime picking area for his family in the years that followed. As recounted previously, Heiltsuk people also transplanted berry bushes (Cyril Carpenter, Pauline Waterfall, personal communication to NT, 2002). One Gitga'at woman suggested "borrowing a root" from a rare, yellow-berried variety of a red elderberry (*Sambucus racemosa*) bush growing away from their territory, in order to start one growing in Hartley Bay (Mildred Wilson, personal communication to NT, 2002).

Compton (1993a: 394) noted that springbank clover and Pacific hemlock-parsley, or "wild carrot," were reportedly "obtained from the Nuxalkmc who were said to have brought this plant to Hanaksiala territory. In return, the Nuxalkmc from Kimsquit and Bella Coola, as well as some Coast Tsimshian people from Metlakatla, were known to have returned home with stinging nettle plants from Hanaksiala territory following trading visits." Accounts of the transplanting of estuarine roots have been documented in several other contexts (Deur, this volume; Edwards 1979).

Hesquiaht elder Alice Paul (Turner and Efrat 1982) reported that camas from the Victoria area was once planted in the meadow behind Hesquiat Village. There are records of wapato (*Sagittaria latifolia*) being transplanted from one wetland area to another (see Darby, this volume). It does not grow naturally on Vancouver Island, for example, but was known to have been planted in small lakes on some of the Gulf Islands, and possibly also on Vancouver Island, by Cowichan peoples obtaining the tubers from their mainland relatives in the Fraser Valley (Arvid Charlie, personal communication to NT 1996, 1999).

Material plants, too, are sometimes transplanted. Alice Paul said that cat-tails had been introduced to Village Lake behind Hesquiat in order to have a ready supply of mat-making material. Ditidaht elder and basket-maker Ida Jones made several attempts to transplant American bulrush, or "three-square" (*Schoenoplectus olneyi*; syn. *Scirpus americanus*) to wet places along the San Juan River to give her a ready source of this important foundation material for her wrap-twined trinket baskets, a practice that was apparently rooted in traditional Ditidaht practices (Turner et al. 1983). Gitga'at elder Tina Robinson (personal communication to NT, 2002) also noted that the Kitamaat (Haisla) were known to plant stinging nettle in their yards, for use in making string and medicine.

LANDSCAPE BURNING AND CLEARING

On a larger scale, people managed to create a particular habitat type or successional stage, rather than to increase the production of individual species per se. The use of controlled burning to promote the growth and productivity of culturally important plant communities has been increasingly recognized for the Northwest Coast region, as for western North America in general (Blackburn and Anderson 1993; Boyd 1986, 1999; Deur 1999, 2000; Deur and Turner 1999; French n.d.; Gottesfeld 1994a; Norton 1979a,b; Suttles 1951a,b; Turner 1999) and was perhaps the most common form of community-level resource management.

The camas prairies of western Washington, the Willamette Valley and southern Vancouver Island and Whidbey Island have been particularly noted for their anthropogenic influences (e.g., Boyd 1999; Norton 1979b; Suttles, this volume). One article on colonization of Vancouver Island reported in the mid-1800s, pertaining to the area around western Victoria, in Esquimalt:

> Miles of the ground were burnt and smoky, and miles were still burning. The Indians burn the country in order to [promote] . . . more especially, the roots which they eat. The fire runs along at a great pace, and it is the custom here if you are caught to gallop right through it; the grass being short, the flame is very little; and you are through in a second. . . . [Anonymous 1849: 18–19]

"The roots which they eat" would certainly have included camas bulbs, among others. The fires were evidently within, and contributed to, the perpetuation of a Garry oak (*Quercus garryana*) parkland type landscape, and were low and cool enough to remove understory species without killing tall, mature trees. If repeated in several-year intervals, such fires would have served to maintain the parkland communities so conducive to camas growth. The camas itself, and the other edible-"rooted" perennials, would not have been harmed by the fire since they would have been in their summertime dormant phase. Brenda Beckwith (2002, 2004) speculates that landscape burning was a necessary final procedure in a well-timed and instrumented camas harvest season (see also McDougall et al. in press). The burning of the camas meadows on southern Vancouver Island could have provided valuable nutrients to both the bulbs and camas seeds, but the primary purpose of the use of fire was most likely for the continued maintenance of an open habitat free of encroaching woody plants (e.g., Douglas-fir, *Pseudotsuga menziesii*, and waxberry, *Symphoricarpos albus*). Hence, regular fire indirectly promoted the growth and productivity of camas by eliminating invasive and competitive species.

An early settler in Sooke, W. C. Grant, in a report to Governor James Douglas, complained of the practice of burning around his house by aboriginal people. He evidently understood the purposes of the burning, but did not attach significance or importance to it. He reported the fires were

> kindled promiscuously by the natives both in the wood and the prairie between the months of August and October. Their object is to clear away the thick fern and underwood in order that the roots and fruit on which they in a great measure subsist may grow more freely and be the more easily dug up. [Grant ca. 1848]

Less well known are the traditional burning practices of the Coast Tsimshian, Haida, Haisla, Nuxalk, Kwakwaka'wakw, Nuu-chah-nulth, and other peoples of the Coastal Western Hemlock Zone, where precipitation levels generally preclude natural fires. Although evidence is still limited, use of fire by these peoples is known (Turner 1999, 2004). It is generally restricted to production of small clearings, sometimes on islands, that enhance the growth of berries such as salal, huckleberry, blueberry, and strawberry, though camas sometimes appears to have been present in the anthropogenic clearings of the outer south coast (Deur 1999). Within this zone, for example, Kwakwaka'wakw apparently burned berry bushes, as reflected in the "prayer" to Berries, spoken by a woman before she picks the berries (translated by Boas into English; species not given; even in original Kwak'wala language version a general term for fruit is used):

> I have come, Supernatural Ones, you, Long-Life-Makers, that I may take you, for that is the reason why you have come, brought by your creator, that you may come and satisfy me; you Supernatural Ones; *and this, that you do not blame me for what I do to you when I set fire to you the way it is done by my root (ancestor) who set fire to you in his manner when you get old on the ground that you may bear much fruit* . . . [emphasis added; Boas 1930: 203]

GARDENING STRATEGIES

As noted throughout the above descriptions of plant management, several important root foods—particularly springbank clover, silverweed, rice-root, and lupine—were intensively cultivated in "gardens" located in tidal marshes (Deur, this volume). Individual chiefs and families commonly owned silverweed and clover beds up and down the coast. Many intensively managed root-digging sites have been identified in this region (Arima et al. 1991; Boas 1934; Bouchard and Kennedy 1990; Compton 1993a,b; Sapir 1913–14; Turner and Efrat 1982; Turner et al. 1983). The patches of silverweed, clover and other root vegetables were certainly regarded as gardens. Plots were often marked out with posts, wooden pegs, or poles laid on the ground, or had boundaries determined by natural features such as large rocks on the tidal flats. Families traditionally attempted to expand their plots, incorporating more marsh area into them every year. In some places, this was accomplished through extensive modifications of the estuarine marsh. These gardens were carefully tended and looked after, with the plants being selectively harvested, replanted, weeded, and occasionally transplanted, as discussed above. The constant influx of nutrients from regular tidal inundation and alluvial deposits would preclude the necessity for using extra fertilizers to maintain productivity (Deur 2000, this volume).

TOBACCO CULTIVATION

As suggested in the Introduction of this volume, evidence suggests the Northwest Coast tobacco of the Haida, Tlingit, and probably the Tsimshian, was grown in garden settings and was propagated from "seed" each year (probably seed capsules, containing seed, judging from the explorers' descriptions of the size of the seed). The way the original seed was obtained, from supernatural circumstances, is described in several different Haida narratives (Turner and Taylor 1972). Not surprisingly, the tobacco gardens of the Haida were also weeded and fertilized in the fall by mixing rotten wood into the soil after the tobacco had been harvested (Turner and Taylor 1972).

The Social Context for Plant Management

Although not represented in Table 4.2 as management "techniques," peoples of the Northwest Coast also employed a number of resource management strategies on a larger scale, such as within a traditional territory, which, in turn, influenced species and community productivity and diversity. These included a planned and patterned seasonal round, the rotation of harvesting locales, controlling access to resource patches, and religious ceremonies and moral sanctions, which altered the time or intensity of plant harvests (e.g., Turner 1997a; Turner and Atleo 1998; Turner et al. 2000; Turner et al. Chapter 5, this volume). These principles provided the social context for plant management.

SCHEDULING

Decisions concerning where, when, and what to harvest were dictated by cultural preferences and necessity, but limited by the spatial and temporal availability of plant resources. On one hand, the timing and frequency of collecting were determined by the life cycles of the plants themselves. These life cycles varied between species and according to the specific environment (e.g., aspect, precipitation, elevation) of a harvesting locale. Such natural scheduling constraints had to be balanced with cultural preferences for species at certain growth stages, as well as with scheduling conflicts that might arise when several species were available simultaneously for harvesting in different locales. Each group, and often each family, had its own routine patterns of travel and resource harvest within its traditional territory. Of course, the timing varied with latitude and specific climate and other patterns of productivity.

Starting in early spring, with the flowering of the salmonberry and elderberry bushes indicating the beginning of the growing season, people traveled to particular sites to harvest seaweed, collect herring spawn, gather seabird eggs, and harvest green shoots and early spring crops of wild root vegetables, as well as to hunt (for land and marine mammals and gamebirds) and fish (e.g., for eulachon, halibut, spring salmon). What followed was an ongoing procession of harvesting activities, with berries ripening in succession: salmonberries, strawberries, elderberries, huckleberries, blueberries, thimbleberries, salalberries, currants, and, later in summer and into fall, the mountain huckleberries and blueberries, crabapples, highbush cranberries, bog cranberries, and evergreen huckleberries, depending on the locality. Edible inner bark of hemlock and some other trees was harvested around June. Various root vegetables were harvested in summer or fall, usually at the time they were going to seed and their leaves were dying back. Red-cedar and yellow-cedar bark, and roots of red-cedar and Sitka spruce were usually harvested in early summer, whereas other basket materials like cattail, basket sedges, and tule were harvested in late summer or early fall. Nettle stems, for use in fishnets and lines, mat-making, and cordage were cut around the time the first snow hit the mountains. All during this period, animals were hunted and fish caught and processed. People generally harvested shellfish. such as clams, mussels, abalone, and rock chitons, from fall through to spring, but not usually in summer.

Exact determination of the time to harvest specific resources might be left to the control and judgement of a "chief" or clan/household leader, with samples and advice from family members and the use of seasonal indicators such as the blooming of certain plants, or the presence of a specific migratory bird (Lantz and Turner 2003). People usually went to particular camps that were the centers for many subsistence activities, from plant gathering to salmon or halibut fishing, rendering eulachon grease, hunting seals, or, for the Nuu-chah-nulth, whale hunting. An example is the Gitga'at seaweed camp on Prince Royal Island south of Hartley Bay, where Gitga'at families have gone for countless generations for three or four weeks or more around May to harvest and process their seaweed, halibut, spring salmon, rockfish, chitons, seal, seabird eggs, and other spring foods.

Scheduling of harvesting and peoples' presence in particular areas was itself a management strategy, one that would tend to limit overharvesting because of time constraints and diversification of resource use. The necessity of moving on to other harvesting locales as salmon or other resources became available would itself confine the extent of harvesting a particular resource in any one location.

CONCEPTS OF OWNERSHIP OF PLANT RESOURCES

Considerable variation is found in concepts of ownership of land and resources along the coast, but virtually everywhere there were areas of recognized proprietary and territorial jurisdiction, as discussed by Suttles in this volume. Whether recognized at the community level or at the level of clan, family, or individual, the right to harvest and to control the harvest of other people at highly valued places and for high-value resources were widely established. Such proprietorship resulted in intensive monitoring, harvesting, and managing of sites and resources, and, we would argue, ultimately led to sustainable resource use.

Among the Kwakwaka'wakw, for example, the clans, or lineages from a single ancestor would own and have jurisdiction over an entire territory. This jurisdiction was held by a clan even if it moved to another village. Then, within the clan, individual families would have their own root-digging plots or crabapple trees. When asked if outside people would have to ask permission to use a clan's traditional territory, Daisy Sewid-Smith and Chief Adam Dick said that people very seldom crossed the boundaries, since the clan's territory was widely recognized and respected. Adam Dick added, "They don't do that, come across, 'cause they know they're owned. They will respect each other." Daisy Sewid-Smith pointed out that if people needed particular resources from another clan's territory, they would trade for them in exchange some product they had readily available to them.

Plant resources for which ownership was often claimed included root digging grounds, seaweed gathering sites, productive berry patches, and crabapple trees. Pacific crabapple trees in Haisla and Hanaksiala territory, for example, belonged to a number of different families from several different tribes (Compton 1993a; McIlwraith 1948, 1: 133), and the harvest from individual trees or "crabapple gardens" was controlled by families or chiefs. Olson (1940: 182) reported that "It is customary . . . at crabapple time for the ranking chief to test the berries and to give the word when they are ripe enough. No one may gather them before this." He also stated that controlled berry grounds never were "owned" by women. This was not the case on Haida Gwaii, however, where high-ranked women at Skidegate were said to own highbush cranberry patches behind the village (Emma Wilson, personal communication to NT, 1970). Highbush cranberry patches were also owned by Hanaksiala and Tsimshian chiefs (Compton 1993a). Gitga'at Eagle chief Ernie Hill Jr. described how his grandfather, Ambrose Robinson, who was in the Blackfish Clan, gained the rights to a highbush cranberry patch across from Hartley Bay. The rights were bequeathed to him by an elderly man of the Eagle clan in recognition of an act of kindness. Then, after his grandfather's death, the rights to this patch reverted to Ernie Hill Jr. as current Chief of the Eagle clan (Ernie Hill Jr., personal communication to NT, 2002).

Adam Dick and Daisy Sewid-Smith elaborated further on ownership of resources:

> Everybody had their own berry patches, just like everybody had their own clam beds. Things like [salal patches], Yeah, salmonberries and all that, all kinds of berries, wild crabapple, you just don't go [out and pick] . . . There's a certain places that a certain family goes, especially that wild crabapples. Our family used to go over here. And the other families go over here. They got markers too, for *celxw* [crabapples]. Oh, yes, they have pegs, you put pegs all around the tree. Especially the . . . wild crabapples. You just pick up those little sticks and you just peg, put it around the [tree]. Any kind of sticks. If you can get cedar, that would be good. Anything that's pegged, you know it's someone's [Chief Adam Dick, personal communication to NT, 1997].

Not all berry patches were under exclusive control, however. When asked about Kwakwaka'wakw ownership of the bog cranberry beds at Kingcome, Adam Dick said that they were so extensive, anyone could go there to pick.

As noted in Suttles (this volume), some camas beds, particularly those on small offshore islands around the Saanich Peninsula, were owned by certain Coast Salish families (Cheryl Bryce, personal communication to NT 2000; Elsie Claxton, personal communication to NT, 1996; Suttles 1951a, this volume). Nuu-chah-nulth and Ditidaht people sometimes traveled to Victoria to dig camas bulbs, and they would always ask permission from the Coast Salish people to do this (Turner et al. 1983).

Ownership brought with it responsibilities. These included the obligation to look after the resources in order to ensure their continuous production into the future, and to look after the needs of the tribe or community by sharing the proceeds with them. With his first harvest of berries, for example, a Nuu-chah-nulth chief gave a feast to his people (Drucker 1951: 252). The Nuu-chah-nulth system of chiefs' proprietorship over resources and their responsibilities to maintain them and to share them with their people is called *hahuulhi*. Chiefs' areas of *hahuulhi* are still recognized today, and may become once more an important element in management and decision-making along the west coast of Vancouver Island (Scientific Panel for Sustainable Forest Practices in Clayoquot Sound 1995; Foreword, this volume). Ernie Hill Jr. (personal communication to NT, 2002) explained the situation with his grandfather's high-bush cranberry patch: When the berries were ripe, his grandfather would take a whole group of Gitga'at people—anyone who wanted to go—over to pick them. They would spend the night there, and there was a dugout canoe kept there to help them access the berries along the shore of a lake. Each person who was picking gave the first bucket they picked to Ambrose Robinson, as the "owner" chief. After that, they were permitted to pick any quantity of berries they wished for their own use. Ambrose Robinson, on his part, used the berries he received to host a feast for the village, as part of his obligations as a chief.

CEREMONIAL RECOGNITION OF RESOURCES

The ownership and rights to use certain areas and resources were validated at many opportunities, through public recognition at feasts and potlatches and other ceremonial occasions. Ceremonies, such as the "first salmon" ceremony of many groups and the sacred cedar-bark ceremonial of the Kwakwaka'wakw (Sewid-Smith et al. 1998) were also a means of recognizing the importance of resource sustainability, and of formally showing respect for the other life-forms that people depend upon. These ceremonies, while perhaps not directly connected with anthropogenic plant communities, certainly influenced and constrained peoples' attitudes and behavior toward their resources, and they form part of the major context of resource use that determined how people related to their lands (Turner 1997a).

One example of how people's ceremonial practices determined specific harvesting is in the Chehalis "First-Fruits Ceremony," as described by Hill-Tout (1978: 116):

> Another of these ceremonies was kept in connection with the *satske*, or young succulent suckers of the wild raspberry ["*Rubus nutkanus*"; thimbleberry, *Rubus parviflorus*—Figure 4.8] which the Indians of this region eat in large quantities, both cooked and raw. When cooked, I am told they eat like asparagus. The time for gathering these was left to the judgment and determination of the chief. When ready to gather, he would direct his wife or daughter to pick a bunch and bring them to him; and then, the people all being assembled, a ceremony similar to that connected with the salmon ceremony would take place. After the ceremony anyone might pick as much as he liked. A similar ceremony took place later in the summer, when the berries of this plant were ripe.

Hill-Tout continues, expressing the opinion that such ceremonies were:

> intended to placate the spirits of the fish, or the plant, or the fruit . . . in order that a plentiful supply of the same might be vouchsafed to them. The ceremony was not so much a thanksgiving as a performance to ensure a plentiful supply of the particular object desired; for if these ceremonies were not properly and reverently carried out there was danger of giving offense to the spirits of the objects and being deprived of them. . . . For it must be remembered that . . . the salmon, or the deer, or the berry, or the root, was not merely a fish or an animal or a fruit, in our sense of these things, but something more. The Indian's view of the universe was essentially an anthropopathic one. . . .

4.8. Thimbleberry (Photo by N. J. Turner)

Another indication of peoples' deep respect for the plant resources they used is reflected in the words people used to address the spirit of the plants before they harvested them. Important examples are seen in Franz Boas's recorded Kwakwaka'wakw "prayers" (more accurately translated as "words of praise"). The first is an address to berries by a highbush cranberry picker, "Prayer of a Woman in Charge of Berry Picking in Knights Inlet":

> I come, One-Prayed-to, I try to come to you, means of mercy to me, that I may eat, that I may keep alive for a long time, you, Chief of the Upper World; you Life-Owner. Pray, let me come next year to stand again at the place where I am standing to pray to you. [Boas 1930: 203]

The second is offered to a young red-cedar:

> Look at me, friend! I come to ask for your dress. For you have come to take pity on us; For there is nothing for which you cannot be used. For you are really willing to give us your dress, I come to beg you for this, Long-Life Maker. For I am going to make a basket for Lily-roots [*Erythronium revolutum*] out of you. I pray, friend, not to feel angry on account of what I am going to do to you. And I beg you, friend, to tell our friends about what I ask of you! Take care, friend! Keep sickness away from me, so that I may not be killed by sickness or in war, O friend! [recorded by Boas 1921: 619]

Peoples' reverent, yet pragmatic, attitudes toward the plants they used would encourage resource conservation and sustainability, both of which are promoted in the concept of "keeping it living." Cultural constraints against taking too much or wasting resources, talked about by many elders as part of their early training (see Scientific Panel for Sustainable Forest Practices in Clayoquot Sound 1995; Turner et al. 2000), are closely related to this concept.

The Making of an Anthropogenic Landscape

The ethnobotanical evidence presented in the preceding discussion, when framed in terms of models of people–plant interactions, demonstrates that peoples of the Northwest Coast were active managers of their plant resources, employing a variety of strategies and practices, in a variety of habitats, to ensure a reliable, predictable supply of culturally important plant products. These practices, in turn, shaped the landscape of the Northwest, creating a series of anthropogenic landscapes. Table 4.3 serves to summarize and explain this point. It outlines the wild-plant management strategies of Indigenous peoples and the ecological effects of those strategies on the landscape at three spatial scales or levels of biological organization (Peacock 1998). It suggests the use of a wide range of cultivation methods, guided by management activities designed to regulate the timing, intensity, and frequency of harvest, that created anthropogenic disturbances within populations of culturally valued species in the habitats (communities) where they occurred. These disturbances have had both intentional and incidental impacts, and acted to increase the productivity, distribution, and predictability of key species, and to maintain habitats and conditions conducive to their growth. This had the effect of creating mosaics of productive communities on the landscape, increasing not only the diversity of species, but ultimately the productivity of the landscape.

Anthropogenic plant communities may be identified in at least eight environmental zones within coastal British Columbia, as summarized in Table 4.4:

    low elevation meadows
    rainshadow (Douglas-fir) forest
    coastal rainforest (and associated clearings)
    montane forests (and associated clearings)
    freshwater marshes and swamps
    freshwater bogs and fens
    tidal wetlands
    human occupation sites

Each of the habitat types identified in table 4.4 is discussed briefly below.

(text continues on page 140)

TABLE 4.3. INDIGENOUS PLANT MANAGEMENT STRATEGIES
AND THEIR IMPACTS ON THE PRODUCTIVITY AND
AVAILABILITY OF PLANT RESOURCES

Use of Horticultural Methods

selective harvesting, digging, and replanting;
tilling and weeding; sowing and transplanting;
pruning and coppicing; burning

Guided by Management Activities

scheduling of seasonal rounds; rotation
of harvesting locales; controlled access;
religion/moral sanctions;

Regulates

the scale, frequency and intensity
of anthropogenic disturbance

Scale of Application	*Results in:*
Species Level	*Increased productivity of selected species through:* Altered age structure, density, distribution, genetic structure, longevity, range, and yield of species with cultural utility
Community Level	*Increased habitat diversity through:* Altered community dynamics, size, and types; altered species associations, composition, diversity, richness, and vertical structure; creates openings or ecotones; halts successional sequences, creating productive seres
Landscape Level	*Increased heterogeniety of the landscape* Creates a mosaic of productive plant communities across landscape with both structural and compositional diversity

NET RESULT

Increased productivity and availability of culturally significant
plant resources in an anthropogenic landscape

(from Peacock 1998)

TABLE 4.4 MAJOR HABITAT TYPES OF BRITISH COLUMBIA'S NORTHWEST COAST, WITH REPRESENTATIVE CULTURALLY
IMPORTANT PLANT SPECIES AND ASSOCIATED MANAGEMENT PRACTICES

Habitat type	Examples of culturally important plants	Associated management
Meadow: low elevation	*Achillea millefolium* (yarrow); *Camassia* spp. (edible blue camas); *Fragaria* spp. (wild strawberry); *Fritillaria lanceolata* (chocolate lily); *Lilium columbianum* (tiger lily); *Pteridium aquilinum* (bracken fern); grass species	Burning; selective harvesting; tilling; weeding; ownership
Forest: rainshadow	*Abies grandis* (grand fir); *Acer macrophyllum* (broad leaf maple); *Allium cernuum* (nodding onion); *Alnus rubra* (red alder); *Arbutus menziesii* (arbutus); *Asarum caudatum* (wild ginger); *Fragaria* spp. (wild strawberry); *Heracleum lanatum* (cow-parsnip); *Holodiscus discolor* (oceanspray); *Lilium columbianum* (tiger lily); *Mahonia aquifolium* (Oregon-grape); *Perideridia gairdneri* (wild caraway); *Polystichum munitum* (sword fern); *Prunus emarginata* (bitter cherry); *Pseudotsuga menziesii* (Douglas-fir); *Pteridium aquilinum* (bracken	Burning; selective harvesting; pruning; ownership

	fern); *Rhamnus purshiana* (cascara); *Rubus parviflorus* (thimbleberry); *Salix* spp. (willow); *Shepherdia canadensis* (soapberry); *Symphoricarpos albus* (waxberry); *Taxus brevifolia* (Pacific yew); *Thuja plicata* (western red-cedar); *Vaccinium parvifolium* (red huckleberry)	Limited (patch) burning; selective harvesting; pruning; fertilizing; transplanting; ownership
Forest: rainforest (including small clearings within forest)	*Abies amabilis* (silver fir); *Alectoria sarmentosa* (old man's beard lichen); *Alnus rubra* (red alder); *Blechnum spicant* (deer fern); *Chamaecyparis nootkatensis* (yellow-cedar); *Dryopteris expansa* (spiny wood fern); *Echinodontium tinctorium* (Indian paint fungus); *Epilobium angustifolium* (fireweed); *Erythronium revolutum* (pink fawn lily); *Gaultheria shallon* (salal); *Heracleum lanatum* (cow-parsnip); *Maianthemum dilatatum* (wild lily-of-the-valley); *Oplopanax horridum* (devil's-club); *Picea sitchensis* (Sitka spruce); *Polystichum munitum* (sword fern); *Prunus emarginata* (bitter cherry); *Taxus brevifolia* (Pacific yew); *Tsuga heterophylla* (western hemlock); *Vaccinium alaskaense, Vaccinium ovalifolium* (blueberries); *Vaccinium parvifolium* (red huckleberry); *Thuja plicata* (western red-cedar)	

TABLE 4.4. (continued)

Habitat type	Examples of culturally important plants	Associated management
Forest: montane (including small clearings)	*Chamaecyparis nootkatensis* (yellow-cedar); *Epilobium angustifolium* (fireweed); *Oplopanax horridum* (devil's club); *Shepherdia canadensis* (soapberry); *Vaccinium alaskaense* (Alaska blueberry); *Vaccinium deliciosum* (Cascade bilberry); *Vaccinium membranaceum* (black huckleberry); *Vaccinium ovalifolium* (blueberry); *Vaccinium parvifolium* (red huckleberry); *Veratrum viride* (false hellebore)	Burning; selective harvesting; pruning
Freshwater wetlands: marshes, swamps, alluvial floodplains	*Alnus rubra* (red alder); *Carex obnupta* (basket sedge); *Equisetum telmateia* (giant horsetail); *Lysichiton americanum* (skunk-cabbage); *Populus balsamifera* ssp. *trichocarpa* (cottonwood); *Pyrus fusca* (Pacific crabapple); *Ribes bracteosum* (gray currant); *Rubus spectabilis* (salmonberry); *Sagittaria latifolia* (wapato); *Salix* spp. (willow); *Schoenoplectus acutus* (tule); *Typha latifolia* (cattail); *Veratrum viride* (false hellebore); *Viburnum edule* (highbush cranberry)	Selective harvesting; pruning; ownership

Habitat	Species	Management
Freshwater wetlands: fens, bogs	*Chamaecyparis nootkatensis* (yellow-cedar); *Ledum groenlandicum* (Labrador tea: beverage); *Nuphar polysepalum* (yellow pond-lily); *Pinus contorta* (lodgepole pine); *Sphagnum* spp.; *Vaccinium oxycoccos* (bog cranberry); *Vaccinium uliginosum* (bog blueberry)	Selective harvesting; some ownership
Tidal wetlands: salt marshes, tidal floodplains	*Conioselinum pacificum* ("wild carrot"); *Fritillaria camschatcensis* (rice-root); *Lupinus nootkatensis* (Nootka lupine); *Potentilla anserina* ssp. *pacifica* (Pacific silverweed); *Pyrus fusca* (Pacific crabapple); *Rumex occidentalis* (western dock); *Schoenoplectus olneyi* ("three-square"); *Trifolium wormskjoldii* (springbank clover); *Triglochin maritimum* (arrow-grass); *Viburnum edule* (highbush cranberry)	Intensive tilling, selective harvesting, transplanting; replanting propagules, weeding, ownership
Human occupation sites	*Achillea millefolium* (yarrow); *Nicotiana* sp. (Haida tobacco; leaves chewed); *Rubus* spp. (thimbleberry, salmonberry); *Plantago major* (broad-leaved plantain); *Rumex occidentalis* (western dock); *Urtica dioica* (stinging nettle); *Pyrus fusca* (Pacific crabapple)	General use: fertilizing; pruning; clearing; selective harvesting; accidental/possibly intentional introduction

Perhaps the best-known anthropogenic plant communities of the Northwest Coast in British Columbia are the Garry oak parkland areas such as those occurring extensively around the Victoria region of southern Vancouver Island. The most intensively targeted plant resources in this community are the edible bulbs of blue camas, which were extremely important to the Straits Salish peoples of this region (Beckwith 2004; Lutz 1995; McDougall et al. in press; Suttles 1951a,b, this volume; Turner 1995, 1999; Turner and Bell 1971; Turner and Kuhnlein 1983). Camas beds were burned over, weeded, cleared, selectively harvested, and sometimes intentionally seeded, and were also sometimes owned by individuals or families. Other culturally important plants that would have benefited from cultivation practices for camas include chocolate lily, Hooker's onion, false onion, bracken fern, and tiger lily. Some camas meadows were also situated within the wetter forest ecosystems along the Northwest Coast, where burning or tidal inundation maintained an open landscape.

RAINSHADOW (DOUGLAS-FIR) FORESTS

As well as Garry oak parkland, clearings and small openings in the relatively dry conifer woods on the southeast coast of Vancouver Island and the Gulf Islands were also burned over, as discussed earlier, in order to maintain production of large patches of fruit-bearing shrubs (salal; thimbleberry—both sprouts and berries; red huckleberry; trailing blackberry—*Rubus ursinus*; and blackcap—*Rubus leucodermis*), as well as wild strawberries (*Fragaria* spp.) (Figure 4.9) and possibly hazelnut (*Corylus cornuta*), which was known to have been burned to enhance its productivity in other regions (Boyd 1986, 1999; Turner 1999). The burning extended along the west coast of Vancouver Island and into the transition region with the Coastal Western Hemlock zone. Burning may have enhanced the growth of Douglas-fir (*Pseudotsuga menziesii*), as a fire climax species, and its associated communities in areas where climax western hemlock forests would be expected in the absence of frequent fire.

COASTAL RAINFOREST

Intentional burning and other practices also modified habitats and enhanced resources within the moist western hemlock forests in various places along the coast. On the west coast of Vancouver Island, areas of Clayoquot Sound were reported to have been burned by Nuu-chah-nulth people to stimulate the growth of berries (George Louie, personal communication to Randy Bouchard, 1991). The burning of this area was confirmed by Stanley Sam (Ahousaht/Clayoquot), who noted that Alaska blueberries (*Vaccinium alaskaense*), red huckleberries, and salal berries grew particularly well after an area had been burned (Bouchard and Kennedy 1990).

As a boy, Willie Hans, a Nuxalk elder (personal communication to NT,

4.9. Wild strawberries (Photo by N. J. Turner)

1981), asked his father why there were so many burned trees around the Bella Coola valley, and was told that they burned to encourage the berries, especially raspberries (*Rubus idaeus*) and blackcaps. The late Felicity Walkus, who was originally from South Bentinck Arm, recalled that they "burned lots" around that area, and the late Dr. Margaret Siwallace, born at Kimsquit, said they also used to burn there to promote berry growth (both personal communications to NT, 1981). The Haisla and Haida also burned areas within their respective territories, around Kitimaat, and on Haida Gwaii, to encourage the growth of berries such as salal and red huckleberries (Lopatin 1945: 14; Turner 1999). Ernie Hill Jr. of the Gitga'at Nation reported (personal communication to NT, 2002) that there was an island in a long inlet near Kitkatla, where his father was raised, that people used to burn over completely to enhance the growth of blueberries (*Vaccinium ovalifolium*), and that the Tsimshian name for this island translated as "on-burned-down-for-blueberries." As previously discussed, the Kwakwaka'wakw people also burned berry bushes. Burning also allowed for the creation of prairies of bracken fern—the rhizome of which was roasted, ground into a flour, and baked into a bread—as well as the maintenance of small camas prairies far from the natural range of peak camas productivity. Once burning ceased, the coastal forest has reoccupied these prairies, and in some cases largely eliminated once abundant species of culturally preferred plants (Deur 1999).

Since considerable harvesting of trees and wood took place (especially red-cedar), with tree roots used for cordage and basketry, tree bark for construction, weaving, and medicine, branches for material and medicine, and a variety of other useful and culturally important plants, these forests, even those not burned over, can certainly be described as anthropogenic. Peoples' visible impacts on the forests were not as obvious as the impacts of logging and other activities of the incoming Europeans, but they were nonetheless present. They can be seen in the "culturally modified" trees, stumps, and logs throughout the coastal forests (British Columbia Ministry of Forests 1997; Garrick 1998). The gathering of cedar and spruce roots from the forest floor, and from rotten logs in the forest where the penetrating roots grow long and straight, was an activity that would have impacted the forest plant communities, but since it was done carefully and selectively, the impacts would be very subtle.

MONTANE FOREST

Contrary to common opinion, coastal peoples did venture into the mountains, and did use both plant and animal resources from montane forests (see Lepofsky et al., this volume). Yellow-cedar bark and many types of berries were particularly sought (R. Y. Smith 1997). For example, Duff (1952: 73) notes that for the Stó:lō (Halq'emelem) people of the Fraser Valley, "To obtain most of the berries they preserved for winter use, parties of women and a few men went into the mountains for several days during the late summer [for huckleberries and blueberries of various kinds (*Vaccinium* spp.)]. Then he adds, "Sometimes patches of these berries were burned to improve their future yield."

This indicates at least one management practice for this Coast Salish group (see also Lepofsky et al., this volume). Their northern neighbors, the Pemberton Stl'atl'imx, or Lil'wat are well known for their practice of burning mountainsides (Peacock and Turner 2000; Turner 1999). Such burning for berry enhancement has also been reported in montane areas at the far south end of the region, where such methods were employed to increase output of huckleberries and blueberries (*Vaccinium* spp.) (e.g., Deur 2002c; LaLande and Pullen 1999). The effect of such burning on the production of berries, as well as of certain root vegetables, by clearing out brush from under the trees, was said to be major; conversely, since these practices have been stopped because of forestry regulations, these food plants are widely considered to be less productive, or to even be disappearing altogether (Minore 1972; Peacock and Turner 2000; Turner 1999). Other details and impacts of human interactions with montane ecosystems will require further investigation.

Many culturally important plants, mostly herbaceous perennials, were harvested from marshes, swamps, riverbanks, and lakeshores. Harvesting was often intensive. A relatively small tule mat, for example, of about 1 m square, might require 100 to 150 individual tule stalks. Considering the immense numbers and large size of such mats used traditionally by Nuu-chah-nulth and Coast Salish peoples especially, one would expect that thousands of tule stalks would have been gathered every year from designated harvesting areas. Some people maintained that thinning out the leaves or stems through harvesting actually allowed the plants to grow better the following year. Wapato, with its edible corms, was also harvested in considerable quantities within freshwater wetland environments, including tidally influenced freshwater wetlands (Darby, this volume; Spurgeon 2001). Other than this harvesting-related human influence, through which the habitats of these plants can be collectively identified as anthropogenic communities, there is little evidence of major human impacts, perhaps with the exception of wapato, which was harvested intensively, and Pacific crabapple, which was owned, pruned, and tended in specific places. Wapato swamps, once common and extensive in the Fraser Valley,[3] have been almost entirely eliminated from many sites by draining for agricultural purposes, and perhaps by a reduction in human management of wapato resources (see Darby, this volume; Carlson 2001). The pruning of Pacific crabapple trees has been noted, as has family ownership of these trees, and of highbush cranberries, which can often be found in the same wetland sites. Some people refer to sites with crabapple trees as "crabapple gardens," a significant appellation.

FRESHWATER BOGS AND FENS

These are specialized habitats that are wet but poorly drained and acidic, so that organic matter builds up without decomposing. Specific plants, adapted to these conditions, include sphagnum moss (used for diapering), Labrador-tea (*Ledum groenlandicum*), bog cranberry (*Vaccinium oxycoccos*), and bog blueberry (*Vaccinium uliginosum*). Often these habitats form extensive meadows from which people would harvest large quantities of plants, as well as using them for hunting the geese, swans, and ducks that light in these areas in large numbers to feed during their fall migrations.

One such meadow occurs at Kingcome Inlet, as recalled by Adam Dick:

It's a *big* field. As far as you can see, it's just flat. With *qiqelis* [bog cranberries]. And . . . swans, were there . . . all the birds, ducks . . . they eat that. You pole up there [by canoe]. It's a long ways from the village. Then we tent, put a tent there, on the edge of that meadow. . . . Anybody could go . . . from the village. They can go and pick, all the . . . acres and acres of [cranberries]. . . .

Village Lake behind Hesquiat was a similarly important bog cranberry site for the Hesquiaht Nuu-chah-nulth. Bouchard and Kennedy (1990: 64–65) record a place called *tl'aaxaktis* ("cleared area") from Alice Paul, as "a narrow area between the northeast end of Village Lake and the beach at Hesquiat Harbour. In former times, this area was kept cleared in order to provide easy access to Village Lake for purposes of obtaining bog cranberries." Wiiknit is another meadow west of Hesquiat village where bog cranberries were also obtained. It was "a flat piece of land with no trees on it, of about 50 acres" in 1911 according to Father Charles Moser (Bouchard and Kennedy 1990). Similar muskeg areas occur near Fort Rupert, Bella Bella, in places of the Bella Coola and Kitlope valleys, Hartley Bay, in the vicinity of Prince Rupert, and on Haida Gwaii, especially the northeastern part, and all were known to and used by local peoples for harvesting plant resources. As well as harvesting large quantities of berries, people would have selectively harvested the leaves of Labrador-tea and various other plant products, including a range of medicinal plants. The ultimate effects of harvesting have not been determined, but they must have been significant.

TIDAL WETLANDS

Aside from the Garry oak meadow communities, there is perhaps the most extensive evidence of the cultivation and enhancement of the root-vegetable communities of the river estuaries and tidal flats all along the coast, producing intensively managed species such as springbank clover, Pacific silverweed, Nootka lupine, rice-root, and wild carrot (*Conioselinum pacificum*). These root vegetables were harvested in significant quantities, being served at major feasts and family gatherings, and also sometimes being dried and stored for winter (Boas 1921, 1930; Bouchard and Kennedy 1990; Compton 1993a,b; Edwards 1979; Turner 1995; Turner and Kuhnlein 1982, 1983; Turner et al. 1983). Notably, blue camas (specifically *Camassia quamash*) also grows prolifically in tidal marshes at the mouth of the Somass River in Port Alberni, where it was cultivated along with the other species, as well as a type of wild onion (apparently *Allium geyeri*, a rare species that is known to occur there—H. Roemer, personal communication to NT, 1999) (Arima et al. 1991). Bracken fern (*Pteridium aquilinum*), Pacific crabapple, and red elderberry (*Sambucus racemosa*) also occur in moist areas around the periphery of tidal marshes, as do tule, cattail, and various sedges.

The intensity of cultivation practices in this habitat is evident from many sources and on many parts of the coast (Deur, this volume). One tidal wetlands site in Clayoquot Sound is named after these practices. This site is called *shishp'ika* ("cultivated") because of the way people used to look after the beds of Pacific silverweed here (Hesquiaht Elder Alice Paul, cited by Bouchard and Kennedy 1990: 43, place name #11; see also Drucker n.d.: 1935–36, 23, 12).

Village and camp sites and their immediate vicinities were intensively modified by humans. Stands of red alder (*Alnus rubra*) often signify the presence of an old village, according to Nuu-chah-nulth elder Roy Haiyupis (personal communication to NT, 1994)[4] This is because they come in after an area is cleared. Although alders were not specifically tended, they were, and are, culturally valued. They are trees that are successful cohabitors with humans. Other kinds of plants, too, are known for their ability to occupy heavily trampled and impacted sites, such as villages.[5] Stinging nettle (*Urtica dioica*) patches are almost always found in the rich moist soils around West Coast villages. Since this plant was required in large quantities as a cordage material everywhere, it was not only welcomed, but encouraged through transplanting or fertilizing in some instances (Leslie M. Johnson, personal communication 1999; Compton 1993a). It is possible that nettle, commonly found on village refuse areas many kilometers from its natural range, was introduced both intentionally and accidentally to places such as middens and village-associated clearings, where it became established on the long term (D. Deur, personal communication 1998). From personal experience, we know that stinging nettle is easily introduced into gardens, and will regenerate and spread copiously.

Drawings and photographs of coastal village sites from a century or more ago invariably show the houses set in rows within a clearing. Haida historian Captain Gold (personal communication to NT, 2000) explained that people cut the trees around the villages both for fuel and for house construction. However, clearing had another purpose as well: berry bushes, cow-parsnip, western dock (*Rumex occidentalis*), and other culturally important plants thrived in the more open environment of a village. As well, the stumps of the cut trees were especially valuable as growing substrates for berry species such as salal, red huckleberry, blueberry, and trailing currant.[6] Areas around some Haida and Nuu-chah-nulth villages were also formerly burned to enhance production of berries and other useful plants (Turner 1999, 2004).

Discussion

The culturally important plants of the Northwest Coast, with the exception of the former Northwest Coast tobacco, are all perennial species, either herbaceous perennials or woody perennials (trees and shrubs). This is the key to the management practices presented here—and to their lack of recognition in the past. These plants are not limited to seed production for propagation and maintenance of their populations. They have meristematic tissues in their rootstocks, rhizomes, roots, and buds that allow them to grow vegetatively, as well as from seed. The management practices are reliant on the broad capacity of these perennials for regrowth and regeneration. This, then, is the "Perennial Paradox" referred to in our title. In many of our past assessments of the lack of agriculture on the Northwest Coast, our attentions have been focused on European-style agriculture: sowing the seeds of annual species—the lettuce, beans, cabbages, and carrots of the classic Mr. McGregor's garden—in neat rows. For perennials, different strategies are needed, and these are not necessarily easily identified as management or cultivation. We contend that Northwest Coast peoples *were* active managers of their plant resources, who did not simply "adapt to" the environment but actively modified it to ensure reliable, predictable and abundant supplies of key resources for food, materials and medicine.

At the species population level, plant management was practiced through harvesting strategies dictated by both cultural and biological factors. Harvesting had both intentional and incidental but frequently positive effects on the productivity of targeted species. By selecting individuals at certain life cycles, or according to age and size, indigenous peoples thinned the populations, decreasing intraspecies competition. Weeding also decreased competition between desired and undesired species, giving the culturally important plants a competitive advantage. Pruning and coppicing of herbaceous plants, trees, and shrubs encouraged the growth of new shoots, leaving the roots or rhizomes intact, as did the burning of selected individuals such as hazelnut bushes. The intentional replanting of "roots" and other propagules was also an important factor in maintaining population productivity.

Incidental impacts of harvesting practices included localized soil disturbances from digging and tilling. In addition, the accidental detachment of portions of taproots, tubers, corms, and bulbs would enable vegetative reproduction of the species (as well as intentional replanting). Further, harvesting of some species was done at a time when seeds were in production, and the activities associated with harvesting—digging, tilling, turning over the turf—would help to distribute seeds and propagules. Not only did people actually seem to have enhanced the growth and abundance of particular species through time, they may have, in some instances, extended the range of particular species through purposeful or accidental transport and replanting (Deur 1999; Turner and Loewen 1998).

The use of controlled fires to manage plant communities is well documented for Indigenous peoples throughout the world (e.g., Anderson 1993b; Boyd 1986, 1999; Day 1953; Deur 1999; Gottesfeld 1994a,b; Lewis 1973, 1977, 1978, 1982; Lewis and Ferguson 1988; Timbrook et al. 1982; White 1980). It is not surprising, then, that this was one of the more important management tools of the Northwest Coast peoples, who burned habitats at specified times of the year and at regular intervals to enhance the productivity of roots and berries. In some cases, the continued productivity of these habitats was ensured through alternating harvest locales.

Finally, while these plant management techniques had economic motives, they were embedded in a larger decision-making system structured by religious and moral ideologies and concepts of social reciprocity. These principles arguably guided people's interactions with the environment and ensured careful, considerate use of plant resources.

These plant management principles and practices represent a continuum of people–plant interactions, with differing levels of intensity and differing impacts on landscape. At the foraging end of the scale, plant harvesting activities had little or no impact on the plant populations or the landscape. In contrast, activities associated with plant cultivation and domestication structured the species composition and diversity and thus had impacts at the population, community, and landscape levels, altering the natural (that is, successional) state, and creating anthropogenic landscapes.

It is important to stress that we believe that most of the plant management practices as described in this chapter are long-standing, and that they existed in some form prior to the coming of Europeans to the Northwest Coast. The intensive harvesting of springbank clover rhizomes, for example, was observed at Tahsis by Archibald Menzies (Newcombe 1923) at the time of Vancouver's circumnavigation of Vancouver Island. The fact that there are words in the aboriginal languages like the Kwak'wala *t'ekilakw* ("manufactured soil"), and the Nuu-chah-nulth *tlh'ayaqak*, is an indication of the antiquity of the gardening of tidal flat areas (see Deur, this volume). There are also terms corresponding with "weeding" and "pruning," as well as terms for the posts and pegs used to delineate harvesting areas owned by families or clans. Although much of this discussion has been based on the recollections of contemporary elders, there is no reason to doubt Adam Dick's contentions that the tidal flats at Kingcome Inlet, for example, were under intensive cultivation for countless previous generations. Hopefully, further research, including soil analysis, will be brought to bear on the question of the antiquity of plant management (Deur and Turner 1999).

Many questions still remain to be answered regarding plant cultivation and anthropogenic plant communities on the Northwest Coast. If Northwest Coast peoples' traditional plant production represents an intermediate stage between foraging and farming, what would the ultimate result of such activities have been, had the people been allowed to develop uninfluenced from the outside? Would they have intensified plant production to the point of dramatic genetic change from wild species? Already there was recognition of special types or variants of different foods and materials. For example, a special type of "sweet crabapple" from "a different kind of crabapple tree" is recognized by some Haisla people, one said to grow, among other places, at Kildala, in Haisla territory, and at the head of the Kitlope River. These are present in a ratio of perhaps 1 to 50 of the more common type (Compton 1993a: 269). The neighboring Gitga'at recognize and distinguish by name at least five different crabapple varieties (Marjorie Hill, Mildred Wilson, Annetta Robinson, personal communication to NT, 2002). Highly productive patches of good-tasting salal, highbush cranberry, red huckleberry, and other fruits are recognized and sought after in many areas. Special varieties of yellow-cedar (*Chamaecyparis nootkatensis*) and springbank clover are recognized and named in Ditidaht (Turner et al. 1983). Do these reflect the beginnings of human selection? Or has human use and influence already changed the genetic composition of coastal plants?

Many proto-domesticates could have been lost soon after Europeans arrived, just as the Northwest Coast tobacco was lost (see also Deur 1999, 2000).

Another major question needing intensive investigation is the relationship among plant and animal resources and human interactions. What role do animals play in anthropogenic plant communities? Here we have focused on plant resources, but in reality these are inextricably linked not only to the people who use them but to animals as well. Some plant–animal relationships known to aboriginal peoples are discussed elsewhere (Turner 1997b), but much is yet to be learned. For example, in traditional narratives on many parts of the coast, Canada goose in her human form was said to have revealed the value of silverweed, and other root vegetables to people (cf Bach 1992b: 97, cited in Compton 1993a; Edwards 1979). Mallard ducks are also associated with roots such as clover and silverweed. The phenomenon of bears pruning highbush cranberries, thereby increasing the berries' productivity, has already been mentioned. Also, burning meadows and forests is said to provide forage for deer, as well as enhancing food-plant production for people; thus hunting and plant gathering can be undertaken at the same sites. Bog cranberry meadows and other culturally important wetlands are frequented for hunting of swans, geese, and ducks in many areas, and what influence these birds may have on the cultural plant communities is not known.

Conclusions

By reviewing the nature of people–plant interactions on the Northwest Coast and situating these activities in the context of ecological-evolutionary models of plant-food production, we see that the "affluent foragers" of the region have much more in common with "farmers" than previously supposed. The ethnobotanical evidence presented in this chapter reveals not only that plants were significant contributors to *all* aspects of traditional economies, but that First Nations peoples actively managed those resources through strategies more commonly considered as "horticultural." By mimicking natural forms of intermediate disturbance regimes at differing spatial scales, these practices created and maintained a mosaic of productive and diverse habitats across the landscape. It was these anthropogenic landscapes that Europeans first encountered and mistook as "natural." Clearly, these data challenge the prevailing wisdom that Northwest Coast peoples achieved a high degree of cultural complexity *without* plant food production simply by gathering the abundant, natural resources of the landscape.

The final word in this story goes to Kwakwa̱ka̱'wakw historian Daisy Sewid-Smith and Hereditary Chief and elder Adam Dick. When asked if the "grandfather" (*gagemp*) segment of the rice-root (Figure 4.10) that was removed and placed back in the ground was removed simply as a means of cleaning off the edible portion of the root, or if his ancestors recognized that it might actually grow into another plant, Adam Dick was adamant:

4.10. Rice-root flowers (Photo by N. J. Turner)

ADAM DICK: That's what it does [re-grows]. That's why we were picking them off! That's why . . . [it was replanted]. That's going to grow. That's going to grow the next season.

NANCY TURNER: I see. So, it's just like gardening, really?

ADAM DICK: It *is* gardening!

DAISY SEWID-SMITH: But, see, people—this is what I'm saying—people didn't believe that we did this. They think that Nature just grows on its own. But our people felt to get more harvest, and bigger berries, they did these things. Same thing a farmer does.

Notes

We gratefully acknowledge the Elders and other knowledgeable people who have contributed information and insights in this work. We especially thank the following people: Chief Adam Dick (*Kwaxistala*), Dr. Daisy Sewid-Smith (*Mayanilth*), and Kim Recalma-Clutesi (*Ogwilogwa*) (Kwakwaka'wakw); Cyril Carpenter, Jennifer Carpenter, and Pauline Waterfall (Heiltsuk); Helen Clifton, Chief Johnny Clifton, Ernie Hill Jr., Marjorie Hill, Mildred Wilson, Annetta Robinson, Tina Robinson (Gitga'at–Tsimshian); Bea Wilson (Haisla); Florence Davidson, Barbara Wilson, Captain Gold, and Emma Wilson (Haida); Dr. Margaret Siwallace, Felicity Walkus, Edward Tallio, and Willie Hans (Nuxalk); Baptiste Ritchie (Stl'atl'imx–Lil'wat); Dr. Richard Atleo (Chief *Umeek*), Chief Earl Maquinna George, Gloria Frank, Lena Jumbo, Arlene Paul, Stanley Sam, Roy Haiyupis, Brandy Lauder (Nuu-chah-nulth), Larry Paul, Alice Paul, Ruth Tom, and Rocky Amos (Nuu-chah-nulth–Hesquiaht); John Thomas, Ida Jones, and Chief Charlie Jones (*Queesto*) (Ditidaht); Tom Sampson,

Elsie Claxton, Christopher Paul, Mary George, Agnes George, and Cheryl Bryce (Straits Salish); Percy and Dolores Louie (Chemainus); Arvid Charlie (Cowichan/ Hulq'umi'num), and Dr. Mary Thomas (Secwepemc).

Many others helped us in providing information, analysis, and editorial criticism. These include: Dr. Douglas Deur, Marguerite Babcock, M.Ed. (who interviewed Saanich Elder Christopher Paul and other Tsartlip elders about camas in 1967), Denis St. Clair, Randy Bouchard and Dr. Dorothy Kennedy, R. Andrew Reed, Dr. Wayne Suttles, Dr. Brian Compton, Dr. Richard Ford, Dr. Eugene Anderson, Dr. Reese Halter, Alison Davis, Dr. Hans Roemer, Dr. Marilyn Walker, Dr. George Nicholas, Dr. Leslie M. Johnson, Juliet Craig, Dawn Loewen, Dr. Brenda Beckwith, Trevor Lantz, Ann Garibaldi, Wendy Cocksedge, Robin Smith, Jim Jones, Grant Thomas Edwards, Dr. Kat Anderson, and Dr. Kay Fowler.

This research was supported, in part from Social Sciences and Humanities Research Council of Canada Faculty Research Grants (especially # 410-2000-1166) to Nancy Turner and colleagues, and a Social Sciences and Humanities Research Council of Canada Doctoral Fellowship to Sandra Peacock. The Global Forest Foundation has also supported graduate work of Trevor Lantz and Brenda Beckwith, cited here. Recent research (2001–02) with members of the Gitga'at Nation was undertaken through the Coasts Under Stress Major Collaborative Research Initiative grant funding jointly by the federal granting councils (Rosemary Ommer, P.I.).

The botanical names for some of the plants mentioned in the epigraph are springbank clover (*Trifolium wormskjoldii*), Pacific silverweed (*Potentilla anserina* ssp. *pacifica*), Nootka lupine (*Lupinus nootkatensis*) and rice-root (*Fritillaria camschatcensis*). Deur (this volume) provides a detailed overview of traditional uses and management of these estuarine species.

Unless otherwise noted, all information in this paper from Chief Adam Dick and Dr. Daisy Sewid-Smith is from taped interviews with Nancy Turner in November 1996 and November 1997, and with Nancy Turner and Douglas Deur in August of 1998 and 1999. We acknowledge with gratitude their important contributions to the understanding of traditional plant cultivation.

1. The actual number of languages varies according to methods used to classify them; some are considered either as separate, or as dialects of a common language, depending upon different linguistic parameters.

2. This plant is sometimes also called "chocolate lily," a name usually applied to its relative, *F. lanceolata*.

3. For example, an entry in the Fort Langley Journal (p. 32) notes that on October 5, 1828, a large number of people had gathered at the "Forks" (where the Pitt and Fraser rivers join) to procure " . . . Wappatoes a root from under water in pools and marshes and held by them in great estimation as an article of food: The name they give it here is Scous or rather Skous. . . . " The journal also notes that (wild) onions were growing close to the river.

4. Alders are highly valued for their wood, as a source of fuel for smoking fish, for carving, for their bark (which is used as a red dye), and as a medicine (Turner 1998; Turner and Hebda 1990).

5. Broad-leaved plantain (*Plantago major*), an important medicine plant, is known as "village skunk-cabbage" in Haida, because it is always found around people's paths and yards (Turner 2004).

6. Today on Haida Gwaii, heavy browsing by introduced deer around villages such as Sgang Gwaay llnagaay has eliminated the productive berry bushes from all except the high stump locations (Turner 2004).

Chapter 5

"A Fine Line between Two Nations"

*Ownership Patterns for Plant Resources
among Northwest Coast Indigenous Peoples*

NANCY J. TURNER, ROBIN SMITH,

AND JAMES T. JONES

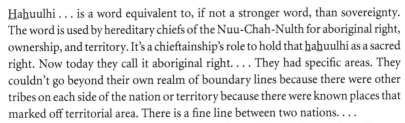

Hahuulhi . . . is a word equivalent to, if not a stronger word, than sovereignty. The word is used by hereditary chiefs of the Nuu-Chah-Nulth for aboriginal right, ownership, and territory. It's a chieftainship's role to hold that hahuulhi as a sacred right. Now today they call it aboriginal right. . . . They had specific areas. They couldn't go beyond their own realm of boundary lines because there were other tribes on each side of the nation or territory because there were known places that marked off territorial area. There is a fine line between two nations. . . .

—Ahousaht Hereditary Chief Earl Maquinna George,
personal communication to NT, May 10, 1996

R ecently, we were discussing concepts of land ownership with Gabe Bartle-man, a Saanich (Wsánech) Elder. In our conversation, we commented that the Saanich people once must have "owned" all the land on the Saanich Peninsula, which lies around and north of the present city of Victoria. He corrected us, saying: "No, we didn't own the land; we just lived on it and used it and looked after it. There's a difference!" Although Gabe Bartleman did not feel that the Euro-Canadian concept of "ownership" was appropriate for describing Saanich land tenure, there were clearly mechanisms for maintaining control and authority over traditional territories. Kevin Paul (Paul et al. 1994: 4), in *The Caretakers: The Re-emergence of the Saanich Indian Map*, elaborated on this point: "The range of my people's movements during their summer activities determined the 'boundary' of our territory; you would not likely have found other peoples using this land." These comments suggest that factors such as occupation, control of access, and rights to use are all parts of the ways in which cultural groups define and exercise rights to lands and resources.

As suggested by the statements made by Ahousaht Chief Earl Maquinna George and Saanich Elder Gabe Bartleman, quoted above, among the First Peoples of the Northwest Coast, there were and are multiple concepts of land and resource ownership. It is important to note that the sources we utilized in writing this paper are taken from a range of time periods—memories and teachings of contemporary indigenous peoples, and ethnographic descriptions from the late nineteenth century to the present day. Traditional patterns of land and resource ownership have been severely disrupted by European colonization and the associated imposition of new economic and political structures. Still, as the words of contemporary elders show, the traditional concepts have been passed down orally and indicate a very strong correlation with accounts of ownership protocols found in early ethnographic descriptions.[1]

The various traditional concepts of ownership (as well as their contemporary expressions) range from the general recognition of communal territory defined by seasonal movements, as in the Saanich example, to authority over specific resource sites and locations. In addition, these ownership rights might be held by individuals, by culturally defined kinship groups (such as households, lineages, or clans), or by larger village and ethnic groupings. Also, ownership rights and responsibilities may be limited to certain resources (e.g., fish, clams, berries, roots) at particular locations, without ownership or control by the same authority of other resources at the same location. Imbued in all types of ownership traditions developed by Northwest Coast peoples are concepts of stewardship, in which an individual's rights to use the land and its resources are contingent upon their sustainable management, and the sharing of resources with other group members.

In this chapter, we explore patterns of land and resource ownership on the Northwest Coast, and reflect on how these patterns influence and are influenced by resource management and conservation strategies of Northwest Coast peoples. Following this discussion, we will consider the ways in which European concepts of ownership and property have differed from and have disrupted traditional patterns of ownership on the Northwest Coast, and the subsequent implications for plant management and conservation.

Patterns of Land Tenure and Ownership

There is a notable lack of discussion in the ethnographic literature concerning land ownership and stewardship patterns among "hunter-gatherer" societies (Richard Daly, personal communication to NT, 1997; Daly and Vast 2001). This silence has had significant implications for Aboriginal peoples who have entered the era of modern treaty negotiations and land claims. The widespread, though highly erroneous, assumption that indigenous peoples on the Northwest Coast were foraging societies that did not have clearly defined ideas of property, ownership, or responsibility for stewardship has worked against the interests of Aboriginal peoples attempting to secure title to traditional territories in British Columbia.

The rapidly growing literature on traditional ecological and Indigenous knowledge has contributed to a more finely tuned appreciation for the ways in which Indigenous peoples in many areas of the world conserve and manage their resources (see for example: Anderson 1996; Berkes 1989, 1999; Brody 2000; Freeman and Carbyn 1988; Inglis 1993; Johnson 1992; Williams and Baines 1993). As we begin to reassess human–environment interactions in light of growing ecological crises, it is important to look more closely at some of the cultural institutions and strategies that have contributed to the development of sustainable societies in particular environments. At the heart of these strategies for responsible resource use and management are culturally defined prescriptions of authority over land and resources.

In general, all lands and waters along the Northwest Coast of North America have been occupied, traversed, and managed by groups of Aboriginal peoples for thousands of years. In some cases, harvesting locations were frequented by more than one people, especially at the outer limits of territories or in places where affinal and other types of social relationships allowed overlapping use. However, primary rights to use were almost always recognized and maintained by traditional protocols, and were sometimes resolved through outright conflicts. Land and resource use was not at all casual or random, nor were lands and resources "freely accessible to all" as many early European settlers would have liked to believe. As pointed out by Gunther (1927: 213), "Early travelers along the coast very frequently mention meeting a band of 'wandering savages.' To strangers, it was not always apparent that these wanderers usually had a very definite destination and that they were moving at a specified time set by the ripening of the berries or a run of salmon."

First Peoples on the Northwest Coast traditionally engaged in annual rounds of activities to take advantage of resources that were available in their home territories at certain times of the year in particular locations. It should be noted, however, that settlement patterns were more sedentary than for many groups classified as "hunter-gatherers"; the seasonal round was generally regular, and was often anchored by time spent at permanent winter villages. Much emphasis has been placed on the importance of seafood, especially salmon, in the scheme of resource-based activities on the Northwest Coast. While the economic and cultural importance of salmon as an abundant and predictable— although spatially and temporally limited—resource is undeniable, plant resources have also contributed significantly to the physical, intellectual, and spiritual life of aboriginal peoples (see Ames, Deur, and Turner and Peacock, this volume).

Seasonal movements took advantage of culturally significant plants growing in different habitats and locations at different times of the year. These plant resources and gathering sites were regulated by traditional patterns of ownership and control. Certain habitats, such as open meadows and prairies containing camas bulbs (*Camassia* spp.; *C. quamash*, *C. leichtlinii*), and estuaries and tidal flats containing silverweed (*Potentilla anserina* spp. *pacifica*), springbank clover (*Trifolium wormskjoldii*), rice-root (*Fritillaria camschatcensis*) and

Nootka lupine (*Lupinus nootkatensis*), were especially prized and cared for, while other habitats were frequented for secondary or backup resources in times of scarcity. These plant resources continue to play an important role in the lives of contemporary Aboriginal peoples on the Northwest Coast and elsewhere (see Turner and Peacock, Suttles, Deur, this volume).

Obviously, there are many cultural differences among the various groups and communities with respect to ownership of land and resources. Since traditional patterns of ownership have been severely disrupted by European colonization, and by the subsequent loss of so many people through the disease epidemics of the past two hundred years (Duff 1964: 58–60), these patterns are sometimes difficult to discern. But they do persist, in the memories of Elders, in ongoing traditions that have been sustained despite colonization, and in historical writings and ethnographies of anthropologists and others. In the following section, we have attempted to characterize some of these patterns of ownership for different language groups of the Northwest Coast as drawn from these sources, especially as they pertain to plant resource use of coastal British Columbia.

Specific Land and Resource Ownership Patterns in Northwest Coast Societies

In the context of this chapter, ownership may be defined as the formal recognition of rights to control access to lands and resources; these rights may be dormant or actively exercised (Richardson 1982: 96). Cove (1992: 235) suggests that ownership of property, or "title," refers to a collection of rights including the right to use or enjoy, the right to restrict access, and the right to include or alienate. In addition, among peoples of the Northwest Coast, ownership rights invariably entail some reciprocal responsibilities on the part of the owner, to the land and to other group members. There is significant variation among Northwest Coast cultural groups in the extent to which these rights are defined and exercised, and in the specific places and resources to which access is restricted.

SALISHAN PEOPLES

For the Coast Salish peoples (with traditional territories on southern Vancouver Island, the lower mainland of British Columbia, and northwestern Washington), the general pattern of traditional ownership was based on recognized territorial boundaries defined by occupation or use, and control of a few key resources, either by a family, household, or village group (Richardson 1982: 100). This is reflected in the statements of contemporary Saanich people, as discussed in the introduction to this chapter.

Traditionally, among Central Coast Salish peoples (comprising the speakers of Squamish, Halkomelem, Nooksack, Northern Straits and Clallam languages), private property in the form of goods and privileges was important (Barnett 1955: 250; Kennedy 1993: 69; Suttles 1974: 324).[2] Property—such as canoes, weapons, and slaves—was individually owned, while family groups owned names, ceremonial prerogatives, and associated material goods such as masks and dancing regalia. The display of such wealth was an important social function in Salish society. However, according to Kennedy (1993: 69), "ownership of land and resources . . . was conceptually different," and did not generally involve exclusive ownership of the land. Kennedy (1993: 69–70) points out, however, that this did not preclude a strong connection with tribal territory. This is reflected, for example, in the traditional names given to general territories and specific locations. The suffix /-ulh/, meaning "of or belonging to," is used to identify a group's territory. The large number of places that were specifically named, and the teachings and stories explaining the origins of various features in the landscape attest to the long history and intimate connection with particular areas.

Within tribal territories, ownership of land and resources was, in general, inclusive for all group members. However, it is possible to find a number of exceptions to this pattern, providing important insights into the variation in resource ownership and control. As part of the annual round of activities, household groups would go to different locations at different times of the year to exploit specific resources. Many of these specific sites for seasonal resource gathering were owned and controlled by families, and sometimes by individuals (Barnett 1955: 241).

Among the Central Coast Salish, some of the resource sites that were owned by family groups included camas beds, "wild carrot" patches (probably Pacific hemlock parsley—*Conioselinum pacificum*, and/or wild caraway—*Perideridia gairdneri*), bracken fern (*Pteridium aquilinum*) rhizome sites, wapato (*Sagittaria latifolia*) patches, bog cranberry (*Vaccinium oxycoccos*) gathering areas, horse clam and butter clam beds, sites for duck nets and sturgeon fishing, dipnet locations and certain fishing streams (Richardson 1982). Below, we will focus on case studies of plant ownership among two Central Coast Salish peoples: the Katzie of the lower Fraser River valley on the mainland of British Columbia, and the Saanich of Vancouver Island. Ownership patterns among a third Salishan group, the Nuxalk of the central British Columbia (BC) coast, are also discussed. The Nuxalk are geographically separated from other Salishan speakers to the south, and their ownership practices reflect their particular geographic and cultural location on the Northwest Coast.

Katzie (Halq'emeylem) of the Lower Fraser Valley. The traditional territory of the Katzie contains large areas of low, seasonally flooded lands, which provide excellent growing conditions for wetland plant species. Two of these plants, bog cranberry (*Vaccinium oxycoccos*) and wapato (*Sagittaria latifolia*), were extremely important in the traditional diet of the Katzie people, and valuable as trading goods, both with Aboriginal and, later, non-Aboriginal groups.

According to Suttles (1955: 26), "Katzie territory was famous for these [plants]," and in the fall harvesting season they attracted people from a number of other tribes.

Although many areas were open to anyone for gathering, certain choice locations were owned by individual families. For example, according to Suttles's consultant, Simon Pierre, the bog south of the Alouette belonged to all the Katzie; however, other important cranberry areas north of Sturgeon Slough and along Widgeon Creek belonged specifically to Simon Pierre's father's family from Pitt Meadows. Outsiders had to seek permission from the owners before gathering cranberries in these locations, although apparently they were rarely denied permission, and no tribute was required (Suttles 1955: 26).[3]

Use by outsiders was closely monitored by the family who owned an area. For example, it was the responsibility of Simon Pierre's father to watch the cranberry bogs and ensure that no one picked the berries until they were fully ripe (Suttles 1955: 26). Suttles attributes this practice to economic motives; the cranberries were sold to European settlers at New Westminster, and if green berries were brought in the price would drop. However, regulating the timing of the harvest was a common practice among a number of Salishan groups on the Northwest Coast and adjacent Plateau regions. This was often achieved through holding first-fruit ceremonies (for example, see Hill-Tout's descriptions in Maud 1978; and in Turner and Peacock, this volume). This practice likely served a variety of functions including, but not limited to, economic considerations. As Suttles points out, ownership of (or identification with) a rich cranberry bog might have been its own reward in invoking reciprocity from other families who were given permission to gather there.

A similar situation was found with wapato patches in Katzie territory, some of which were owned by the entire tribe, and some of which belonged to certain families. For example, Suttles's consultant, Simon Pierre, gave the traditional place names for nine wapato patches belonging to his father's father's family. Even within the areas open to all members of the group, specific locations might be temporarily claimed by family groups. This was true for the area near the head of Sturgeon Slough, where families could establish seasonal claims to wapato plots. These areas (which might be several hundred feet long) would be cleared of other growth so that the wapato tubers might be more easily gathered (Suttles 1955: 27). Wapato productivity in these areas over the following years would most likely have been enhanced through these weeding and tending activities (see Darby, this volume; Spurgeon 2001).

Saanich of Vancouver Island. As described by Saanich Elder Gabe Bartleman in the introduction to this chapter, Saanich peoples of Vancouver Island may not have always considered themselves "owners" of the land in a general sense. Rather, they had socially imposed responsibilities to look after and carefully use the resources of the land, and their rights to do so were generally recognized and respected by other cultural groups. Within tribal territories, however, certain important locations were recognized as falling under more restricted authority. Saanich Elder Elsie Claxton (personal communication to NT, 1996) described some of the locations where ownership rights were more specifically defined. For example, the West Saanich peoples traditionally owned a small island off Mill Bay, while *Lauwelnaw* [Mount Newton], and Sidney Spit belong to the Tsawout (East Saanich) people. Both of these places are well known for their locally abundant plant and animal resources.

In addition, Saanich peoples of Vancouver Island developed protocols for delegating authority over certain important resource-gathering sites. In particular, specific locations for gathering camas fell under these protocols. Camas grows in open meadow areas on Vancouver Island and on ledges of rocky slopes on the Gulf and San Juan Islands. Suttles (1974: 60) suggests that in general the camas beds were open to everybody; however, some locations were privately owned and inherited among the Saanich and Samish peoples.[4] Examples of specific camas bed locations that were privately owned included Mandarte Island in Haro Strait, which was owned by three people, and an islet south of Sidney Island, which was owned by a single person (Elsie Claxton, personal communication to NT, 1996; Suttles 1974: 60). An interesting comment made by Elsie Claxton was that although an island was owned, anyone could go there to get camas bulbs or seagull eggs. This suggests that ownership did not necessarily preclude the sharing of resources among group members or even with outsiders.

For both the Katzie and the Saanich, gathering sites for particular plant species were owned and cared for by certain families and individuals. This was the case with bog cranberry and wapato patches among the Katzie, and camas beds among the Saanich. These ownership patterns allowed careful monitoring of plant populations through long-term and consistent use, and regulation of harvesting activities. For example, by requiring outsiders to seek permission to harvest bog cranberries, the family that owned the patch was able to keep track of and, if necessary, limit how many people were harvesting from that particular location, and ensure that the timing of the harvest was appropriate. Ownership protocols also facilitated plant management techniques such as weeding and tending patches, and landscape burning. In turn, the labor invested in the gathering areas would have encouraged the maintenance of usufructuary rights. For example, one of Suttles's consultants stated that she was responsible for burning the camas area that belonged to her grandmother once they had finished harvesting for the season (Suttles 1974: 60). Other people were allowed to harvest from this site, but responsibility for managing the resource site through landscape burning was clearly delegated to ensure that the resource would be available in following years.

Nuxalk of the Central Coast. The Nuxalk, who speak a Salishan language, inhabit the Bella Coola River valley as well as adjacent inlets and river valleys along the central coast of British Columbia. As a geographically isolated Salish-speaking population, their culture and their territory are distinguished, somewhat, from those characteristic of the primary concentrations of Salish-speaking peoples along the Puget Sound and Strait of Georgia, some distance to the south. Among the Nuxalk, concepts of property are more strictly defined than among the Salish-speaking people to the south, and bear stronger similarities to those of neighboring populations along the central mainland coast. The Nuxalk have developed a distinctive system of land tenure based on oral traditions establishing their ancestral rights to the land. McIlwraith (1948: 130), who worked among the Nuxalk in the early twentieth century, found that "great difficulty was experienced in obtaining information about land tenure." The people of the Bella Coola valley were sensitive about the topic, and resented the fact that white settlers had encroached on their traditional territories without proper permission or compensation.

According to oral tradition, it was the original ancestral families that claimed the tracts of land now owned by different villages. A supernatural being, *Älhquntä*, created the first men and women, who were then sent down to populate the earth. They came in small groups of two, three, four or more; usually brothers and sisters but sometimes husbands and wives. These groups settled in the different parts of the Bella Coola valley. They brought with them tools, houses, clothing, knowledge of ceremonies and prerogatives, names, and plant and animal species (McIlwraith 1948: 117).

Each of these groups of first people formed a village community and claimed the hunting grounds in the vicinity. Areas that were particularly favored for settlements included places along the river where salmon and eulachon could be caught, and places where wild berries were abundant nearby. These were clearly defined tracts of land, which consisted of small valleys off the main watershed of the Bella Coola River (McIlwraith 1948: 118). In these small valleys, the first people released the animal and plant species they had brought with them. Among the people of Kimsquit, there is a tradition that a few years after the first settlements were made, *Älhquntä* came to mark and allot the territory that was the property of each village community (McIlwraith 1948: 131). These original village communities formed the ancestral families from which the Nuxalk trace their descent. It is by having names from, or being a member of, these ancestral families that a person obtains the rights to use certain territories.

WAKASHAN-SPEAKING PEOPLES

Among the peoples speaking Wakashan languages on Vancouver Island and the central and north coast of British Columbia (Nuu-chah-nulth, Kwakwa̱-ka̱'wakw, Heiltsuk, Oweekeno, and Haisla), resources were primarily managed through restricted kin groups who owned land and resources in traditional territories (Richardson 1982: 102). Among the different Wakashan-speaking peoples, as with the Nuxalk, ownership rights to lands and resources were clearly defined, although the social group that regulated access varied from group to group. These ownership rights remain a vital element of the contemporary cultures of these peoples.

Haisla. The Haisla (including the Kitamaat, Hanaksiala, and Kemano), who speak a Wakashan language, are the traditional inhabitants of the territory immediately north of the Nuxalk. McIlwraith (1948: 133) notes that among the Hanaksiala of the Kitlope valley, crabapple trees were the property of specific ancestral families. When Indian Reserve Commissioner Peter O'Reilly was defining reserve lands in British Columbia in the late nineteenth century, he did not include the crabapple groves in the land allotted to the people of the Kitlope. Subsequently, they wrote a letter to the "Chief Commissioner of Lands and Works " requesting that the important crabapple groves be added to their reserve lands (Compton 1993a). This omission points to the colonial disregard for the traditional uses of wild plant resources among First Peoples. Primary attention was given by the European newcomers to fishing stations for peoples on the coast. The use of wild plants was not considered to constitute "cultivation," and therefore plant-gathering areas were not often specifically included in reserve lands.

Nuu-chah-nulth-aht. Among the Nuu-chah-nulth on the west coast of Vancouver Island, ownership concepts are based on a complex and highly developed system of chieftainships, reflecting hereditary lineages (Beaglehole 1967: 306; Bouchard and Kennedy 1990; Drucker 1951: 247). The contemporary Nuu-chah-nulth people comprise fourteen distinct First Nations, including the Ditidaht and Pacheedaht, although the population and composition of these groups have changed over time. Among these peoples, the Chiefs' proprietary rights to and authority over territory and resources are known as ḥaḥuulhi. Historically, ḥaḥuulhi authority was passed on as a hereditary privilege. Rivers, fishing areas, hunting areas, and plant harvesting areas were all considered private property. In the past, it would be unthinkable to violate a Chief's ḥaḥuulhi.

All areas, from the peaks and ridges of the Vancouver Island mountain range and the river valleys down to the beaches and out to the distant halibut fishing banks, were privately controlled by the Chiefs and their representatives. According to Peter Webster (1983: 17),

> The land fought for in inter-tribal wars long ago, belonged to hereditary chiefs and extended to the west as far as the horizon and to the east deep into the mountainous forests on the mainland of Vancouver Island.

James Adams of Ahousaht described ḥaḥuulhi more specifically as extending as far inland as salmon go up the streams and as far up the mountains as the people would go to obtain cedar. It extended as far out to sea as a person could go and still see the mountains (Bouchard and Kennedy 1990: 20).

The meaning of the word ḥaḥuulhi is addressed by Chief Earl Maquinna George at the beginning of this chapter. As described by Earl Maquinna George, sub-Chiefs and others could gain usufructuary rights to land and resources, but would always be required to pay tribute to the head Chief, usually in the form of a portion of the harvest. In exchange for these rights, the user was also expected to perform certain management functions. For example, when a Chief grants a sub-Chief rights to a salmon stream, it is called "looking after the river." As part of this agreement, the sub-Chief "ensures the stream is kept clear, determines who and when others can fish, and collects a tribute from those who do" in exchange for rights to the first run of salmon (Bouchard and Kennedy 1990: 21). Chum salmon streams in particular were closely monitored to prevent unauthorized use. Dr. George Louie of Ahousaht and Luke Swan of Manhousaht noted that some coho fisheries were also owned, as were trap locations at sockeye fisheries. These Elders were able to list the names and trap locations at *hisnit* (place name # 148 in Bouchard and Kennedy 1990) that were owned by high-ranking people. Other types of resource gathering sites that were traditionally owned included particularly good gooseneck barnacle gathering places; specific rocks that were good locations for hunting sea lions and seals; and bear trails. A chief's property also included any salvage found on a beach within his ḥaḥuulhi territory.

Some examples of plant-gathering areas that were considered private property included berry patches, root vegetable plots, and stands of western red-cedar. Concerning the ownership of berry-picking grounds, Manhousaht Elder Luke Swan stated that the first crop was picked for the Chief (Bouchard and Kennedy 1990: 23). With this first harvest, the Chief gave a feast to his people (Drucker 1951: 252). Certain plant-harvesting areas, as well as falling under the general jurisdiction of a Chief's ḥaḥuulhi territory, would be managed by certain individuals and families who would be granted limited ownership rights that were resource specific. For example, Hesquiaht Elder Alice Paul stated that although two individuals held the right to hunt geese on Village Lake behind Hesquiaht village, they did not have the right to restrict other uses of this same lake, for example, harvesting the bog cranberries that grew around its shore.

Plant gathering sites where important root foods such as Pacific silverweed (Figure 5.1), springbank clover, and northern rice-root were found in abundance were frequently owned and cared for by specific families (Turner and Efrat 1982). As mentioned in other chapters of this volume (Deur, Turner and Peacock, this volume), these sites were often maintained intensively, and were commonly termed "gardens" by both Indigenous and European peoples (Figure 5.2). Hesquiaht Elder Alice Paul explained that the owners of the silverweed gardens were possessive of them because of the care taken to cultivate and tend the plots. Arima et al. (1991: 190) recorded that "A place where one has rights to collect roots or berries is called *tl'ayaqak*. Often four cedar stakes . . . were used to mark the boundaries of such places, and were replaced periodically to avoid rotting." This practice was documented in Bouchard and Kennedy (1990) in their study of Aboriginal land use in the Clayoquot Sound area.[5] Jessie Webster of Ahousaht noted that at *wa7uus* poles and rocks were laid on the ground to identify the boundaries of each silverweed plot, areas that were considered to be private property.

5.1. Pacific silverweed (Photo by N. J. Turner)

Many Europeans who arrived on the west coast of Vancouver Island commented on the well-defined concepts of property and ownership among the various Nuu-chah-nulth nations. For example, Captain James Cook remarked, "Here I must observe that I have no where met with Indians who had such highly developed notions of every thing the Country produced being their exclusive property as these" (Beaglehole 1967: 306). The reports of early explorers such as Cook are significant, particularly in light of contemporary land claims, in documenting territorial ownership of land and resources preceding contact with Europeans. Cook's observation about the ownership of territories and resources among the Nuu-chah-nulth was echoed over 150 years later by anthropologist Philip Drucker (1951: 247), who worked among the Nuu-chah-nulth in the 1930s:

> The Nootkans [Nuu-chah-nulth] carried the concept of ownership to an
> incredible extreme. Not only rivers and fishing stations close at hand, but
> the waters of the sea for miles offshore, the land, houses, carvings on a house-
> post ... names, songs, dances, medicines, and rituals, all were privately
> owned.

Drucker elaborates on this system of ownership, discussing it in terms of privileges that can be divided into two broad categories: economic and ceremonial. He goes on to outline how these privileges can be acquired (through inheritance, or as a reward for bravery) and the various rights they confer (for example, rights to fish in certain locations, harvesting rights for berry and root patches, and salvage rights along the beach).

5.2. E. S. Curtis, *The Berry Picker*, about 1915. The woman is Jessie Sye of Hesquiaht, picking evergreen huckleberries. (British Columbia Archives, D-08315)

However, these descriptions may not fully reflect the meaning of concepts such as "property" and "ownership" within the context of Nuu-chah-nulth culture. These non-Aboriginal terms tend to emphasize rights of exclusive ownership, reflecting European attitudes toward natural resources as commodities to be exploited for the benefit of the owner. The Nuu-chah-nulth concept of authority over lands and resources, as embodied in the term hahuulhi and described by contemporary Nuu-chah-nulth people, recognizes these exclusive rights of ownership; however, it also recognizes the many responsibilities associated with these rights. These responsibilities included looking after and managing the land and resources, and sharing resources with other members of the group. According to Ahousaht Elder Roy Haiyupis (quoted in the Scientific Panel for Sustainable Forest Practices in Clayoquot Sound 1995: 9),

> *Ha hoolthe* [hahuulhi] . . . indicates . . . that the hereditary chiefs have the responsibility to take care of the forests, the land, and the sea within his *ha hoolthe,* and a responsibility to look after his *mus chum* or tribal members.

This description highlights the reciprocal rights and responsibilities that were embodied within traditional concepts of ownership and stewardship. It also demonstrates an abiding concern for other members of the group going beyond the individual self-interest that tends to be emphasized in classic European notions of ownership and property. It is important to look beyond the individual as the unit of analysis when exploring traditional systems of property ownership. Often, the focus on individuals obscures the mechanisms and practices that were developed at family, lineage, or clan levels to manage and own property.

Kwakwaka'wakw. The Kwakwaka'wakw are a Wakashan-speaking people with traditional territories on the north end of Vancouver Island, on the adjacent mainland of British Columbia, and on the islands in between. Like their Nuu-chah-nulth neighbors, the Kwakwaka'wakw have carefully defined ownership traditions based on descent groups. According to Boas (1966: 36), each Kwakwaka'wakw winter village group is a separate unit bearing the name of the location the group inhabits.

Groups of extended families within each village form clans,[6] which are considered to be "of one kind" (Boas 1966: 35). It is the clan which acts as the corporate (property owning) group in Kwakwaka'wakw society. A clan is made up of one or several extended families who all trace descent back to a common ancestor (Galois 1994). According to Daisy Sewid-Smith, Kwakwaka'wakw historian and language specialist, these clans would own and have jurisdiction over an entire territory, within which member families would have their own root-digging plots, berry patches, and crabapple trees (personal communications to NT, 1996, 1997).

As the fundamental unit in Kwakwaka'wakw society, each clan has its own traditions, crests, ceremonial privileges, family names, rights to own houses in the winter village, seasonal occupation sites, and fishing, hunting and gathering areas. Kwakwa̱ka'wakw land ownership, like the ḥaḥuulhi of the Nuu-chah-nulth peoples, involves rights based on inheritance and the ability to restrict access to certain resources and resource areas. For example, clans owned and controlled hunting grounds (especially areas for mountain goat hunting), rivers, eulachon fishing areas, salmon fishing sites, berry patches, eelgrass (*Zostera marina*) beds, root gardens, and clam beds. Chief Adam Dick's ancestors, for many generations back, had the position of river guardian for the eulachon at Kingcome Inlet. These individuals would keep a close eye on the river, and ensure that the eulachon were spawning well before they allowed people to begin catching these fish (Chief Adam Dick, personal communication to NT, 1997).

According to Daisy Sewid-Smith (personal communication to NT, 1996) and Boas (1966: 42), the ancestors to whom descent in each clan is traced originally came down from the sky, out of the sea, or from underground to appear at specific locations where they would take off their animal masks and become people. These would be the locations of the original village sites, which Boas (1966: 47) suggests each clan represents. Daisy Sewid-Smith explained the origins of rights to gathering areas (personal communication to NT, 1996):

> Well, it would more be a clan ownership . . . and then the clan families would have their own plots . . . we are made out of what I call confederated clans. And your plots [for plant resource harvesting] would be where your beginnings were . . . so they would have their area in their beginning, this is where their area would be.

Chief Adam Dick and Daisy Sewid-Smith (personal communication to NT, 1996) explained that people would use pegs to mark off their gathering areas, including tidal flats for digging root vegetables, berry patches, clam beds, and wild crabapple areas. As Daisy Sewid-Smith explained, "Anything that's pegged, you know it's someone's." This observation was also made by Charles James Nowell (*Owadi*) in his autobiography (Ford 1941). *Owadi* described traditional ownership of springbank clover gardens:

> In the olden days, the women had their own clover patches marked with sticks on the four corners. . . . Clover patches are women's property, and it comes down to her daughter. If they have no daughter, the boy gets it, and then when he gets a wife the wife uses it. It is only this way with patches of roots. [Ford 1941: 51–52]

When asked if outside people would need to ask permission to gather resources in a clan's territory, Daisy Sewid-Smith said that people seldom crossed the boundaries of another group's territories, since these were well known and respected. As she pointed out, if people needed particular resources from

5.3. Highbush cranberry (Photo by R. D. Turner)

another clan's territory, they would trade one of their own products for the needed resource (see Deur 2000; Turner and Loewen 1998).

Boas (1966: 36) names eight kinds of berries—crabapples, highbush cranberries (Figure 5.3), salal, bog cranberries, red elderberries, stink currants, salmonberries, and huckleberries—for which harvesting sites were owned. The berry-picking grounds would be closely monitored and guarded by their clan owners against intrusion from outsiders. Boas recorded that if someone from a different clan picked berries without permission, it would result in a fight and often death. A number of important root foods were also regulated by clan ownership rights, including sea milkwort (*Glaux maritima*), "wild carrot," eelgrass, springbank clover, Nootka lupine, rice-root, and silverweed (Boas 1921).

The ultimate basis of authority over resources in Kwakwaka'wakw society was the ancestral creation story that explained the origin of each group, and defined the named and ranked positions within the clan (Weinstein and Morrell 1994: 35–36). The clan also had important political functions, closely related to the responsibilities for resource management and ownership. According to Marty Weinstein and Mike Morrell (1994: 35–36) the head of the clan, along with a council of advisors, would be responsible for three major administrative functions: administering the economic system of the group, managing the group's resources, and ensuring that the group's ceremonial obligations were fulfilled.

HAIDA

The Haida system of land and resource ownership has evolved through continuous occupation of Haida Gwaii (the Queen Charlotte Islands) since time immemorial. A strong ideology of stewardship is incorporated into traditional Haida concepts of ownership. On November 6, 1985, Gwaganad of Haida Gwaii spoke before the Supreme Court of British Columbia, regarding an application by Frank Beban Logging and Western Forest Products Ltd. for an injunction against the Haida protesting logging activities on Lyell Island (South Moresby). In her statement, Gwaganad (1990: 49) expressed the intimate connection between Haida and the land:

> Since the beginning of time—I have been told this through our oral stories—since the beginning of time the Haidas have been on the Queen Charlotte Islands. That was our place, given to us. We were put on these islands as caretakers of this land.

More recently (in March 2002) the Haida First Nation filed a Statement of Claim with the Supreme Court of British Columbia, asserting their title to the lands and surrounding waters of Haida Gwaii.

Among the Haida, tracts of land along the coastline traditionally were owned by different families, or lineages, and were considered private property. Some families did not own land; however, they were able to harvest resources such as berries from the land of another family, provided they paid some form of tribute. The permission of the owning family had to first be sought, and the visiting family was only allowed to harvest once the owners had finished (Swanton 1905: 71). The boundaries of territories belonging to families were strictly observed, and were inherited according to rules of succession (generally from a man to his sister's son); alternately lands could be obtained through trade or gifts. Generally, every family had one or more creeks or portions of creeks where they might fish, although some of the larger salmon streams were used jointly by several families (Harrison 1925: 63).

According to Dawson (cited in Cole and Lochner 1993: 117–18), "the Haida trouble themselves little about the interior country [in terms of intensified ownership, *not* in terms of use], but the coast line, and especially the various rivers and streams, are divided among the different families." It was along the coast that the primary berry patches were located and divided among the different families. Boundaries of these different territories were sometimes indicated by placing poles along the perimeter (Cole and Lochner 1993: 118). Similarly, Charles Newcombe (British Columbia Archives, MS-1077) noted that in the annual construction of fishing weirs, for streams that were used jointly by several families: "the uprights made of hemlock [were] left in place to mark each family's territory." Another type of resource-gathering site owned by families was seabird nesting sites. According to Dawson (cited in Cole and Lochner 1993: 134b), "every lonely and wave-washed rock on which these birds deposit their eggs, is known to the natives, who have even these, apportioned among the families as hereditary property."

Some of the different kinds of property owned by Haida lineages included rights to certain salmon spawning streams, lakes, trapping sites, patches of edible plants (including root vegetables such as silverweed, wild fruits such as highbush cranberry, bog cranberry, and wild crabapple, and shoot vegetables such as fireweed—Figure 5.4), as well as stands of western red-cedar, and tracts of land along the coast line (Blackman 1990: 249). Among the Haida of Skidegate, high-ranked women are said to have owned highbush cranberry patches behind the village at Miller Creek (Ada Yovanovich, personal communication to NT, 1995; Turner 2004).

TSIMSHIAN AND ATHAPASKAN SPEAKING PEOPLES OF THE NORTHWESTERN BRITISH COLUMBIA

Coastal areas in northern British Columbia and the major river valleys extending inland are inhabited by a number of peoples speaking Tsimshian languages, including the Southern Tsimshian, Coast Tsimshian, Nisga'a, and Gitxsan. The Witsuwit'en are an Athapaskan-speaking inland group with traditional territories in the upper drainage of the Skeena River (along the Bulkley River), neighboring on the Gitxsan. The Gitxsan and Witsuwit'en have a long history of interaction and mutual cultural borrowing; they inhabit similar environments, and have developed many shared cultural features and institutions for managing and owning lands and resources (Gottesfeld 1994a).[7]

As with the Haida, among these peoples ownership of traditional lands is based on the matrilineal descent group or clan. The traditional territories of these corporate groups were extensive and contained multiple resources in a wide variety of habitat types (Richardson 1982: 104). In northwestern British Columbia, many resources tend to be in localized patches, and are often avail-

5.4. Fireweed (Photo by R. D. Turner)

able only at very specific times of the year (Gottesfeld 1994c: 447). The matrilineal descent group, commonly referred to as the House, traditionally regulated access to these resources through well-defined systems of land ownership. For the purposes of this chapter, we will focus our discussion on land and resource management among the Gitxsan and Witsuwit'en. However, the general patterns discussed also apply, with some variation, to the Tsimshian and Nisga'a peoples.

Gitxsan and Witsuwit'en. Gottesfeld (1994a: 171–72) describes the lands inhabited by the Gitxsan and Witsuwit'en along the Skeena and Bulkley River valleys as diverse in topography, climate, and vegetation. This area represents a transition zone between the Northwest Coast and the Boreal Interior of the province. Among the Gitxsan and Witsuwit'en, resource use is regulated through the division of land into owned territories or sites. Ownership of these lands and their resources is primarily through the House group, which is called Wilp among the Gitxsan and Yikh among the Witsuwit'en; lands and resources are administered by the Chief on behalf of the group.[8] Each particular territory generally corresponds to a watershed along the Skeena River and its tributaries, with boundaries running along the height of the land in the drainage area, and including all the habitat types at different altitudes within these zones (Cove 1992: 234; Gottesfeld 1994c: 448).

Access to these territories was, and still is, strictly controlled, as was the harvesting of resources. Nominally, it would be the head of the House who regulated access to resources, including areas for hunting mountain goat, beaver, and other fur-bearing animals, as well as berry patches and fishing sites. Outsiders wishing to pass through these territories were required to seek permission from the head of the House group. In the past, punishment for not following this protocol could be death. In some cases, however, House groups might own specific fishing sites (particularly along the major rivers such as the Skeena or the Kispiox) and berry patches within the general territory of a different House group. In addition, berry patches that were in close proximity to village sites were generally used in common by all residents of the village. In the past, the head of the House, as the nominal owner of these berry patches, would be acknowledged by small public gifts after the harvest (Gottesfeld 1994a: 176; 1994c: 448).

Gottesfeld (1994c: 451) suggests that the territorial system of ownership by House groups represented by Chiefs had important implications for the conservation of plant resources such as root vegetables and wild berries. She suggests that House restrictions on harvesting rights would ensure that knowledgeable elder women would decide which wood fern (*Dryopteris expansa*) rhizome or rice-root bulb patch would be harvested at any given time. These patches had to be carefully monitored, since it took several years for populations to recover for sufficient biomass to be harvested again. In addition, the owner of a territory would decide when to burn the berry patches for optimal production (Gottesfeld 1994c: 451). Ownership of territories among the

Gitxsan and Witsuwit'en provided the cultural framework to coordinate, monitor, and regulate important management activities such as selective harvesting and burning. Similar regulatory functions were performed by the owners of salmon spawning streams and fishing sites, and are still in effect today in portions of Gitxsan and Witsuwit'en territory (Gottesfeld 1994c: 454).

Traditional concepts of ownership among the Gitxsan, and the histories that define and validate ownership patterns, are explicitly expressed in the publication entitled *Histories, Territories, and Laws of the Kitwancool* (Duff 1959).[9] In the village of Kitwancool, one of the primary Gitxsan villages along the Skeena River, there are two matrilineal clans, the wolf clan and the frog clan. Within these clans, matrilineal Houses have their own territories and traditions. The totem poles represent the histories of each clan's migration to different places, and how they claimed ownership:

> They had a ceremony and put their power on that river and land, which meant that it belonged to them as they had found it first . . . they built a permanent village here and put their mark on the river, thus claiming ownership of it. . . . Once more they moved on, leaving their power and mark which made this country theirs, and returned to their former village, Ks-gay-gai-net. The reason they were traveling so much was that they were making their map, and on each piece of land when they stopped they had left their mark and power, making it theirs. [Chief Gam-gak-muk, in Duff 1959: 24–25]

Finally, the Chiefs made their permanent village at Gitanyow (Kitwancool) where they set their totem poles in the ground. This is an act of great significance because the poles gave them "the right of ownership of all the lands, mountains, lakes, and streams they had passed through or camped over or built villages in as their own" (Duff 1959: 25). It was the ceremonial act of placing a totem pole in the ground that in effect placed the "mark" of the Chief on the land by symbolically rooting the history of the House and its crests in that place. Duff (1959: 12) describes the totem poles as "the legal deed to the land." Similarly, Cove (1992: 236) describes the totem poles as the physical connection between the House and its territories. The power of these poles persists, and the people of Gitanyow have these histories behind them to prove how they own their lands, and the totem poles that visually represent these claims. Laws concerning ownership of territories are very strict among the Gitanyow, as with other Aboriginal peoples on the North Coast, since the ownership of land is connected with supernatural powers and the origins of their people:

> These laws go back thousands of years and have been handed down from one generation to another, and they must be held and protected at all costs by the people owning these lands. These laws are the constitutional laws, going back many thousands of years and are in full force to-day and forever. [Duff 1959: 36]

According to Cove (1992: 235), a House gained title to its territory by merging its essence with the land itself. That essence is based on the supernatural powers acquired by House ancestors through encounters with supernatural beings at particular locations. These physical locations, as recorded in oral tradition, essentially became part of the House itself through these encounters. Thus the House and the territory belonging to the House are inseparable.

The Gitxsan and Witsuwit'en reaffirmed the continuing vitality of their laws in the Supreme Court of British Columbia, and eventually in the Supreme Court of Canada, with a land title action against the provincial crown (*Delgamuukw v. British Columbia*). The people who spoke at the hearings for this case eloquently expressed their laws and claims to ownership:

> My name is Delgam Uukw. I am a Gitksan Chief and a plaintiff in this case. My House owns territories in the Upper Kispiox Valley and the Upper Nass Valley. Each Gitksan plaintiff's House owns similar territories. Together, the Gitksan and Wit'suwet'in Chiefs own and govern the 22,000 square miles of Gitksan and Wit'suwet'in territory.
>
> For us, the ownership of territory is a marriage of the Chief and the land. Each Chief has an ancestor who encountered and acknowledged the life of the land. From such encounters come power. The land, the plants, the animals and the people all have spirit—they all must be shown respect. That is the basis of our law. . . . By following the law, the power flows from the land to the people through the Chief; by using the wealth of the territory, the House feasts its Chief so he can properly fulfill the law. This cycle has been repeated on my land for thousands of years. [Gisday Wa and Delgam Uukw 1989: 7–8]

Delgam Uukw (Gisday Wa and Delgam Uukw 1989: 7–8) further stated that the Europeans have never wanted to know the histories of the Gitxsan, nor did they respect Gitxsan laws and ownership rights. This court case has been a major factor in prompting the recognition of Aboriginal title in British Columbia. Although originally denied by the presiding trial judge, a decision on appeal at the Supreme Court of Canada has recognized the existence of Aboriginal title in British Columbia, and the importance of oral histories in establishing claims to title (*Delgamuukw v. British Columbia*). Until very recently, however, there has been a long-standing reluctance on the part of Euro-Canadians to recognize and respect traditional Aboriginal ownership rights and stewardship imperatives in British Columbia and elsewhere. As described in the following section, the implications of this reluctance have been significant, resulting in the disruption of traditional ownership and management practices for plants and other resources.

Historical Changes in Aboriginal Land Tenure and Ownership

Northwest Coast First Peoples' concepts of land and resource ownership differed substantially from those of the European newcomers to the Northwest

Coast. Seen through European eyes, neither the Aboriginal peoples' use of the land, nor their ownership of it was considered valid or legitimate, perhaps because it was so different from their own. In most cases, the newcomers recognized only large, permanent settlements and highly visible agricultural modification as criteria for land ownership. This demonstrates the important connections between land ownership and the question of cultivation and land management. The perception that Northwest Coast Indigenous peoples were foragers who did not modify or manage their resources provided the rationalization for dispossessing the Indigenous populations of their lands (Deur 1997, 1998, 2002a,b; Deur and Turner, introduction to this volume). In turn, the finely tuned mechanisms for managing and conserving resources, which were embedded in traditions and institutions for ownership, were disrupted when traditional territories were taken over by European systems of land use and ownership.

The primacy given to European-style agriculture and the lack of understanding of traditional ownership and resource use are reflected in the fourteen treaties signed on Vancouver Island between 1850 and 1854. These agreements are known as the Douglas Treaties, since they were negotiated under the authority of James Douglas, Chief Factor of Fort Victoria for the Hudson's Bay Company (1849–1858) and Governor of the Colony of Vancouver Island (1851–1864). Nine treaties were signed at Fort Victoria in 1850, two were signed at Fort Rupert in 1851, two were signed on the Saanich Peninsula in 1852, and one was signed at Nanaimo in 1854. There is little variation in the format and terms of the treaties, with the exception of the Nanaimo treaty, which has no text, but which is assumed to be based on the same terms as the other treaties since it is described as a "similar conveyance." In each treaty, the boundaries of the traditional territory to be surrendered are given in detail, and the following conditions are specified:

> The condition of our understanding of this sale is this, that our village sites and enclosed fields are to be kept for our own use, for the use of our children, and for those who may follow after us; and the land shall be properly surveyed hereafter. It is understood, however, that the land itself, with these small exceptions, becomes the entire property of the white people forever; it is also understood that we are at liberty to hunt over the unoccupied lands, and to carry on our fisheries as formerly. [British Columbia 1875: 5][10]

Wilson Duff (1969) points out that these early treaties reflect more of the European concepts of land and property than the realities of the Aboriginal peoples with whom they were negotiated. According to Duff (1969: 3), "A treaty, of the kind discussed here, is a white man's certificate of a transaction . . . to read a treaty is to understand the white man's conception (or at least his rationalization) of the situation as it was and of the transaction that took place." In fact, the Douglas Treaties may be regarded as primarily reflecting the aims and attitudes of the Governor and Committee of the Hudson's Bay Company (and likely by extension the British Colonial Office), since the actual text of the treaties was sent from the HBC Governor and Committee in London to James Douglas on Vancouver Island after the original negotiations took place at Fort Victoria in 1850 (Tennant 1990: 19).

Nevertheless, these early treaties contain an implicit recognition that Aboriginal peoples exercised some form of ownership over the land that had to be extinguished by colonial authorities (Fisher 1992: 67; Tennant 1990: 20). This recognition, however, was short-lived, as later administrators would effectively deny the existence of Aboriginal title in British Columbia, and deny the applicability of the Royal Proclamation of 1763, which asserted the need to negotiate with Aboriginal peoples for the surrender of lands. The result was a system whereby reserves were allotted by executive action, rather than by negotiation (Bartlett 1990). The unresolved issues of Aboriginal title to land in British Columbia are a legacy that is only now being addressed, albeit halt-ingly, through modern-day treaty negotiations.

The emphasis on cultivation and enclosure of lands as a prerequisite for title in the Douglas Treaties and the colonial correspondence reflects a broader philosophical attitude toward the land and land use, one that was shared by many Europeans at the time (Deur and Turner, Introduction, this volume). The relationship between private ownership and "cultivation" of land was deeply embedded in European cultural conditions, and was upheld with almost religious conviction, as reflected in enlightenment philosopher John Locke's (1988: 291) words, originally written in 1690:

> God gave the World to Men in Common; but since he gave it them for their benefit, and the greatest Conveniences of Life they were capable to draw from it, it cannot be supposed that he meant it should always remain com-mon and uncultivated. He gave it to the use of the Industrious and Rational (and Labour was to be his Title to it).

One of the most revealing examples of European attitudes towards Indigen-ous occupants of the land is in the book by Gilbert Malcolm Sproat (1987), *The Nootka: Scenes and Studies of Savage Life*, originally published in 1868. Sproat (1834–1913) was a pioneer businessman who was in charge of the daily operations of the sawmill, townsite, farming, and trading activities at Alberni on Vancouver Island, and later served as a Commissioner on the Joint Com-mission on Indian Reserves, which was appointed in 1876 (Duff 1964: 94; Lillard 1987: xvi). In his book, written before his involvement with the Joint Commission, Sproat records his early opinions on the question of the "supe-rior" newcomers' taking over of the land, in an effort to rationalize what he and others are doing:

> My own notion is that the particular circumstances which make the delib-erate intrusion of a superior people into another country lawful or expe-dient are connected to some extent with the use which the dispossessed or

conquered people have made of the soil, and with their general behavior as a nation. *For instance we might justify our occupation of Vancouver Island by the fact of all the land lying waste without prospect of improvement. . . .* Any extreme act, such as a general confiscation of cultivated land . . . would be quite unjustifiable. [Sproat 1987: 8; emphasis ours]

The justification of Sproat and his fellow European and Canadian newcomers for taking over the Nuu-chah-nulth lands was based upon the premise that the land was not cultivated, and therefore not really "occupied," in any "civilized sense." It is interesting to note that, later in his career, as he became more intimately aware of the circumstances of Aboriginal peoples in British Columbia through his work on the Joint Commission on Indian Reserves, Sproat became one of the few European voices advocating for a just resolution of Aboriginal land grievances, or at the very least the assignment of larger Indian Reserves, in British Columbia. He became increasingly critical of the provincial government and their refusal to recognize Aboriginal land rights, and eventually was pressured to resign from his position in 1880 (Fisher 1992: 188–98).

In many ways, the problem was more acute on the coast than in the interior. Europeans assumed that the coastal peoples were mainly fishers who required only enough land for their winter houses and for small summer camps, since they spent their time on the ocean and at focused salmon-fishing points at river estuaries and similar sites. As a result, the coastal reserves established were—and still are—generally small and scattered, with little provision for the extensive areas required for the seasonal rounds people undertook to access all the resources they required to live.

Although not recognized by early European settlers,[11] nor later by anthropologists, Northwest Coast Aboriginal peoples were not passive occupants of their territories (Deur and Turner, Introduction, this volume). They did cultivate their lands and resources, as many chapters in this volume demonstrate. However, their management practices were often subtle and not obvious to the newcomers, whose views were shaped by notions of domesticated environments of the European style, such as ploughed fields, grazing areas for livestock, and formal, fenced and rowed gardens.

Discussion

In this chapter, we have described the patterns of ownership for plant and other resources among various First Nations along the Northwest Coast. Each group developed unique protocols for ownership tailored to its social and natural environment. The specific traditions and protocols that governed resource ownership and management among the various peoples on the Northwest Coast are likely related to both environmental and cultural factors. According to Richardson (1982: 107), along the Northwest Coast there may be a south-to-north gradient of decreasing resource abundance, with resources becoming increasingly patchy and localized in northern parts of the region. Richardson (1982) suggests that there is a relationship between ownership patterns and resource abundance or distribution, with a tendency for community control and open access in areas of greater resource abundance, and more restricted kin-group ownership in areas where resources are patchy or scarce. More intensive analysis and comparison of the relationships between resource diversity, resource abundance, and social structures would need to be undertaken to confirm such a relationship (Deur 1999, 2000).[12]

Also important, however, are the cultural values placed on the various plant resources. For example, certain types of berries were highly prized and given as gifts at feasts. In particular, Pacific crabapples and highbush cranberries were highly prestigious foods that were used as gifts to chiefs at potlatches and as bridal gifts. Unusually long silverweed roots, too, were particularly valued, and were usually eaten only by chiefs in Kwakwaka'wakw society (Boas 1921; Turner 1995).

Other plant resources served as valuable trade items that would be exchanged among Aboriginal groups, and later with the European newcomers. Evidently, these tradable plant resources, more than others, were subject to conventions of ownership and tenure. This was the case with ownership of certain choice camas beds among the Saanich and other Salish peoples of southern Vancouver Island. Camas was traded from Salish peoples of southern Vancouver Island to a number of other coastal groups, including the Nuu-chah-nulth, Kwakwaka'wakw, Ditidaht, Mainland Comox, and Halq'emeylem (Turner and Loewen 1998).

As described in this chapter, there are a number of ways in which ownership rights facilitate the management of resources. Ownership protocols that granted certain families or groups rights to visit particular locations from year to year facilitated the long-term observation and management of plant resources. This is true for the southern parts of the coast, with inclusive ownership and customary patterns of use, as well as in northern areas where exclusive rights were clearly defined and protocols in place for enforcing these rights.

Many of the ownership patterns described in this chapter include provisions for sharing resources, and for allowing other groups the rights to use resources. Excessive hoarding of resources is widely recognized among Northwest Coast peoples as a negative trait, and regarded with scorn. There are many sanctions against such behavior, which are reinforced in traditional narratives and teachings. Conversely, those individuals and families who share their food and other resources with their neighbors, or even with strangers, are highly regarded (Turner and Atleo 1998).

Of course, ownership rules are not enough to ensure sustainable use of resources. There must also be an accompanying ideology that fosters respectful use and conservation, as well as an understanding of strategies to enhance resource productivity and to prevent overexploitation. Along the Northwest Coast, the use of plant resources was guided by concepts of stewardship and respectful attitudes to all life-forms. This is reflected in historical sources and oral traditions, as well as in the words of contemporary Aboriginal peoples. For instance, the life force of plant resources is explicitly recognized in the words of praise offered by Kwakwaka'wakw gatherers and others to plants (see Turner and Peacock, this volume). The concept of interdependence among life forms is eloquently expressed in the Nuu-chah-nulth phrase *hishuk ish ts'awalk*, which means "all things are one." This sentiment was recently reaffirmed in the work of Nuu-chah-nulth people with the Scientific Panel for Sustainable Forest Practices in Clayoquot Sound (1995). Contemporary Aboriginal peoples are affirming their rights as caretakers, stewards, and owners of their traditional territories in the courts and at the negotiating table.

Conclusion

Since the British Columbia Treaty Process was initiated in 1991, many First Nations in British Columbia have entered into treaty negotiations with the provincial and federal governments. Throughout this process, differing understandings of ownership rights (and responsibilities) have been at the heart of debate and negotiations. People with a background in the British system of land tenure hold a certain concept of ownership that usually requires absolute (and written) title to a given piece of property, which then allows the owner a high degree of control over what is done to the land and resources on that property. Theoretically, this avoids "the tragedy of the commons" as described by Hardin (1968). According to Hardin, in the absence of a regulating authority, individuals with access to a common but limited resource will maximize their own benefits at the expense of the group as a whole, and ultimately lead to the degradation of the common resource.

However, Aboriginal peoples in British Columbia have never existed in an environment of "open access" to resources or in the absence of culturally prescribed regulation of resources. As described in this chapter, ownership and rights of use were clearly defined through social structures and protocols, whether through communal territories with known boundaries in combination with ownership of key resources, or through strictly defined descent groups owning extensive tracts of land and multiple resources from which others are forcibly excluded.

Through developing protocols for ownership and management of plant resources, Aboriginal peoples on the Northwest Coast have conserved and sustained populations of culturally important plants, and important resource-gathering sites. In combination with respectful attitudes towards the land, and an ethic of sharing within groups, Aboriginal peoples on the Northwest Coast have developed ways of living sustainably in local environments, as demonstrated by their successful long-term inhabitation of particular locales. Abundant resources and populations of culturally important plants have been maintained, and even enhanced, by careful use and management through traditional ownership protocols. These ownership patterns, although disrupted through colonization and population loss, are still entrenched in the traditions and laws of the different peoples, along with the underlying worldviews and philosophies that have guided resource use in traditional territories since time immemorial.

Notes

We gratefully acknowledge the contributions of the many Elders and other aboriginal specialists quoted or cited in this chapter: Gabe Bartleman, Elsie Claxton, and Kevin Paul (Saanich); Chief *Umeek* (Dr. Richard Atleo), Chief Earl Maquinna George, Roy Haiyupis, Stanley Sam, Gertrude Frank, Lena Jumbo, James Adam (Ahousaht); Luke Swan (Manhousaht); Alice Paul, Larry Paul, Ruth Tom, Rocky Amos (Hesquiaht); John Thomas, Chief Charlie Jones (*Queesto*) and Ida Jones (Dididaht); Dr. Margaret Siwallace (Nuxalk), Cyril and Jennifer Carpenter, Pauline Waterfall (Heiltsuk); Chief Adam Dick (*Kwaxistala*), Kim Recalma-Clutesi (*Ogwilogwa*), John Macko, and Dr. Daisy Sewid-Smith (*Mayanilth*) (Kwakwaka'wakw); Barbara Wilson (*kii7lljuus*), Nonnie Florence Davidson, George Young, Emma and Solomon Wilson; Ada Yovanovich (Haida); Chief Johnny and Helen Clifton, and Ernie Hill Jr. (Gitga'at, Tsimshian). We are also grateful to the following people for help, advice and information: Dr. Brenda Beckwith, Randy Bouchard, Juliet Craig, Dr. Richard Daly, Dr. Douglas Deur, Dr. Marianne Ignace, Chief Ron Ignace, Dr. Dorothy Kennedy, Dawn Loewen, Dr. Sandra Peacock, R. Andrew Reed, and Dr. Wayne Suttles.

1. Another factor that needs to be taken into account is that ownership rights and traditions have come under increasing scrutiny as Aboriginal peoples in British Columbia have entered into modern-day land rights negotiations and litigation. This circumstance may have brought into sharper focus certain features of traditional ownership patterns, as Aboriginal peoples have been compelled to establish their claims according to European standards of ownership and property rights. Of course, "traditions" are dynamic and evolve and are reinterpreted through social and cultural change. To expect otherwise would be to assume a static, time-bound view of Aboriginal cultures.

2. This statement is largely based on ethnographic descriptions dating from the early to mid twentieth century, and probably applies to the time period from the late nineteenth to early twentieth century.

3. It appears from the ethnographic source that this information, collected in the mid twentieth century, would have applied to the consultant's parent's generation, that is, back to the early twentieth century.

4. For a description of management strategies for camas beds (including burning, tilling, and weeding practices) see: Suttles (1974) and this volume; Beckwith 2004; Turner and Peacock, this volume.

5. The practices described here probably would have been operative into the early twentieth century.

6. Boas (1966) refers to these groups as numayma; however, Daisy Sewid-Smith, a Kwakwaka'wakw language specialist and historian, prefers the term "clans."

7. In 1968, the Gitxsan established a regional tribal council for purposes of asserting sovereignty over traditional territories (Inglis et al. 1990). A number of Babine (Carrier) villages were included in this organization, which was called the Gitksan-Wet'suwet'en Tribal Council.

8. A more thorough examination of property rights and ownership among the Gitxsan and Witsuwit'en is provided by Daly and Vast (2001).

9. This publication grew out of a request by the people of Kitwancool to record their histories and laws for future generations at a time when totem poles were being removed from their village and transferred to the British Columbia Provincial Museum (now Royal British Columbia Museum). The stories of these totem poles were told by the chiefs of Kitwancool to Mrs. Constance Cox; the material was then edited by Wilson Duff for publication.

10. The original manuscript treaties are held at the BC Archives, MS-0772.

11. Even with camas (*Camassia* spp.), which was indisputably managed by peoples of southeastern Vancouver Island (Babcock, ca. 1967; Beckwith 2004; Suttles, this volume), Sproat (1868 [1987]: 42) comments, "They have never attempted to increase the production of camas by any kind of cultivation." Yet, concerning Pacific crab-apple (*Pyrus fusca*), he noted, "The natives are as careful of their crab-apples as we are of our orchards; and it is a sure sign of their losing heart before intruding whites when, in the neighborhood of settlements, they sullenly cut down their crabapple trees, in order to gather the fruit for the last time without trouble, as the tree lies upon the ground." (op. cit., p. 43).

12. Douglas Deur (1999, 2000) has argued that high population concentrations around productive resource sites on the Northwest Coast (particularly estuarine environments and/or salmon fisheries) increased dependence on a finite range of terrestrial resource sites, and thus contributed to the intensification of ownership patterns and land tenure. In turn, this situation encouraged the intensification of resource harvesting and management activities at these sites.

PART II

Case Studies

6.1. Flowers of common blue camas (Photo by R. D. Turner)

6.2. Camas bulbs (Photo by R. D. Turner)

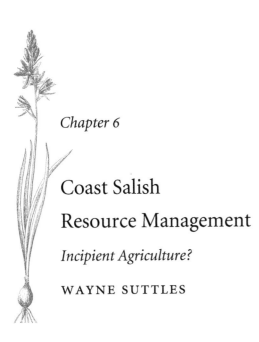

Chapter 6

Coast Salish
Resource Management

Incipient Agriculture?

WAYNE SUTTLES

A half century or more ago, in a part of the Northwest Coast split by the International Boundary, older Coast Salish people told ethnographers that their forebears owned and tended beds of edible native "roots," harvesting them in ways that ensured their reproduction and taking measures to increase the yield.

In 1952, on the southern shore of Vancouver Island west of Victoria, Mary and Agnes George of the Sooke Band described to me how the Sooke people had harvested camas (*Camassia* spp.) (Figures 6.1 and 6.2). At several places where the plants grew thick, women had their own plots, marked off with stakes so that no one else would dig there. A woman defended her rights to her plot. After her death, a relative inherited it. Women dug the bulbs in the spring, while the flowers were in bloom. Using digging sticks (I assume), they pried up a piece of sod, described as solid camas bulbs. "There was no grass then," I was told, "because the patches were cared for." They turned a strip of sod over, pulled off the bulbs from the under side, and then replaced it with the flowers on it. Their hands got purple from the flowers. When they had finished, they leveled the ground and covered it with seaweed. Later, when it was all dry, they burned it over. This was said to make the bulbs bigger the next year.

The bulbs were carried in openwork cedar-root or spruce-root baskets to where they were steamed. Women only did the digging—the men would have been fishing, but the men helped with steaming the camas. This was done in pits used year after year. They set stones in the bottom of the pit and built a fire. When the stones were hot, they removed the ashes and put in the bulbs surrounded by kelp, salal (*Gaultheria shallon*), grand fir (*Abies grandis*) boughs, and sword ferns (*Polystichum munitum*), leaving a stick to make an opening to pour water through. The camas had to steam for a day and half to be cooked properly (Suttles 1947–52).

Early in the twentieth century, Franz Boas (1921: 189; 1934: 37) and others had reported that Kwakiutl (Kwakwa̱ka'wakw) families owned beds of silverweed and springbank clover (see Deur, Turner and Peacock, this volume). For the Coast Salish, one of the first descriptions of harvesting practices suggesting cultivation was recorded from Lummi consultants in 1928–29 by Bernhard Stern (1934: 42–43). One of the first references to ownership of plots was recorded by Marian Smith in 1938, from a Nooksack man, Jack Jimmy. In 1950 Smith wrote:

> Near Goshen there was a small prairie where many "Indian carrots" [possibly *Perideridia gairdneri*] grew. These were a prized food and were dug in March. Each family owned its own plot, about forty feet square, and the plots were marked off from each other by shallow ditches. Trespass on another's plot was a serious offense and caused "big fights." The roots were dug, the edible portion broken off, and the sprout replanted. The soil was kept loose and "easy to dig." [M. W. Smith 1950: 336–37]

In 1942 June Collins got similar information from Upper Skagit people. In the mid-1940s and later I got similar statements from members of several other Coast Salish groups, and Paul Fetzer got more from Nooksack sources.[1]

These practices sound like a kind of gardening. Smith (1950: 336–37) suggested: "In an otherwise purely food-gathering economy, the Nooksack seem to have already foreshadowed simple agricultural procedures." Collins (1974: 55) refers to the Upper Skagits' "semidomestication" of tiger lilies (*Lilium columbianum*) and "wild carrots." Writing of camas on southeastern Vancouver Island, Nancy Turner (1975: 81) has called Native practices "semi-agricultural." Helen Norton (1979a: 392) has used the term "proto-horticulture."

But were these practices aboriginal? Accounts of them were recorded a century and a half after the Native people first met Europeans and over a century after they were introduced to cultivation by the Hudson's Bay Company, were given potatoes (*Solanum tuberosum*), and began raising them throughout the region. Smith (1950: 337) suggested of the Nooksack, "Their almost immediate adoption of potato cultivation proves the ease with which they could make the transition [from tending native plants to simple agriculture]." In a paper on the spread of potato cultivation (Suttles 1951a), I also commented on how easily the potato had been adopted and argued that several prerequisites for agriculture were already in place.[2] But couldn't it have been the other way around—the Native people began cultivating native plants *after* they learned to cultivate potatoes? To answer this I will review the data and try to weigh the evidence, concluding with some thoughts about the relevance of Native social organization.

The Data

THE PEOPLES

Statements about the ownership or tending of beds of plants with edible "roots" (in the general sense) come from an area that includes southeastern Vancouver Island and the mainland from the Fraser to the Skagit and from speakers of four Salishan languages—Northern Straits, Nooksack, Lushootseed, and Halkomelem. The Northern Straits groups (the Sooke, Songhees, Saanich, Semiahmoo, Lummi, and Samish) are saltwater people. The Nooksack and the two Lushootseed groups for whom we have data (the Nuwhaha, in the Samish Valley, and the Upper Skagit) are inland river people, and the two Halkomelem groups from whom we have data (the Musqueam and Katzie) are on the Lower Fraser River. These groups form a continuous area.

It seems likely that the adjacent Halkomelem to the north on Vancouver Island (the Cowichan and Nanaimo) and perhaps a few more of the Northern Lushootseed to the south (as the Skagit and Stillaguamish) may have followed these practices, but we have no clear indication of this. For some neighboring Coast Salish groups we have explicit denials of ownership of root beds.[3]

THE PLANTS

Eleven species of native plants were said to have grown in plots owned by individuals or families somewhere in this area. These plants included several members of the lily family: camas (probably both *Camassia quamash* and *C. leichtlinii*), two kinds of "wild onions" (probably nodding onion, *Allium cernuum*, and Hooker's onion, *A. acuminatum*), chocolate lily or rice-root (*Fritillaria lanceolata*), and tiger lily (*Lilium columbianum*); "wild carrots" (probably the umbelliferous, wild caraway *Perideridia gairdneri*); bracken fern (*Pteridium aquilinum*); wapato (*Sagittaria latifolia*); and the bog cranberry (*Vaccinium oxycoccos*). (Identifications are from Turner 1995.)

Camas was evidently the most important food plant for the Northern Straits peoples and for many of their neighbors. It grew abundantly in dry open areas on the shores of southeastern Vancouver Island and the San Juan Islands (Figure 6.3), and on the mainland in open areas known locally as "prairies." The islands seem to have been preferred; the soil over rocks was thinner and the camas easier to dig. But the work was not easy. In 1868 the botanist Robert Brown wrote of camas:

> The gathering is nearly wholly done by women and children, who use a sharp-pointed stick for the purpose; and it is surprising to see the aptitude with which the root is dug out. A botanist, who has attempted the same feat with his spade, will appreciate their skill. [Brown 1868: 378–79]

6.3. Camas meadow, Beacon Hill Park, Victoria (Photo by R. D. Turner)

The Northern Straits people generally harvested camas in late May, near the end of its blooming season. According to Brown:

> In Vancouver Island this plant comes into flower about the middle or end of April, and remains in flower until June, when, just as it is fading, the roots are in a condition to be gathered. Until that time it is watery and unpalatable. [Brown 1868: 378–79]

Nancy Turner suggests (personal communication 1997) that some people may have dug later in the year, when the bulbs were more mature. This seems reasonable, but I have no evidence for it from the Northern Straits.[4]

The bulbs were evidently dug in considerable quantities. But ethnographic data on just how much could be dug are contradictory. In the 1930s Jenness (1934–35: 7) was told by a Saanich that during the three-week season an energetic family could fill ten or twelve bags. In 1967 a Saanich source said that his family group "collected four or five potato sacks full" per session (Marguerite Babcock, quoted in Turner and Bell (1971: 75).

As my Sooke sources said, the bulbs were usually steamed in an earth oven for a day and a half. Turner and Kuhnlein (1983: 214) report that the bulbs are rich in inulin, which is converted into sweet and digestible fructose through this prolonged steaming. After steaming, the bulbs could be dried and stored for winter use. In the Fort Langley Journal, the entry for 21 December 1828 reported Cowichans from Vancouver Island passing on their way up the Fraser to their summer fishing sites and notes:

> Their Canoes are loaded with Kamas which I believe is procured in abundance on the Island—with it and the Salmon they left *in Cache* last fall they propose living well for the Winter. [Maclachlan 1998: 90]

On upriver prairies, beds of "wild carrots," chocolate lilies, tiger lilies, and other bulbs were owned by the inland groups, the Nooksack, Nuwhaha, and Upper Skagit. Wild carrots were harvested in early spring, in March according to Smith's Nooksack source. Turner's Saanich source said that "plants were marked in the summer and dug in the following spring before the leaves came up." They were also cooked by steaming in pits. Presumably other "roots" growing with camas or carrots would have been harvested at the same time. Again, we have no clear statements on quantities.[5]

Bracken fern was probably used everywhere; the roots (rhizomes) were singed and then pounded to convert the edible material into a dough that was eaten in that form or baked as a "bread."[6] Wapato grows in standing fresh water; it was absent over much of the region, but it was especially abundant in ponds in Katzie territory in the lower Fraser Valley, to which people from a wide area came for the harvest. The bog cranberry was also especially abundant in the Lower Fraser Valley.

OWNERSHIP

It appears that throughout the Salish-speaking region where ownership of root plots is reported, those owned were only some—though possibly the best— plots; there were always other places where any member of the local group could dig. Several sources said that if someone found a good bed earlier in the year, she could mark it as hers to harvest later. Everywhere the network of intermarriage and obligations to kin and affines enabled people from outside the local group to share in its resources. Thus when one source said plots were owned and another said people from all over could come and dig, these statements are not necessarily contradictory. Everywhere women did the harvesting, but statements from Saanich and Nooksack sources indicate that men too might own plots. Several sources identified owners by name. Several indicated that people from different villages might own adjacent plots at a productive site.

The ownership of beds of camas was mentioned by Sooke (as quoted above), Saanich (Jenness 1934–35: 52–53; Suttles 1951a), and Samish (Suttles 1951a) sources. My Saanich source, Louie Pelkey, said a plot was marked with rocks. He identified by name three men who owned the northern, middle, and southern portions of one small island; the three were related but lived in two different villages. And he identified the woman who owned a smaller islet. Stern (1934: 42–43) said Lummi women returned to the same places yearly "but they do not have exclusive rights to these plots," which could simply refer to the usual sharing of access to owned resources. My Sooke sources said the Songhees "had plots" on the islands in Oak Bay, but my Songhees sources did not mention ownership. Julius Charles, my oldest source on the Semiahmoo and Lummi, did not mention ownership of camas beds by those peoples, although he was aware of the ownership of (wild) carrot beds by Nooksack women. Annie Lyons, one of the two older Samish I interviewed, said that an islet off Lopez Island was her grandmother's camas plot, but the other, Charley Edwards, did not think camas beds were owned, although he described his grandmother's role as owner of a horse-clam bed.

Probable evidence of ownership of camas beds was observed in Northern Straits territory in 1860 by William J. Warren and Dr. C. B. R. Kennerly, members of a party from the North West Boundary Commission exploring the San Juan Islands (Warren 1860: 117–20). On San Juan Island the party camped on what may have been Garrison Bay, where they saw the ruins of a large plank house. On the hills nearby Warren reported, "We saw in different places cobble stones placed in lines about 100 feet long, arranged in this position probably by the Indians who have been at the place, though for what purpose we could not conjecture." The following day (8 February 1860) the party went over to Stuart Island, probably to Reid Harbor, where they saw deserted houses, which they assumed were used during fishing season. Warren and Kennerly "climbed up the steep mountain immediately behind the camp following a well-worn Indian trail." They found "a fine growth of grass even to the summit of the highest ridge" and observed that the south slope of the mountain "had been dug up a great deal by Indians gathering Kamass roots." And they found "more of those lines of cobble stones that we had seen on San Juan Id." The ruin on Garrison Bay was probably the village of the Klalákamish people, who were claimed as ancestors by Lummi, Saanich, and Songhees people. The houses on Stuart would have belonged to Saanich people from Louie Pelkey's village on Saanichton Bay (Tsawout). Gibbs (1863: 39) identified the Klalákamish as "a band of Lummis now extinct," but people claiming Klalákamish ancestry could well have been returning each year to harvest camas.[7]

Smith's Nooksack source said a plot was about forty feet square and marked by shallow ditches. The Nooksack people Fetzer and I interviewed all said that individuals or families from different Nooksack villages owned plots of "carrots" and other plants, probably camas, wild onions, and chocolate lilies, on the prairie near Goshen. Josephine George said that the plots were strips about four feet wide, the corners marked with rocks; her paternal grandfather had inherited four such plots, which his wife harvested. If a man died without heirs, she said, the person who disposed of the body properly inherited the plot. Lottie Tom thought the plots had been marked with rocks and lines drawn in the earth—perhaps the "ditches" mentioned by Smith's source. Agnes James and George Swaniset thought there were stakes at the corners, and Mr. Swaniset speculated that the stakes may have been marked to show which family they belonged to.

Collins (1974: 55–57) reported that the people she identifies as "Upper Skagit" maintained beds of tiger lilies and wild carrots on Jarman and Warner prairies on the Samish River and Sauk Prairie on a tributary of the Skagit. The first two prairies were in the territory of a group better identified as the Nuwhaha. Collins states:

> Land was divided into individual plots in these prairies, marked by sticks at the four corners. A daughter inherited the right to obtain roots from one plot from her mother. It was not necessary for women to live in the village . . . located at the prairie . . . in order to harvest a crop. Use rights were based on descent; during the late summer women with such rights came from widely distant villages to the plot of their mother. . . . [On] Sauk prairie, each plot was three to four acres in size. [Collins 1974: 55]

A Nuwhaha, George Bob, told me that each woman had her own strip, fenced off with tall cedar poles tied with cedar rope.

Although bracken fern roots were widely used in the region, the only report of ownership comes from a Musqueam source, Christine Charles, who told me that Musqueam families owned beds of bracken-ferns near Point Grey (Suttles 1956–62).[8] A Katzie source, Simon Pierre, said some wapato ponds and some cranberry bogs were owned by Katzie families (Suttles 1955).

GARDENING-LIKE PRACTICES

If loosening the soil while digging and burning afterward count as "tending," then tending was practiced widely in Western North America. The harvesting process no doubt loosened the soil, which would have increased production. Several Native people mentioned this, and George Gibbs (1855: 39), referring to the effect of American settlement in Western Washington, wrote, "The kamas . . . improves very much by cultivation, and it is said to attain the size of a hen's egg in land that has been ploughed." Burning after a harvest of several kinds of plants may have been practiced for millennia (Boyd 1999; Norton 1979b; Turner 1999).

But in this part of the Coast Salish region we have scattered reports of other, more agricultural tending practices. Weeding after digging is mentioned for the Upper Skagit. Fertilizing with seaweed was mentioned by my Sooke source. Lifting and turning the sod in order to remove the larger bulbs and replacing it with the smaller bulbs, which would amount to replanting, is reported for the Sooke and the Lummi. Transplanting was mentioned by Nooksack and Nuwhaha sources; women, they said, dug bulbs they happened upon and planted them in their plots. South of this region, some Lushootseed people may have transplanted wapato (Haeberlin and Gunther 1930: 21). Reseeding is mentioned by two sources; Stern (1934: 42–43) says that when digging camas the Lummi "crush the soil directly afterwards and plant the seeds broken from the stems," and Smith's statement quoted above indicates the Nooksack replanted carrot sprouts. Can these practices have been truly aboriginal?

The Case against Aboriginal Origin

Arguing against aboriginal practice we might consider the history of contact in the region, the statements of early observers, the weakness of the testimony, and possible biases of the ethnographers who collected the data, and a problem with the data.

1. In this region, European influence was felt at least as early as the 1770s. The first recorded contact was in the 1790s. Continuous contact began in 1827 when the Hudson's Bay Company built Fort Langley on the Fraser River about 25 miles (40 km) upstream from its mouth. During its first three years, the traders at Fort Langley began growing potatoes and other crops, and they hired local Coast Salish people as farm laborers. Very quickly the local Coast Salish began growing their own potatoes, and this practice rapidly spread throughout the region (Suttles 1951a). Thus a model for owning and tending beds of native root crops was available from around 1830. White settlers arrived in the 1850s and soon their livestock were destroying native plants everywhere except on offshore islets. By then some of the Coast Salish were growing seed crops, and by 1880s a good many were full-time or part-time farmers (Knight 1996: 166–78; Suttles 1954). Native knowledge of aboriginal culture has to have filtered through this experience. Moreover, beginning in the 1890s, litigation over Native rights has probably colored Native perceptions of Native culture.

2. There is nothing (that I know of) in the earlier literature on this region to suggest incipient agriculture. Sproat ([1868] 1987: 43) wrote, "They have never attempted to increase the production of camas by any kind of cultivation." Brown (1873–76 (1): 50) wrote, "Of agriculture they are quite ignorant" and "they have no aboriginal plant which they cultivate."[9] Nor did the earlier ethnographers, Wilson (1866), Boas (1891, 1894), and Hill-Tout (1907), record anything about ownership, though admittedly their reports are brief.

3. Native accounts of ownership and tending were all recorded toward the middle of the twentieth century. They come from probably no more than a dozen people, most of whom had not seen for themselves all that they described.

4. The data were collected mainly by anthropologists who were working with the objectives and assumptions of Boasian ethnography. We were attempting to "reconstruct" Native culture as it had been before contact disrupted it, and we assumed that we could do this by interviewing a few of the oldest and most knowledgeable members (key "informants") of each Native group. We tended to assume that Native culture was static and that acculturation was simply loss. Ignoring contemporary life and what later came to be called "ethnohistory," we often failed to consider possible causes of change

in ongoing Native culture. We could have missed clues that might have told us that what looked like aboriginal cultivation was simply a response to introduced cultivation.

5. It is doubtful that people were reseeding in the way suggested by Stern and Smith. At the season when camas bulbs were generally said to have been harvested the seeds would not have been mature. Likewise, if "carrots" were dug in early spring, the previous year's seeds would likely have been dispersed and not available for reseeding. We may also ask how much people really knew about plant reproduction. Camas sometimes reproduces by splitting or multiplication of bulbs as well as by seeds, and women must have seen that little bulbs become big bulbs. But it is not certain that they understood reproduction by seeds before the introduction of seed crops. Such knowledge was not required for potato cultivation; you simply plant a part of the tuber. Tobacco was not grown in this region and was probably unknown; the sowing of tobacco seeds could not have provided a model.[10]

The Case for Aboriginal Origin

However, there are also good arguments for aboriginal origins of cultivation.

1. When the data were gathered, traditional if not purely aboriginal culture was not really so far in the past. The Coast Salish people I interviewed in the late 1940s and early '50s had been born in the 1860s and '70s. They knew a good deal about traditional life from their own experiences and had learned more from their parents and grandparents, who had lived before the great changes that occurred in the 1840s and '50s. Some gave genealogies that went back to the eighteenth century. There had clearly been continuity in some aspects of Native culture, especially in social relations, values, and knowledge of the environment. In spite of our assumptions and our ignorance, I believe we recorded much that really reflected culture practices as they existed in precontact times.

2. Some features of plant tending were not a part of introduced cultivation. The turning over of strips of sod is not (as far as I know) a European practice, nor is burning after the harvest.

3. The distribution of the ownership of plots—over a nearly continuous area, absent to the south, close in the north to another area of plot ownership—suggests a precontact status, though it might be argued that Kwakiutl (Kwakwaka'wakw) practices too were the result of long familiarity with potato cultivation. But potato cultivation would have had to have had a very rapid influence on Native practices to account for the existence of "those lines of cobble stones" on San Juan and Stuart islands in 1860, only thirty-three years after Fort Langley was founded.

4. An even better argument in favor of aboriginal status is that plant-bed ownership and tending fits well into the Native pattern of "resource management" in this region. The people of the region had an intimate knowledge

of many, though certainly not all, features of their environment, which enabled them to be at the right place and time to harvest its resources. And their technology was sufficient to get and preserve enough of those resources to support what was, for a food-gathering economy, a relatively large population, though not so effective as to exhaust their resources. Some of their practices, such as periodically lifting sections of a salmon weir to let the fish go upstream, were conservationist in effect though probably social in intent—letting the folks upstream get their share. Others may have been ritual in intent, like the custom, shared by much of Western North America, of forbidding the taking of a resource before a ceremony was performed to honor the species. For salmon, this allowed some escapement of the earliest fish.

But it seems to me that the most effective means of resource management was a social organization that adjusted people to resources. The principal features of this organization were two balanced principles. On the one hand, there was a strong identification of people with place. Each social group had a home territory and was identified with it by myths and traditions, ceremonies, and even distinct features of speech, and its most productive sites were identified with families or individual owners. On the other hand, there was a wide social network, based on bilateral kinship and intervillage marriage, with obligations to kin and affines leading to a controlled flow of food, goods, and people.

The control given by ownership varied with the resource. A number of kinds of resource sites were owned—sites for taking salmon, sites for netting waterfowl, and shellfish beds, as well as beds of plants. For some resources the owner had important functions. The clearest example is the Northern Straits reef net. This is a complex device that required the cooperation of several families, through the sockeye salmon season, under the leadership of someone with special knowledge and organizational skill. The owner, or a "captain" appointed by him, had to engage a crew and supervise the making and setting of the net and the catching and dividing of the fish. When members of his crew had enough fish, he could allow others to work on a day-by-day basis for a share of the catch. In this instance the function of the owner or captain was by no means simply symbolic. He was responsible for directing as complex an operation as existed anywhere on the Northwest Coast and had considerable control over the disposition of the resource. Nor was this a recent development; underwater archeology (Easton 1985) indicates reef netting was practiced well before first contact.

For other resources, such as wapato ponds, the owners may have had no function other than to give permission to harvest and may have seldom, if ever, denied permission, but their presence established their families and their villages as the hosts, who expected reciprocity from the guests. Some owners exercised what may have been largely a symbolic function, as when a Samish woman supervised digging in her clam bed in order, it was said, to prevent anyone from leaving sharp broken shells there. But such a symbolic function, in a root bed, might have led to cultivation.

The terms "owner" and "ownership" have been questioned by some of my colleagues. Doesn't the site or resource really belong to a kin group, and isn't the so-called "owner" simply a steward for his or her kin group? My answer is: This may be the case in societies with permanent, discrete kin groups like those in the matrilineal north, but in bilateral societies like that of the Coast Salish, there are no enduring kin groups with clear boundaries. Descendants of a famous ancestor formed a kin group, but in each successive generation it overlapped with other such kin groups and diminished in people's consciousness. In this kind of society, control of resources was better achieved by the transmission of ownership from individual to individual. Of course an owner had obligations to his kin, but so do I. The difference between ownership among the Coast Salish and among ourselves is in the degree to which this obligation extends and the extent to which it is mediated by formalized social and ceremonial structures.

Conclusions

Although the ethnographic data suggesting the cultivation of native plants were recorded long after contact with European cultivation and the introduction of the potato among the Native people, the pattern of resource management in aboriginal culture makes it likely that most of the practices were aboriginal. But I think it would be rash to suppose that all practices reported for the late nineteenth century must have gone back to the aboriginal past. Considering the innovations that continue to be made in some surviving features of Native culture, as in ceremonial life, it would be strange if innovations were not made in plant tending. The use of seaweed as fertilizer may be one.

In my potato paper (Suttles 1951b), I argued that the easy acceptance of potato cultivation was made possible by several features of Native culture, and I suggested that looking at how this happened might tell us something about the beginnings of agriculture elsewhere. What made the adoption of potato cultivation easy were the knowledge and the technology for using native plants, and a semi-sedentary life that allowed women to return year after year to the same sites for enough time to tend and harvest native plants. Property rights were probably also important, ensuring that the investment of labor give a return to the investor. In an aboriginal setting property rights may even have promoted cultivation. The owner invests labor in a plot, at first simply to justify a claim, then sees an increase in the yield and in time expands the practice.

Was a knowledge of seed reproduction necessary? As I mentioned earlier it is not clear that the Native people understood that plants grow from seeds. In Downriver Halkomelem, there seems to be no word for "seed." A word sometimes identified as "seed" turned out to mean "bud" or "tip of stem." Different parts of several plants are identified as "male" and "female," with the flowering and seed-bearing part commonly identified as the "male," suggesting that the seeds were not identified as offspring, though perhaps they were identified, as in an older English, with semen. The Northern Lushootseed word for "seed" is composed of root and two suffixes—"bury" + "–activity" + "–plant" (Bates et al. 1994: 350), suggesting the word was coined when planting was introduced. But the words for "seed" in English, Latin, and Greek are all ultimately derived from a root meaning "to sow" (Buck 1949: 505), suggesting that they too were coined when sowing was introduced. Perhaps in the Old World as well, the understanding followed the action.

Notes

1. Paul Fetzer was a graduate student in anthropology whose linguistic and ethnographic work with the Nooksack was cut short by cancer in 1952.

2. When I wrote this paper, M. W. Smith's had not yet been published and I had not seen it, but Smith (1950) deserves the credit for having first made these suggestions.

3. In his culture element list for the groups around Georgia Strait, Barnett (1939: 268) lists (no. 2232) "Plots of roots owned" as positive for every group but the Squamish. But in his monograph (Barnett 1955: 252) he indicates that his Saanich, Cowichan, Nanaimo, and Musqueam informants did not remember "family hunting- and gathering-land rights." Much earlier, Hill-Tout denied ownership of plant resources by the Squamish (Hill-Tout 1900: 491; Maud 1978: 2: 50) and for two Upriver Halkomelem groups, the Chehalis and the Scowlitz (Hill-Tout 1904: 330; Maud 1978: 3: 116). Elmendorf (1960: 268–69) says that Twana families did not claim any sites for clamming, hunting, or fishing, and we can probably assume they did not claim sites for gathering plant foods either. White (1980: 20) writes that on Whidbey Island the Skagit (not to be confused with the Upper Skagit) "cared for the nettle in a manner closely resembling cultivation." His evidence seems to be no more than an affirmative answer to a leading question in a lawsuit. I know of no other reports of nettles, very common around settlements and used mainly as a source of fibers for twine, being tended. But Compton (1993) reports the tending and transplanting of nettles by the Haisla (Douglas Deur and Nancy Turner, personal communication 1998).

4. Gibbs (1877: 193) said: "The kamas season is in the latter part of May and June, and then as well in the fall when the sunflower is dug. . . . " But the reference to "the sunflower" (probably *Balsamorhiza*) suggests he may have been describing people outside of our region.

5. Collins (1974: 55) wrote: "Digging in one of these plots all day, a strong woman, according to Alice Campbell, could get ten roots; a weak one six." Norton (1979a: 185–86) points out that this cannot be true, and estimates that a normal woman should be able to get ten roots in as many minutes. Collins's statement is especially strange, coming shortly before her statement that "each plot was from three to four acres in size." Presumably what was intended was "ten baskets of roots."

6. In my dissertation (Suttles 1951a: 62–63) I wrote: "The fact that both informants said the "bread" was called *saplél*, the Chinook Jargon word for bread, suggests that baking it was a recent practice. However, Gunther [1973: 15] says Reagan has described the same thing for the Quileute." The "however" was meant to question the inference that the practice was recent. But Turner (1975: 58) and Norton (1979a: 384–85) have assumed that I intended to argue that it must have been recent. I do not believe I did then; I certainly do not now. Gibbs (1877: 193), writing on Western Washington from observations in the 1850s, reported "From the fern, they make a species of flour which is baked into bread." The Chinook Jargon term probably has a Native source in "*shapallel*," identified as "a cake made from dried roots," in use

on the Lower Columbia at the time of Lewis and Clark (Thwaites 1969, IV 260, 262; cited in Hajda 1984: 235). These cakes were an important trade good at this time, and were probably made of the bracken fern (*Pteridium aquilinum*) (Deur and Turner, personal communication 1998).

7. I will have more to say on the Klalákamish in a work now in progress.

8. Turner (1975: 58) mentions ownership of fern beds on southeastern Vancouver Island. But this appears to be based on based on Turner and Bell (1971: 69), who cite M. R. Mitchell (1968), who cites Suttles (1960). But my statement there must have been misleading; it was intended simply to indicate what resources were owned somewhere in the region. I was thinking about the Musqueam. As far as I know there is no evidence of the ownership of bracken fern beds anywhere else.

9. But Brown's bias is revealed in his next statement: "Of late years, in the vicinity of most villages, they have begun to grow a few potatoes, but, though a plentiful supply of these would add materially to their comfort, their utter laziness prevents them from scratching over anything but a mere scrap of ground."

10. However, there is another possible model. In 1868 G. M. Sproat reported that some inland hunters, whether Nuu-chah-nulth or Salish is not clear, transplanted salmon eggs. This statement appears in a footnote: "It is a common practice among the few tribes whose hunters go far inland, at certain seasons, to transport the ova of salmon in boxes filled with damp moss, from the rivers to lakes, or to other streams" (Sproat 1987: 148). Earl Maquinna George verified this practice, discussing the transferring of eggs from the Fraser River to Esperanza Inlet, while Bouchard and Kennedy (1990) record the transplanting of smolts between specific sites on Clayoquot Sound (Deur and Turner, personal communication 1998).

Chapter 7

The Intensification of Wapato (*Sagittaria latifolia*) by the Chinookan People of the Lower Columbia River

MELISSA DARBY

In the Lower Columbia River region of the Northwest Coast, wapato (*Sagittaria latifolia* Willd.; Alismataceae), a tuberous plant of aquatic and semiaquatic habitats, formerly grew prolifically in immense, homogeneous "fields," commonly hundreds of acres in extent (Figures 7.1, 7.2, and 7.3). This plant is still widespread in some localities. The tubers were available to local Chinookan indigenous peoples in much greater quantities than they required, even though they were living within the most densely populated region north of Mexico. This situation is what I call the Bountiful Factor.

Nevertheless, the people of the region were not passive exploiters of this bounty. They lived a semi-sedentary lifestyle; their winter villages were clustered around the wapato-filled wetlands. Scattered on the shorelines in the vicinity of the villages were small canoes that were chiefly used to gather wapato tubers (Clark in Thwaites 1969: 237). The abundance of this food, I believe, led to the establishment or enhancement of various strategies to intensify the resource, and, in turn, resulted in a level of social complexity more commonly seen in agricultural groups. I describe various strategies that are indicative of wapato intensification; these include: (1) control of the resource in the form of ownership of resource extraction sites; (2) a pattern of village placement adjacent to wapato fields; (3) establishment of an extensive trade and exchange network to handle the surplus; (4) development of specialized extraction tools including the shovelnose canoe; and (5) control of predators.

Explorers and settlers noted that the starchy tubers of wapato were an essential food staple and trade commodity for numerous riverine peoples in the Northwest, particularly the Chinookan people, who lived in the broad freshwater tidal zone of the Lower Columbia River (Swan 1857; Lewis and Clark in Thwaites 1969). Wapato tubers could be harvested from September through

7.1. Wapato field, Cunningham Lake, Sauvie Island, Columbia River, Oregon
(Photo by Mark Nebecker, 2002)

mid-May in the region, though the main harvests were traditionally carried out between late October and March. The collectors, chiefly women, harvested the tubers by wading into shallow water and stirring up the soft mud with their feet. This activity released the tubers, which then floated to the surface of the water where they were gathered together, sometimes by means of a curved stick, and thrown into a small canoe or a container nearby.

The tuber was widely noted for its resemblance in taste and texture to a white potato (*Solanum tuberosum*), though it is typically smaller, averaging between 35 and 55 mm in length and weighing an average of 13 grams. When growing in dense, monocultural plots, it is an efficient food to gather, and it stores well either fresh or dried (Darby 1996). Nutritionally it compares well with the potato, though wapato typically contains more iron (Keely 1980). Steamed or baked in hot ashes, wapato tubers cook in about ten minutes.

Wapato wetlands occupied so much of the Lower Columbia Valley that Captains William Clark and Meriwether Lewis named this area "Wapato Valley" in 1805. These fields were called the "great water gardens of the Indians" by Ben Hur Lampman, a twentieth-century newspaper man and poet who interviewed some of the early farmers of the area (Lampman 1946: 161). Lewis and Clark also wrote that the people in this region had permanent houses and that

7.2. Wapato (*Sagittaria latifolia*) (Photo by R. D. Turner)

they "hunt some roots in their own villages . . . " suggesting that villages were closely juxtaposed with root grounds (Clark, in Jackson 1962: 500). Maps of village locations compiled by Saleeby and Pettigrew (1983) show that of the thirty villages historically documented in Wapato Valley, all were located directly adjacent to *Sagittaria latifolia* habitat, or within one-eighth of a mile.[1] Among the neighboring Tualatin Kalapuyans, this pattern is seen as well (Zenk

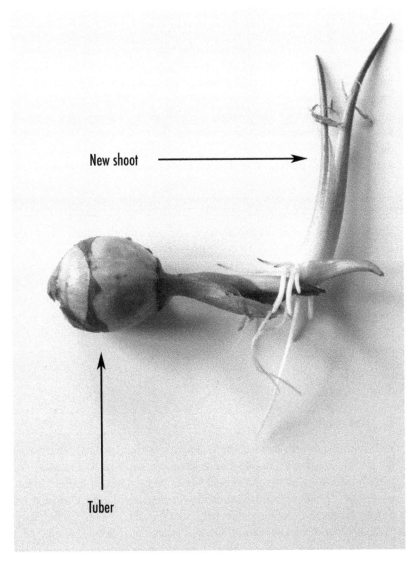

New shoot

Tuber

7.3. Wapato tuber (Photo by Ann Garibaldi)

1994: 155). Villages were clustered in locations where wapato fields were known to be extensive, and the largest villages were located near the largest wapato wetlands.

Ostensibly, people stepped into and claimed a small portion of the very productive ecological niche that waterfowl had occupied before humans arrived in the region. *Sagittaria latifolia* and waterfowl co-evolved over thousands of years in a close symbiotic relationship. Ducks and swans harvest wapato by stirring up the mud in shallow water with their bills and feet, releasing the wapato tubers and associated plant parts, which float to the surface and are eaten by these birds. Uneaten or partially eaten tubers and rhizomes eventually sink to the bottom and establish new plants, sometimes after floating a considerable distance. Human harvesting would have had similar effects.

In the late eighteenth and early nineteenth centuries, Wapato Valley was an ecologically complex and productive environment that provided the region's human inhabitants with numerous types of food, with many resources (most notably salmon runs) varying considerably over time and space. The Lower Columbia region fits the model put forward by D. R. Harris (1977) of an emergent stable agricultural system, characterized by an ecosystem with high species and pattern diversity, intensive management of some resources within the ecosystem, and plant ecology that was conducive to intensification. *Sagittaria latifolia* would have lent itself particularly well to intensification because it was prolific, has a long harvest season, is cost-effective to harvest, stores well fresh but could also be dried, and takes only a little fuel and a short time to cook.

This chapter is divided into three sections. In the first, I present botanical and ecological data relevant to the productivity and intensifiability of this species. In the second, I present a picture of wapato use drawn from ethnographic and ethnohistoric accounts and oral tradition. The final section describes some intensification strategies that were applied to wapato in this context.

The Lower Columbia River

GEOGRAPHY

The Columbia River's large discharge and low gradient created extensive wetlands in the meander floodplain of its lower reaches, which were also subject to daily tidal fluctuations and annual floods. Wapato was ubiquitous in slackwater bays, freshwater tidal mudflats, on marshy islands, and in myriad ponds, lakes, and sloughs, especially on the large, marshy island named "Wapato Island" by Lewis and Clark and today called Sauvie Island.

At the time of European contact, Chinookan-speaking peoples inhabited the Lower Columbia River, from The Dalles to the Pacific Ocean. Chinookan peoples also occupied the Willamette River drainage from Willamette Falls to Multnomah Channel, and from the mouth of the Columbia River north along the coast to Willapa Bay, and south to Tillamook Head. Wapato Valley is the broad, tidally influenced freshwater zone in the Lower Columbia River Valley, beginning at the mouth of the Columbia River gorge near the Sandy River confluence, and extending westward to the Kalama River valley. The Coast Range hems Wapato Valley in on the west, and the foothills of the Cascade Mountains form its eastern boundary. This same region is known today as the Portland Basin.

As they were for other inhabitants of the Northwest Coast cultural area, salmon were a dietary staple of the Chinookan peoples, one particularly abundant within their territories. Eulachon (*Thaleichthys pacificus*) and sturgeon (*Acipenser transmontanus*) were also important food fish. Though these peoples relied heavily on fish, they also hunted waterfowl, seal, elk, deer, and bear, as well as smaller mammals, including beaver, muskrat, and raccoon. The Lower Columbia is still a major stop for waterfowl on the Pacific Flyway, and supports abundant migratory bird populations as well as significant numbers of resident species. A variety of plants, in addition to wapato, were also vitally important in the production of foods, medicines, and materials (see Turner and Peacock, this volume).

The Lower Columbia River had one of the densest human populations north of Mexico. According to a newspaper account announcing the return of the Corps of Discovery in 1806, Lewis and Clark reported the "Indians to be as numerous on the Columbia River, which empties into the Pacific, as the whites in any part of the United States" ("A Boston Newspaper" reprinted in Betts 1985). Governor George Simpson of the Hudson's Bay Company wrote that "The population on the banks of the Columbia River is much greater than in any part of North America that I have visited as from the upper Lake to the Coast it may be said that the shores are actually lined with Indian Lodges" (Simpson 1968: 94). He attributed this to the abundance of "provision" yielded by the river, and further noted "the whole of the Interior population flock to its banks at the Fishing Season" (1968: 94). As this chapter demonstrates, wapato was an essential component of a diet that allowed such a dense population to persist and flourish on the lower Columbia River.

Sagittaria latifolia within the Lower Columbia Ecosystem

Sagittaria latifolia is a species of aquatic plant of the water plantain Family (Alismataceae). *Sagittaria* species are widespread over the Northern Hemisphere and are considered cosmopolitan in distribution. In North America the genus is found from Mexico north to British Columbia, and east to Nova Scotia (Hitchcock and Cronquist 1973). *Sagittaria latifolia* grows in shallow water and on the margins of lakes between seasonal high and low water lines, as well as in slow streams and perennial wetlands. In a dense, naturally occurring wetland wapato patch, there are approximately thirty plants to the square meter, and, as with many wetland species, it can grow as a solid monoculture. Another species with smaller tubers, *Sagittaria cuneata,* also occurs along the Columbia east of the Cascade Mountains and was also used as a root vegetable within its range.[2]

The overwintering part of *Sagittaria* is a tuber. Reproduction is primarily vegetative, with seedlings rarely occurring in undisturbed mature communities. Tubers are formed on the ends of long white rhizomes, which are horizontally creeping subterranean stems. These rhizomes, attached to the base of the plant, are about one centimeter in diameter, and are often as long as 70 centimeters. The distal end has a pointed tip, which helps it to extend through the mud. During the spring and summer, daughter plants grow from the distal ends of the rhizomes, in much the same way a new strawberry plant grows from runners. If a rhizome is broken into several fragments, each fragment is capable of producing shoots above and roots below. By September on the Lower Columbia, the distal ends of these rhizomes cease to produce new plants, but begin to thicken and develop into the starchy overwintering tubers.

A domesticated form of this species, *Sagittaria trifolia* var. *edulis*, is found in Europe and northeast Asia. The leaves of *Sagittaria* spp. are grown in China principally for forage (Hu et al. 1992: 265). The use of the tuber was also widespread, but has been displaced somewhat by the white potato, which was introduced to China primarily by missionaries in the eighteenth and nineteenth centuries (Anderson 1988: 122). *Sagittaria trifolia* var. *edulis* is grown in Japan and referred to as Chinese Arrowhead. It is an important ingredient of traditional dishes in Japan (Tanimoto 1989: 345). Both the young leaves and tubers are edible. The Asian and North American species are almost identical in overall appearance and structure.

Wapato's high productivity can be partially explained by the concept of co-evolution. In fall and spring as many as one million ducks, geese, and swans arrive on the Lower Columbia on their biannual migration along the Pacific Flyway, even today, despite the shrinking of their wetland habitats due to human modifications. Modern aerial waterfowl surveys count as many as 100,000 individuals present on Sauvie Island on any day in November and December (Oregon Department of Fish and Wildlife n.d.).

Wapato is a plant that lends itself to intensification by humans in part because waterfowl became necessary partners in its fitness. Grubbing by swans has been shown to increase the fitness of *Sagittaria* species. The excrement of swans and other waterfowl adds nutrients, especially nitrogen, to the wetland environment, contributing to the productivity of the wetland plants. Waterfowl also contribute to the spreading of the plant by breaking up the rhizomes, allowing ungrazed rhizomes and tubers (which float) to raft to new locations and reestablish themselves. Waterfowl also assist in dispersing the seeds, which stick to the skin on their feet and legs, but also are eaten and subsequently redeposited, enabling the plant to colonize over great distances. Wapato contributes to the fitness of waterfowl by providing large quantities of high-carbohydrate food, one that is predictable and cost-effective, and available exactly when the birds need it the most: during both the fall and spring migrations. Humans, who have occupied this environment for thousands of years, would have had similar effects in enhancing wapato production through thinning, tilling, and seed and tuber dispersal.[3]

Historical and Ethnographic Accounts of Wapato Use

After only forty years of contact with Euro-American explorers and traders, the indigenous societies of the greater Lower Columbia were decimated by introduced diseases (Boyd and Hajda 1987: 309). In part as a result, ethnographic research in the area was not very extensive. However, there is a fairly large body of ethnohistorical literature from explorers and fur trappers who were headquartered on the Lower Columbia. Some of the most descriptive accounts of wapato use are found in the journals of Lewis and Clark, written in 1805 and 1806. In the index of the Thwaites's Lewis and Clark Journals (1969), "wapato" and other versions of the word are cited over ninety times.

ACCOUNTS BY EXPLORERS PRIOR TO LEWIS AND CLARK

Accounts of wapato's importance to the indigenous peoples of the Lower Columbia River appear in descriptions of the first recorded contact between whites and Chinookan people. This first contact occurred when the ship *Columbia Rediviva* crossed the Columbia River bar in May of 1792. On May 18, Boit (1960) described the scene as follows:

> The river abounds with excellent salmon and most other river fish, and the woods with moose [elk] and deer, the skins of which were brought to us in great plenty. The banks produce a ground nut, which is an excellent substitute for bread and potatoes. [Boit 1960: 56]

The "ground nut" that Boit describes is almost certainly wapato. On November 4, 1792, Thomas Manby, the master of the *Chatham*, also made note of a root that appears to have been wapato. While waiting in Bakers Bay for calm water to cross the Columbia River bar, Manby decided to hunt waterfowl in a "swamp" four miles from the cove near the mouths of the Chinook and Wallacut rivers. He met with a small party of Indians who supplied him with salmon and "a basket of roots not much unlike small potatoes and a little inferior to them in taste" (Howay and Elliot 1942: 325).

The *Ruby* crossed the Columbia River bar the first week of June 1795. Captain Charles Bishop wrote an account of this trading voyage, and his words constitute the first explicit mention of wapato on the Lower Columbia:

> As none of us are acquainted with Bottoney, I can offer nothing on that head, but what is discribed in the account of Nootkan Productions, in Cooks voyage we found here, except the Wild Potatoe called by the natives "Wappatoe" which we have seen nowhere else, they in general are of the size of a Pidgeons Egg, and appear to grow like an onion or Turnip, above the surface of the Earth are found in swampy grounds, and when boiled or roasted, eat not unlike potatoes. [Bishop 1927: 274, spelling as in original]

The crew of the *Ruby* subsisted that winter chiefly on salmon, cranberries, wapato, and wild game supplied by the natives.

The Corps of Discovery, led by Captains William Clark and Meriwether Lewis, consisted of thirty men and one aboriginal woman guide known as Sacajawea, with her baby. Sacajawea was almost certainly familiar with the wapato root. *Sagittaria latifolia* is cosmopolitan in distribution over North America, and grows in the areas of Montana where Sacajawea spent her youth. When the vote was taken on where to camp for the winter, she voted to camp near a wapato ground: "Janey [Sacajawea] in favor of a place where there is plenty of Pota [sic]" (Thwaites 1969, 3: 247). She was probably aware that wapato would have provided a good source of winter food for the Corps of Discovery, as well as a good food for her baby when mashed.

Lewis and Clark's first encounter with wapato occurred on November 4, 1805. In Clark's first draft he describes acquiring and eating a "round root near the size of a hens egg" at a village past the mouth of what is now known as the Sandy River. The roots were roasted in the embers until they became soft (Thwaites 1969, 3: 194). In his second draft describing the same day, he noted that the Chinese cultivate the tubers in great quantity and it is called the common arrowhead or "Saggiti folia." The root had "an agreeable taste and answered well in place of bread." The members of the expedition purchased four bushels and divided it among their party (Thwaites 1969, 3: 196–97).

Lewis noted that wapato grew in great profusion on an island farther downstream described as "about 20 miles long and from 5 to 10 in width; the land is high and extreemly fertile and intersected in many parts with ponds which produce great quantities of the sagittaria Sagitifolia [sic], the bulb of which the natives call wappetoe" (Lewis, in Thwaites 1969, 3: 218). They named this island Wappato Island (now Sauvie Island), and they named this part of the Lower Columbia Valley "Wap-pa-too Valley from that root or plants growing Spontaniously in this valley only" (Clark, in Thwaites 1969, 3: 202).

On their way downriver the Corps traded for the root on several occasions, and each time they made new observations about its qualities and uses. They observed that the local population stored wapato in houses under bed platforms. Clark noted that the diet of the Cat-tar-bets (Cathlamet) consisted of fish, elk, and wapato. Yet near the mouth of the Columbia, Clark noted that in this region "the Wapto root is scerce, and highly valued by these people, this root they roste in hot ashes like a potato and the outer skin peals off, tho this is a trouble they seldom perform" (Thwaites 1969, 3: 240).

Against Sacajawea's best advice, the Lewis and Clark party wintered at Fort Clatsop on the south side of the Columbia River, near the river's mouth and far from the wapato fields of Wapato Valley. They occasionally were able to trade for wapato, which was a welcome addition to their diet. On December 31, they purchased one and a half bushels of wapato, for which they were grateful since they had been living on spoiled elk, which was "disagreeable to the smel. as well as the taste" (Clark, in Thwaites 1969, 3: 294). Four days later Lewis wrote, "The hunters were all sent in different directions, and we are now becoming more anxious for their success since our store of wappatoo is all exhausted" (Lewis 1814, 2: 105). On January 10 they were presented with "a basquit of woppetoe" from Wapato Valley from a chief of the Cathlamet people. They were able to obtain wapato several more times during their Fort Clatsop visit.

The journal entries made on the return voyage through Wapato Valley indicate that wapato was a high-priority provision for their return trek over the mountains. On their second day of travel, Monday, March 24, 1806, the Corps of Discovery arrived at the Cathlahmah (Kathlamet) village where they purchased some wapato and a dog for the two sick and weak men of their group (Moulton 1983, 12: 10). The Corps entered Wapato Valley on March 27 and traded for wapato with the Cowlitz people. On March 29 they stopped at a village (Cathlapotle) near the mouth of the Lewis River, and noted an abundance of sturgeon and wapato. They camped near a pond about a mile above the village.

> In this pond the natives inform us they collect great quantities of [w]appato, which the women collect by getting into the water, sometimes to their necks holding by a small canoe and with their feet loosen the wappato or bulb of the root from the bottom from the Fibers, and it imedeately rises to the top of the water. they collect & throw them into the canoe, those deep roots are the largest and best roots. [William Clark, March 29, 1806, in Thwaites 1969, 4: 217, spelling as in original]

The version of this passage in Biddle's edition of the Lewis and Clark Journals, written in 1814, evidently with further clarification by Clark, records that the roots were collected "chiefly by the women" who would remain in the water for several hours even in the "depth of winter" (Lewis 1814: 225).

Lewis wrote that wapato is taken in great quantities from the ponds around Cathlapotle. It was the "principal article of traffic with those Tribes which they despose of to the nativs below in exchange for beeds, cloath, and various articles" (Lewis, in Thwaites 1969, 4: 222). They purchased "a considerable quantity of wappetoes, 12 dogs, and 2 Sea otter skins of these people" (Thwaites 1969, 4: 215). Soon thereafter, the Corps learned that Chinookan peoples carried out seasonal migrations between villages as part of the wapato harvest; the peoples they encountered "inform us that their relations also visit them frequently in the spring to collect this root which is in great quantities on either side of the Columbia" (Thwaites 1969, 4: 226). Indeed, people moved into the valley from east of the Cascade Range mountains "in serch of Subsistence which they find it easy to precure in this fertile Vally" (Clark, in Moulton 1983, 12: 52). Food scarcity appears to have been a strong motivating factor drawing people from east of the Cascades to the wapato grounds. Ordway wrote:

Great numbers of Savages visited the Camp continually Since we have lay at this Camp, who were passing down with their famillys from the country above into the vally of columbia in Search of food. they inform us that the natives above the great falls have no provisions and many are dieing with hunger. This information has been so repeatedly given by different parties of Indians that it does not admit any doubt and is the cause of our delay in this neighborhood for the puropose of procureing as much dryed Elk meat as will last us through the Columbia plains. [Ordway, in Moulton 1983, 9: 286]

The captains realized that they needed to provision themselves for the journey to the Rocky Mountains, since they did not expect to be able to trade for food or find game upstream. Preparing for this journey, the members of the Corps hunted deer and elk, and Clark was able to trade some tobacco for several parcels of wapato.

ACCOUNTS FROM THE FUR TRADE

The land-based fur trade began in 1811 with the establishment of Fort Astoria at the mouth of the Columbia by John Jacob Astor's Pacific Fur Company. Several members of the Astorian enterprise, including Gabriel Franchere, Alexander Ross, and Ross Cox, published accounts of their experiences that mentioned wapato and its use by the Chinookan people.

Alexander Ross described the root as a "favourite article of food at all times of the year" (Ross, in Thwaites 1904a: 109). Cox mentions that wapato "is highly esteemed by the natives, who collect vast quantities of it for their own use and for barder" (Cox 1957: 79). On occasion, wapato served as a trade good, augmenting the diet of Fort Astoria residents. Franchere (in Thwaites 1904b: 278) went on a trading voyage upriver into Wapato Valley in October of 1814, where he purchased over 700 smoked salmon and a "quantity of the Wapto root (so called by the natives), which is found a good substitute for potatoes."

ETHNOGRAPHIC ACCOUNTS

As noted previously, ethnographic materials for the people of the Greater Lower Columbia are quite limited, largely due to the epidemics that eliminated most of the region's indigenous societies well before visits by anthropologists. Fieldwork with the few survivors was largely limited to ethnographic reconstruction on the basis of memory (Hajda 1984: 4). Two such accounts are those of Gatschet (1877) and Spier and Sapir (1930).

Gatschet's account applies to the Tualatin Kalapuyans, neighbors of the Chinookan speakers and fellow participants in a regional network of economic and political interrelationships centered on the Lower Columbia River (Zenk 1976: 5). A published version appearing in Jacobs (1945: 19) mistranslates the Tualatin word *mampdu* "wapato" as "camas" with misleading results (Zenk

1976: 56–57, 85). Zenk (personal communication, 1999) has supplied the following annotated paraphrase of Gatschet's (1877: 29) original field translation
of a Tualatin narrative:

> I know (in) the fall of the year wapato (*mampdu*) were being gathered.
> Women were digging, making pits (*abullum,* four to five foot deep pits in
> the ground for storing wapato tubers), they kept them (there) for the win-
> ter, to preserve them for eating in the winter. In the lake they gathered them
> from underneath the soil, they picked them up, the women gathered them.
> When the lake was overflooded we named it "step in the water" the women
> harvested wapato (*debinfalu*) [translated in Jacobs as 'stepped in the water',
> but explained elsewhere by Gatschet as "to stamp" in water to harvest wapato]
> with a wide stick with a bent end (*aleki*) being used to scoop the floating
> tubers from the water.

According to Spier and Sapir (1930: 183) the Wishram Upper Chinookans collected the small sagittaria root (*Sagittaria cuneata*) from a lake on the south
side of Mt. Adams. "It was gathered in the fall with the aid of a flat stick or by
feeling about with the foot" (Spier and Sapir 1930: 183).

Other ethnographic accounts describe aboriginal use of wapato in other
parts of the Northwest. Haeberlin's Snohomish and Snoqualmie informants,
interviewed at the Tulalip Reservation in 1916 and 1917, stated that wapato was
not only an important food source, but also that "this plant can be easily grown
and transplanted" (Haeberlin and Gunther 1930: 21, citing George Gibbs). An
account of wapato harvesting obtained from the Katzie of the Lower Fraser
River bears special mention because this group lived in an ecosystem similar
to the Lower Columbia, one very favorable to the *Sagittaria latifolia*. The Katzie
gathered the tubers by "wading and treading on the plants, 'dancing' until they
came floating up" (Suttles 1955: 27). They were taken home raw and unwashed,
and would keep for several months. They were baked in hot ashes as they were
needed. One report by McKelvie noted that as many as 5,000 indigenous people
assembled at Pitt River near its Fraser River confluence at the end of salmon
season to collect wapato, which "was considered a delicacy" (reported in Suttles
1955: 227). As along the Columbia, the wetland habitats supporting *Sagittaria*
growth have largely given way to agriculture and the populations of this plant
are greatly diminished in the Fraser Valley (Spurgeon 2001).

WAPATO IN ORAL TRADITIONS

Wapato is mentioned in several narrative texts regarding the mythic past. In
a Klickitat narrative recorded in the mid-nineteenth century by George Gibbs
(1877: 12), wapato and camas are mentioned as the group's "favorite roots."
Boas collected the following Cathlamet tradition about how wapato saved the
people of the myth-age after the salmon failed. Boas entitled the narrative
"Myth of the Salmon" and it begins at a time when the people of mythical

times were dying of hunger. They had only small *Sagittaria*-roots and skunk-cabbage roots and rush roots to eat. In the spring of the year the Salmon went up the river. They went some distance. Then the Skunk-cabbage said: "At last my brother's son has arrived. If it had not been for me your people would have been dead long ago" (Boas 1901: 6).

The myth continues: The Salmon found that it was Skunk-cabbage talking, so they gave him gifts and carried him inland and "placed him among willows." They continued their journey, and small Sagittaria root (*Sagittaria cuneata*) addressed the Salmon in the same way, and identified herself as the Salmon's aunt, saying if it weren't for her all their people would be dead. The Salmon gave the small Sagittaria root three "woodchuck" (probably muskrat) blankets and some dentalia, and left her. The same thing happened with the large Sagittaria root (*Sagittaria latifolia*) except that she got five woodchuck (again, probably muskrat) blankets and some dentalia, and they "placed her in the mud." The same theme is repeated with the rush root, which was put in a swamp, and another root that was placed on the bank of the river. The story ends at the Cascades.

Two features of this narrative may provide important hints about the wapato complex of the Lower Columbia: plant foods, including two forms of wapato, saved the myth-age people from starvation; and the salmon, who owed the plants a debt, gave them gifts and placed them in the mud, or among the willows or along the bank. The latter episode suggests that the Chinookans possessed some understanding of the mechanics of transplantation.

Strategies of Intensification

Indicators of resource intensification in the region that appear regularly in the ethnographic literature include the following: (1) control of the resource in the form of ownership of resource extraction sites; (2) the pattern of village placement adjacent to wapato fields; (3) the existence of an extensive trade and exchange network to handle surplus harvests; (4) specialized extraction tools, including the shovelnose canoe, and (5) control of predators.

RESOURCE OWNERSHIP

Ownership and control of resource areas is a fundamental characteristic of resource intensification. Discussing native land tenure in northwest Oregon and western Washington, George Gibbs observed that while some Northwest tribes recognized no individual right except by actual occupancy, among the Chinook and Lower Chehalis (Coast Salish speakers neighboring the Lower Chinook on the north), "the right may have been carried somewhat further" (Dall 1877: 186; Turner, Smith, and Jones, this volume).

While wapato tenure patterns could not be documented on the Columbia River, ethnographic references suggest that this pattern paralleled what was found among the Katzie of the Lower Fraser River. The Katzie lived in an ecosystem similar to the Lower Columbia, and both were very favorable to wapato: "the unusual extent of low, seasonally flooded lands in Katzie territory gave them an unusual abundance of several bog and marsh plants. The two most important of these were the cranberry and the wapato" (Suttles 1955: 26). Some wapato fields belonged to the Katzie tribe and others belonged to families, while others were unclaimed. Suttles's informant could name nine patches that belonged to his father's family. There was also a large area that "belonged to all":

> Here . . . families might establish claims for the season by clearing tracts, several hundred feet long, of other growth so that the roots could be gathered more easily, and by the following year they would have grown up again and become common property. A family who cleared such a tract might camp there in a mat house for a month or longer harvesting the roots. [Suttles 1955: 27]

Katzie territory was famous for cranberries and wapato, and, as on the Columbia, "in the fall outsiders came from a number of other tribes to gather them" (Suttles 1955: 27). Suttles's consultants said that permission was not refused to outsiders to harvest from cranberry bogs in the nineteenth century, nor was tribute extracted. Suttles infers that during this period "ownership of . . . a rich cranberry bog was its own reward in that it permitted the owners to play the role of hosts. A host at one time and place is potentially a guest at another" (Suttles 1955: 27). I suggest that this resource tenure model is applicable to wapato plots on the Lower Columbia River, based on population movements described by Lewis and Clark and others. As Hajda (1984) asserts, on the Lower Columbia, food was collected in areas controlled by people with whom the collectors had immediate consanguineal ties. These kinship ties established enduring reciprocal relations, providing stability to long-term patterns of wapato plot tenure.

TRADE AND EXCHANGE

"A thorough-going occupation with commerce dominated Chinook life" (Ray 1938: 99). They had a currency, a standardization of products, and a trade language—Chinook Jargon. The Jargon is structured around some Nootkan (Nuu-chah-nulth) as well as Chinookan verbs, and contains many Chinookan nouns. Jargon use certainly became widespread in the early nineteenth century as a trade language, spreading with the fur traders throughout the Northwest Coast area. Trading conditions were ideal due to the juxtaposition of a rich habitat that supplied a surplus of goods, and waterways that facilitated transportation within and between tribal territories (Ray 1938).

A variety of goods and services were locally available in the region, resulting in localized specialization and well-established intraregional trade patterns. This trade served to redistribute desirable products from areas where they were produced to areas where they were not produced. For example, from the mouth of the Columbia River a wide range of ocean resources were shipped by canoe to Wapato Valley in exchange for wapato, which was shipped by canoe back to the coast. There were three centers of trade in the vicinity of Wapato Valley: The Dalles, Willamette Falls on the Willamette River, and the mouth of the Columbia River. Captain Clark describes the trading patterns as follows:

> Down to the lower Indians about the falls (the great emporium) & exchange with them for Wappatoe, Pounded fish &c &c. To this market the Indians of the Colombian plains bring mats, silk grass, rushes, root bread called chapelell. The Chiluckitiquaws & Eleeshoots are the carriers between the great falls & the Indians of Wappatoo valley—who bring wappatoo & the fish peculiarly. [Clark, as quoted by Biddle in Jackson 1962: 543]

Referring to the people of Wapato Valley, Meriwether Lewis wrote that wapato was the "principal article of traffic" traded from the people of the Valley to people at the mouth of the River in exchange for beads, cloth, and various articles. At Fort Clatsop, at the mouth of the Columbia, Lewis wrote that the most valuable of all their roots is foreign to this neighborhood. "The nativs of the sea coast and lower part of this river will dispose of their most valuable articles to obtain this root" (Lewis, in Thwaites 1969, 4: 222). This high demand for wapato brought considerable wealth to the residents of Wapato Valley, and increased demands both to control access to this productive resource and to engage in surplus wapato production at a tremendous scale.

There was a standardization of products and basket sizes within the region, indicating a complexity of food production, storage, and trade more often seen in agricultural groups. Dried eulachons strung head to tail were traded by the fathom (Ross 1966) and dried salmon was traded by the crate at The Dalles (Skarsten 1964: 154). There were three graduated package sizes for huckleberries, hazelnuts, and acorns reported for the upriver Chinookan-speakers, the Wishram (Spier and Sapir 1930: 185). They also had a typical basket for storing roots that was a foot in diameter and two feet deep (Spier and Sapir 1930: 192). Lewis and Clark refer to procuring "bags," "parcels," and "basquits" of wapato, as well as "bushels," but it is unclear whether there was a standard basket size for wapato that differed from other root-storing containers.

HARVESTING TECHNOLOGY

Resource intensification is typically accomplished by strategies that facilitate increased rates of gathering (such as using a net to catch fish), or by strategies that increase the production by manipulating the environment (such as maintaining an anthropogenic prairie through periodic burning). The production of specialized tools is an important part of this larger pattern of resource intensification.

One account mentions that releasing wapato from the substrate was sometimes facilitated by a flat stick (Spier and Sapir 1930). There is also a word in a Kalapuyan dialect for a wide stick with a bent end used to gather the floating tubers from the surface of the water (Zenk 1994: 152). By far the most important tool used to facilitate the gathering of wapato, however, was the small shovelnose canoe. Clark drew a picture of a typical shovelnose canoe from the examples he saw near the village of Neerchokioo (Figure 7.4) and described it as follows:

> On the bank at different places I observed small canoes which the women make use of to gather wappato & roots in the Slashes. those canoes are from 10 to 14 feet long and from 18 to 23 inches wide in the widest part tapering from the center to both ends in this form and about 9 inches deep and so light that a woman may with one hand handle them with ease, and they are sufficient to carry a woman an[d] some loading. I think 100 of these canoes were piled up and scattered in different directions in the woods in the vicinity of this house, the pilot informed me that those canoes were the property of the inhabitants of the Grand rapids who used them ocasionally to gather roots. [Clark, in Thwaites 1969, 4: 237]

7.4. Drawing of a Chinook shovelnose canoe from the journals of Lewis and Clark. Courtesy of the American Philosophical Library.

This number of canoes in this small village is interesting because, in the fall when they first visited, Clark wrote that the village consists of "200 men." If he is referring to both men and women (and not children) in this estimate, this implies that about half of the people seasonally occupying the village owned one of these small canoes. Since about half of the village would be women, and wapato was gathered chiefly by women, it would suggest that every woman in the village owned a canoe. It must be noted that Lewis and Clark's population estimates are generally considered rough estimates, and cannot be taken uncritically (Boyd and Hajda 1987).

Arnold (1995) has proposed that advances in water transportation technology affected the degree to which hunter-gatherers became maritime oriented and hierarchically organized. She uses the Northwest Coast as an example, arguing that "advanced water transport technology" (i.e., good canoes) brings with it a number of symbolic and practical ramifications. Among those she explores are the opportunities it made available for intensifying subsistence, communication, networks of exchange, and hierarchical social and economic relationships. Clearly, in the case of Wapato Valley, the presence of such specialized canoe technology for subsistence activities and the trade of surplus foodstuffs produced the social and economic outcomes that Arnold proposes.

SETTLEMENT AND MIGRATION PATTERNS

The ownership of an intensifiable resource brings with it increased stability in food supply, and this has implications with regard to settlement patterns. Throughout the Greater Lower Columbia region, aboriginal settlements appeared in two forms, including villages, containing several houses, and hamlets with only one or two permanent houses. These villages and hamlets in Wapato Valley swelled with visiting friends and relatives during resource collecting times. People came in the fall for elk hunting and wapato seasons, and again in the spring for fish and wapato harvests (Lewis, in Thwaites 1969, 4: 223). Chinookan consultants reported to Lewis and Clark that many houses along the Lower Columbia were only occupied during the wapato harvest.

Various authors have studied settlement patterns in the Greater Lower Columbia region, including Wapato Valley (Boyd and Hajda 1987; Saleeby and Pettigrew 1983). These studies have concluded that the seasonality of village occupation corresponded to the localized availability of resources. "The density and distribution of food in relation to [a] village site was a major factor influencing the length of its occupation during the year" (Saleeby and Pettigrew 1983: 173). Boyd and Hajda (1987: 313) state that patterns of seasonal migration and settlement were contingent upon "spatial and temporal variations in the resource/subsistence base of the lower Columbia drainage."

The length of the wapato season was not known at the time of Boyd and Hajda's study, but, in retrospect, data on wapato seasonality support their conclusions. My research has indicated that wapato is in season for a very long period compared to other root foods in the area. Wapato can be harvested from the middle of September to mid-May, a season of about eight months, during the coldest time of the year when game and vegetable resources are typically scarce (Darby 1996). However, harvest was concentrated during the fall and the spring. Tubers were harvested in the fall when they first became available, before the winter increase in water levels, and again in the spring when the lakes were at low ebb and the water temperature was moderate.

Comparing population estimates recorded at different times by Lewis and Clark, Boyd and Hajda (1987) noted discrepancies that appear to reflect seasonal variation in village populations along the Lower Columbia River. Initial population estimates for the Greater Lower Columbia at the time of the expedition (9,800 Chinookans), were later revised to a figure of 17,000, which included both resident populations and spring visitors to riverine resource sites (Boyd and Hajda 1987: 322). Population increased at the mouths of rivers that had good eulachon and spring Chinook runs; a 167 percent increase in population is inferred from Lewis and Clark's data for the Cowlitz River mouth, a 212 percent increase at the Clackamas River, and an increase of 262 percent at the lower end of Multnomah Channel and across the Columbia at the mouth of the Lewis River (adjacent to the richest wapato grounds).

> Except for the Willamette Falls and Cascades, the village clusters where the increase is more than 200 percent are all on or across from Sauvie Island. The swell of population in this particular area suggests that *resources in addition* to Eulachon and salmon may also have been drawing people to the area and keeping them there. [Boyd and Hajda 1987: 321, emphasis added]

In light of this evidence, one can conclude that wapato harvests were responsible for the proportionately largest number of cases of the transhumance documented by Lewis and Clark.

Located adjacent to the primary wapato wetlands, the villages of Wapato Valley provided the nodes on which this pattern of seasonal migration centered. In 1814, Biddle interviewed William Clark about the seasonal movements of people on the Lower Columbia River. Clark answered that

> Below the falls [The Dalles] they are fixed permanently in houses. Below the great rapids [and therefore in the Wapato Valley area] they hunt some roots *in their own villages* & shoot deer. . . . Below the rapids also stationery go down for the Wappatoe roots. [Clark, in Jackson 1962: 500; brackets and emphases added]

The suggestion that people collect roots in their own villages is significant. "Hunting" roots in one's own village implies that the village was situated next to a root ground, and arguably suggests that the root ground could be considered an integral component of the village, effectively controlled or owned by its inhabitants. Maps of village locations compiled by Saleeby and Pettigrew (1983) show that of the thirty villages historically documented in Wapato Valley, all were located directly adjacent to wapato habitat, and all but one (Cathlapotle) were located within one-eighth of a mile from documented wapato patches. Where wapato fields were known to be extensive, villages were clustered, and the largest villages were adjacent to the largest wetlands.[4]

While much of this area was rich in other resources, particularly salmon and sturgeon, the demographic pattern of winter-occupied villages situated adjacent to wapato grounds can be seen in outlying areas that lacked fish resources. For example, the Tualatin band of the Kalapuya had seventeen ethnohistorically documented winter villages in the Tualatin Valley. Eight were on Wapato Lake, or on wetlands in the vicinity on the lake's tributaries, where fish resources appear to have been scarce. Three other Tualatin villages were located at similar wetland locations (Zenk 1994: 157). This pattern may hold true for portions of the Willamette Valley and Puget Sound, where detailed settlement data have not been forthcoming or have yet to be analyzed relative to resource patterns.

The clustering of villages adjacent to wetlands suggests another strategy of resource intensification that has ethnographic parallels throughout the region (see Deur, and Turner and Peacock, this volume). Such a close juxtaposition of villages and wapato plots allowed the protection of the wapato resource from nonhuman predators. Wapato was eaten by a number of nonhuman species, including waterfowl—especially tundra swans (*Cygnus columbianus*) and various ducks, muskrats (*Onadatra*), and beaver (*Castor*). Control efforts for these predators increased the productivity available to the human residents of Wapato Valley. Muskrats and beavers were paradoxically both major predators of wapato, and—through the production of dams and other impoundments—producers of additional wapato habitat. These mammals also represented potential sources of roots for human consumption through the caches they made in their lodges.

As waterfowl avoid human contact, the presence of humans along the shore likely kept them away from those particular areas, giving the people the benefit of wapato fields that were not heavily preyed upon before harvest. Also, as mentioned previously, among the Katzie people there were areas where families established seasonal claims by clearing tracts several hundred feet long along the shore (Suttles 1955: 27). In addition to demonstrably identifying a location as a claimed site, and subject to some degree of human proprietorship, this clearing would have eliminated protected areas in which wildlife species could find concealment.

Muskrats were once a major predator of *Sagittaria latifolia* on the Lower Columbia. According to Clark and Kroeker, "muskrats are the most significant resident vertebrate consumer of emergent vegetation in many North American wetlands, and their feeding activities may play an important role in vegetation decomposition" (Clark and Kroeker 1993: 1620). They can produce as many as fourteen offspring per litter, and can exceed the carrying capacity of a marsh, resulting in "eat-outs" of emergent vegetation (Clark and Kroeker 1993: 1621). A population density study of muskrats at Delta Marsh in Manitoba, Canada, found that the average density was 0.4/ ha in May and 21.3/ ha in October (Clark and Kroeker 1993: 1625). This study also found that adult muskrats (living in a stable environment) did not lose mass over a typical winter, and juveniles gained weight.

Evidence for the possible management of muskrat populations comes from archaeological data from the Meier Site (35CO5) on the Lower Columbia, which indicate that muskrat was the third most frequently recovered mammalian taxon, after deer (*Odocoileus*) and elk (*Cervus*) (Ames et al. 1995, his Table 3). In order to protect the wapato fields from eat-outs, the people may have intentionally kept the muskrat populations in check.

Muskrats would have provided meat—rich in fat—during the winter months when fish were scarce and large game animals were lean. Ethnohistoric sources for the Lower Columbia suggest that muskrats were also used for their pelts. Therefore, in addition to protecting wapato plots and providing an additional food resource, control of muskrats may have served to provide an important trade item: the muskrat skin robe. Muskrat pelts shed water, are lightweight, and the fur is so dense that muskrats generally lack external parasites such as fleas. Blankets made from muskrat pelts had the same qualities. In their description of the clothing of the Indians at the mouth of the Columbia, Lewis and Clark mention "they procure a roabe from the nativs above [in Waptao Valley], which is made of Skins of a Small animal about the Size of a cat, which is light and durable and highly prized by those people" (Clark, in Thwaites 1969, 3: 242, bracketed terms added). Paul Kane (1967) describes the clothing of the Chinook men as consisting of "a musk-rat skin robe, the size of our ordinary blanket, thrown over the shoulder, without any breech-cloth, moccasins, or leggings" (Kane 1967). Women, he noted, wear the blanket in very severe weather. Blankets made from "woodchuck" or "groundhog" pelts are mentioned in several Chinookan stories taken down by Boas (1901: 51). Woodchucks and groundhogs were not found in Chinook territory, and these stories are probably referring to muskrat skin blankets. "Garden hunting" within the great wapato plots of the Lower Columbia was almost certainly an important source of the pelts mentioned in these tales.

Wapato's Decline

In 1830, some five years after the establishment of Fort Vancouver, a malaria epidemic struck the Lower Columbia region (Boyd 1985). Mortality was high, perhaps as much as 75 percent along parts of the Lower Columbia. Hall Jackson Kelly wrote that by 1834 the Multnomah Indians "who formerly occupied the Wappatoo islands, and the country around the Wallamette [sic] and who numbered 3,000 souls, are all dead, and their villages reduced to desolation" (Powell 1917: 294).

As the human population disappeared from the wapato patches of the Lower Columbia, so too did the profusion of wapato. Wild populations of *Sagittaria latifolia* are no longer abundant in many parts of its former range due to a number of impacts from the resettlement of the region. Among these factors were the filling of wetlands, and the infestation of the Lower Columbia by carp (*Cyprinus carpio*), an herbivorous fish of shallow freshwater sloughs, lakes, and ponds. Open water habitat of *Sagittaria latifolia* became infested by carp after a flood in 1881 washed 3,000 newly hatched fry from the ponds they were being raised in for the domestic fish market into the Columbia River. Thomas Howell, a botanist whose family home was on Sauvie Island, noted in his book *A Flora of Northwest America* (1903: 679) that "this species [*Sagittaria latifolia*] was very abundant along the Lower Columbia river, but is now almost exterminated by the Carp." Carp live largely on a vegetable diet, and thrived in the wetland ecosystem of the Lower Columbia. Likewise, introduced livestock both ate and trampled wapato tubers; cattle are particularly fond of the leaves, and as soon as cattle and pigs were introduced to the region in the early part of the nineteenth century, wapato populations growing along shorelines and shallow, accessible areas were decimated.

Potentially, the elimination of human populations that protected the wapato plots and valued wapato as a staple resource may have contributed to this overall decline in wapato's abundance. No longer was human proprietorship part of wapato's life cycle; there was no longer protection from non-human predation, for example, nor was there the same degree of nondestructive turbation of dense plots that had arguably fostered the rooting of new propagules each season. Surviving Chinooks largely abandoned the use of wapato in the years following these epidemics. Wapato was beginning to be supplanted by the potato in the region as early as 1830, though there are records of use of wapato as a dietary staple into the 1860s. The potato was adopted readily by the Indians of the region in part because the potato fit well into *their* culinary repertoire, and more to the point, this shift from a lifeway defined as "gatherers" to "cultivators" was accomplished so readily because, as Suttles suggests, they were able to fit food-producing into pre-existing patterns (Suttles 1951b: 283).

Accounts from the late nineteenth century illustrate the rapid decimation of the great wapato plots and the impacts of this loss upon the Lower Columbia ecosystem:

> Sportsmen will remember that several years since the wapatoes which grew so luxuriantly in many ponds, lakes and sloughs of Sauvie's Island and in other places along the Columbia and which were the favorite food of the canvas-back duck, totally disappeared. Their loss was mourned by all sportsmen, as with the wapatoes disappeared almost entirely the flight of canvasbacks they used to attract. [Anonymous 1898]

Discussion and Conclusions

In summary, the extensive wetlands of the Greater Lower Columbia produced an abundance of *Sagittaria latifolia* tubers, so many that this tuber became both a localized dietary staple and one of the region's primary articles of trade. This was possible because people stepped into, claimed, and modified a portion of the very productive ecological niche that waterfowl and other marsh animals had occupied before humans arrived in the region. Wapato was first a food for millions of waterfowl before humans arrived on the Lower Columbia. *Sagittaria* spp. and waterfowl co-evolved in the whole of the Northern Hemisphere with this predator–prey relationship, creating a plant that was both productive during the peak migration seasons in the fall and winter, and adapted to heavy predation.

The indigenous subsistence complex described here brings into sharper focus the precontact lifeways of the Chinookan people (and others) of the region within the context of the Northwest Coast Cultural area. Clark's suggestion that villagers "hunt some roots in their own villages" hints at an integral relationship between the villages and root grounds. I suggest here that the winter villages of Wapato Valley were intentionally placed near or adjacent to wapato fields; that some (but not all) of these fields were owned and in some manner protected or enhanced by resident populations; and that the influx of people to Wapato Valley in the fall and spring was due, in large part, to the fact that these were the best seasons to collect wapato. Kin of Wapato Valley villagers, as Hajda (1984) suggests, may have had access to owned plots immediately adjacent to villages; the itinerant, "half starved" people arriving from other portions of the region to harvest wapato, as described by Lewis and Clark, were probably subject to different rules. These itinerant people were quite possibly collecting wapato from unclaimed fields, and lacked access to the prime fields that had been claimed by the Valley's year-round inhabitants.

The peoples of the Lower Columbia River owned fields, harvested the tubers, stored them, traded the surplus, controlled predators, and perhaps used associated animals (muskrats and beaver) as secondary resources. In short, the people employed some strategies to manage the resource that were more commonly seen in agricultural societies. Still, available evidence suggests that wapato production was not, as it is in the case of truly agricultural societies, largely dependent upon human intervention, despite evidence of human protection of the resource.

Wapato was a "risk reducing" resource, having benefits that led to its intensification. The location of winter villages adjacent to wapato fields was advantageous because it meant that, just outside their back doors, the local people had an energy-rich source of food during the coldest months of the year, when many other resources were scarce. This risk reduction is measured in part by the cost effectiveness of the resource to harvest, process, transport, cook, and keep, but also by the length of its harvest season and the nutritional value of wapato tubers. Wapato cooked rapidly without the need for large amounts of fuel or cooking stones, and was easy to preserve. It did not need to be peeled, pounded, or processed (which, regrettably, renders it almost invisible in the archaeological record since fire-cracked rock or ground or chipped stone tools were not part of wapato use patterns).[5] Nutritional analyses show that the tubers, if used as a staple, would have provided aboriginal populations with substantial quantities of carbohydrates, fiber, and trace elements. Moreover, wapato tastes good, and fit well in the culinary repertoire of European and American explorers who wrote about it and briefly enhanced its importance as a regional trade good, because its taste and texture closely resembled the potato.

This case study provides an interesting context with which to study the intensification process, a context in which demographic pressures led through time to expanded control and ownership of more and more fields, but also where wapato was locally abundant and probably exceeded the demands of the local human population. Wapato production on Sauvie Island, the center of the freshwater wetland complex on the Lower Columbia, was sufficiently large that it could have fed considerably more people than lived in the island's villages, and moreover the resource was utilized by a much larger population of nonresident harvesters who acquired tubers through trade. Although the wetland environment was dynamic, production was relatively predictable, providing a stable subsistence base for resident populations and a coveted resource for surrounding peoples. Conventional models of agricultural development based on population stress and food scarcity could not be applied to the case of wapato on the Lower Columbia. Instead, we might seek a better understanding of the development of this traditional pattern of resource use by considering the implications of a nutritious and culturally preferred food source and trade item, characterized by highly localized abundance. It is from this perspective that future research on wapato use and management by the pre-European inhabitants of the Northwest Coast should begin.

Notes

I am grateful to Henry Zenk for providing editorial advice and linguistic information.

1. The only village not to fit this geographic pattern was Cathlapotle, at the mouth of the Lewis River. At Cathlapotle, patches of wapato were located immediately across the Columbia River and at Carty Lake, about a quarter-mile distant.

2. The word "wapato" is a Chinook Jargon word that refers to both the plant and the edible tuber. In historical accounts it is variously spelled as wappatoe, wapatoo, wap'tu, 'pota, and papato, among others. The tuber was often called "Indian Potato" by explorers and settlers for its resemblance in taste and texture to a white potato. David French suggested that the Chinook Jargon word originated as a linguistic borrowing from Kalapuyan; according to this etymology, Chinookans (presumably, those of the Willamette–Columbia junction, where there were apparently many Kalapuyan–Chinookan intermarriages) borrowed a Kalapuyan nominal stem (cf Tualatin Kalapuyan mam-pdu), turning it into a noun by affixing the feminine singular prefix wa- (compare Clackamas Chinookan wa-qat "wapato, potato") (Zenk 1976: 85; personal communication 1999).

3. In order to determine whether wapato availability exceeded the demands of the local population on Sauvie Island, I calculated the production of Sagittaria latifolia tubers on the Island. I first calculated the productivity of wapato available on the island, and then compared this production to that needed for the population estimated to live on the island in the early nineteenth century. The amount suggests that wapato was sufficiently abundant that production exceeded the human demands of villages on this Island. This model is not intended to be representative of the whole Greater Lower Columbia region. As a model, it is an artificial construct. The main purpose is to illustrate the amount of production available in a year, compared to Lewis and Clark's population estimates of people that were on or visiting Sauvie Island. It is important to understand that wapato was an important resource and trade commodity in a region where many groups were linked by systems of exchange, and rights to resources. This model assumes that the people who lived on the island had rights to the resource, and does not address the substantial wapato consumption of nonresidents.

Sauvie Island comprises 24,064 acres (about 9,738 ha) of land and lakes. Water covered much of the island's surface until many of the smaller lakes were drained in the early 1900s. At one time the island had at least seventy-nine named lakes. In order to calculate wapato productivity on Sauvie Island, I used a map produced in 1890 by the Coast and Geodetic Survey that depicted the lakes and wetlands that were present before the Columbia River had been dammed and the island diked. I calculated the area of probable wapato habitat to be 2,985 hectares (7,373 acres). Studies have shown that the belowground biomass of the tubers for *Sagittaria latifolia* reaches approximately 300–400 dry grams per square meter in pure, monocultural stands (Gilbert 1990: 855). Recent studies suggest the productivity could be greater, reaching 1,500 dry grams or 3,000 grams of fresh tubers per square meter (Tanimoto, personal communication 1999). For this study, I used the more conservative estimate from Gilbert. Given that wapato likely occupied most places indicated on the map as shallow ponds, tide flats, or marshy seasonally inundated lands, the total productivity of tubers from the island in 1890 would have been approximately 17,910 metric tons of wapato (Darby 1996).

The equation for estimating the total amount of wapato available to humans on Sauvie Island is as follows: 60 percent of the net belowground primary production (NBPP) is an estimate of the proportion of wapato removed by waterfowl herbivory; subtract 18 percent (NBPP), which is the proportion of wapato not preyed upon, to insure viability; which equals (x), energy available to the human population. The condensed formula is as follows:

$$.60 \, (\text{NBPP}) - .18 \, (\text{NBPP}) = x$$

Thus, with a NBPP of 17,910 the mass available to the human population is 3,940.2 metric tons.

My model borrows two assumptions used by Thoms (1989) for his model of camas intensification. The first assumption is that the caloric requirement for the average member of an average indigenous family in the Pacific Northwest is 2,500 kcal per person per day. The second assumption is that roots provided at least 20 percent of the annual caloric intake of each individual. Using these estimates, the annual caloric intake of a family of five would total 4,562,500 kcal, of which 20 percent, or 912,500 kcal, would come from root foods. Wapato yields 1.8 calories per gram of fresh tuber. The minimum amount of fresh wapato that would be needed to provide 912,500 kcal, would be 0.507 metric tons (507 kilograms). Following Thoms, 125 percent of this amount (figuring for spoilage) is 633 kilograms (0.633 metric tons) per year of wapato. If 0.633 metric tons of wapato would feed a family of five, the estimated annual available wapato harvest would feed over 31,000 people. Lewis and Clark's highest population estimate for the island is 1,810 (Lewis and Clark, in Hajda 1984). In sum, there was a great disparity between the number of people this root could have supported and the estimated number of individuals dwelling on the island.

4. See Darby (1996) for a more detailed analysis of this aspect of Chinookan settlement patterns.

5. Further archaeological study of this subsistence complex could tell us something about plant use by human populations in other locations within the Northern Hemisphere. Although wapato and its use typically leave little or no archaeological record, plant tissue from a *Sagittaria* spp. tuber has been found in a 9,000 year old site of Calowanie, in the central part of the Polish plain (Kubiak-Martens 1996). Dried tissue has also been found in Early Archaic period (8000–6000 years B.P.) sites found in caves adjacent to dry lakebeds in the American Great Basin (Neumann et al. 1989). At the end of the last glaciation, *Sagittaria* spp. were widespread over the Northern Hemisphere, especially—as pollen data suggest—along the edge of the North American ice sheet and in the wetlands created by meltwater (Stuckey 1994: 283; personal communication 1997).

Chapter 8

Documenting Precontact Plant Management on the Northwest Coast

An Example of Prescribed Burning in the Central and Upper Fraser Valley, British Columbia

DANA LEPOFSKY, DOUGLAS HALLETT,

KEN LERTZMAN, ROLF MATHEWES, ALBERT

(SONNY) MCHALSIE, AND KEVIN WASHBROOK

Ethnographic sources document the importance of prescribed burning practices among hunter-gatherers worldwide (e.g., Mills 1986; Pyne 1993). On the Northwest Coast, scattered references indicate that prescribed burning was widespread at the time of European contact and in the early historic era (e.g., Boyd 1986; Gottesfeld 1994a; Norton 1979b; Turner 1999; White 1992; Turner and Peacock, this volume). Controlled fires were set by Coastal First Nations to enhance the growth of early successional plant species, either for direct consumption or as forage for animals to be hunted.

Despite well-documented evidence for prescribed burning in the early historic period, we know little about the precontact development of this practice on the Northwest Coast. However, the widespread geographic distribution of prescribed burning practices, in a diversity of ecosystems, suggests that such practices are well integrated into the traditional ecological knowledge of coastal First Nations. Such extensive and in-depth knowledge most likely results from a long history of prescribed burning on the coast. Documenting burning practices in the archaeological or paleoecological records, however, has proven difficult.

Various kinds of empirical and theoretical arguments have been suggested as evidence of prescribed burning in the precontact era. Schalk (1988), on purely theoretical grounds, proposed that burning on the Olympic Peninsula began as early as the early to mid-Holocene, when the development of closed-canopy forests resulted in a decline in the productivity of important understory plants such as berries and animal forage. Schalk proposed that these early hunter-gatherers began burning in order to open forest patches to increase productivity of game and plant resources. To our knowledge, convincing empirical support for Schalk's model is lacking.

Indirect evidence for prescribed burning in precontact times comes from the existence of plant communities that are known to be fire-maintained, such as the Garry oak (*Quercus garryana*) parkland communities of southeastern Vancouver Island and the Gulf Islands, parts of Puget Sound in Washington, and the Willamette Valley of western Oregon. Early historical documents that cite the prevalence of prescribed burning in these areas (see reviews in Boyd 1986, 1999; Norton 1979b; Turner 1999), combined with modern reductions in the geographical extent of these ecosystems, has led to the conclusion that prescribed burning enhanced the maintenance of these communities in pre-contact times (Agee 1993; Boyd 1986, 1999; Johannesen et al. 1971; Norton 1979b). However, since both natural and cultural fires have been restricted since the early to mid-twentieth century, it is difficult to discern the role that either natural or cultural fires alone would have played in maintaining these communities.

The presence of plants or communities outside of what appears to be their natural range may also provide indirect evidence for controlled burning in the precontact era (Lepofsky et al. 2003). The presence of camas (*Camassia quamash*) in Alaska (Pojar and MacKinnon 1994: 108), in western Vancouver Island (Turner et al. 1983), in the territories of the Tillamook of northern Oregon (Deur 1999), and in Stó:lō territory in the central Fraser Valley (Gould 1942), are examples of such range extensions. The natural range of camas is largely restricted to the Garry oak parkland community described above (Beckwith 2004). The presence of camas well beyond the range of the Garry oak community likely reflects introductions by humans. The creation of suitable habitat for camas in these areas (i.e., more open forests) was probably achieved in part through prescribed burning. The absence or near absence of camas in these areas today may be due to the fact that without prescribed burning forest openings are too small to maintain populations of camas (cf. Deur 1999).

Expansion of the kinds of habitats in which a species can be found within its natural range provides additional indirect evidence for prescribed burning in the precontact period. This is demonstrated on southeastern Vancouver Island, where Garry oak can be found growing on sites that are not typically well suited for it. Garry oak usually grows on shallow soils, particularly on dry, rocky slopes or bluffs, whereas conifers or other hardwoods grow on the wetter sites and deeper soils in this area. Where large Garry oaks do occur on wetter, deeper soils, some other disturbance process has discouraged the encroaching conifers—most likely frequent prescribed burning by First Nations (Roemer 1992).

Some of the occurrences clearly predate contact. For instance, in a grove of oaks in southeastern Vancouver Island growing on an anomalous site deemed to be too wet to support oaks, one dead oak, which had been overtaken by Douglas-fir trees (*Pseudotsuga menziesii*), was dated to 350 years ago. This indicates both that the oaks at this site had established there prior to contact, and that the grove may not be maintained under the current growing conditions (Hans Roemer, personal communication, 1997).

In some studies, the onset of prescribed burning is offered as a post-hoc explanation for shifts in past fire regimes. A pollen record from the Hazelton area of northern British Columbia indicates an increase in lodgepole pine (*Pinus contorta*) pollen after 2200 B.P. The increased prevalence of this tree, which tends to colonize recently burned areas, has been cited as possible evidence of Aboriginal burning of the landscape (Gottesfeld et al. 1991). It is difficult, however, to separate the actual roles of concurrent and potentially causal factors; a shift in climate may have occurred during the same period, for example, leading to an increase in fires (cf. Hallett et al. 2003).

A large roasting pit recovered from an archaeological site in southeastern Vancouver Island and dating to 3900 B.P. (Brown 1997), may be indirect evidence of prescribed burning considerably earlier in the precontact era. Though no systematic paleoethnobotanical analyses were conducted on the contents of the roasting pit, the form of the feature suggests it may have been used to process camas. The archaeological site is situated within the range for the Garry oak parkland ecosystem, but the specific locale of the site does not support camas meadows today. The roasting pit may indicate an expansion of the camas habitat as a result of precontact prescribed burning.

Similarly, in the Willamette Valley, dramatic increases in charcoal concentrations in lake cores may be evidence of the onset of prescribed burning of the Willamette prairie as early as 2700 years ago (P. Schoonmaker, personal communication, 1995). But again, a shift in climate may also be a factor. In at least one case, however, Aboriginal burning appears to be clearly the driving factor in fire and vegetation change. The fire record from fire-scarred ponderosa pine (*Pinus ponderosa*) trees around a western Oregon meadow indicates a dramatic decline in fires from the pre- to post-contact era that is likely due to a decline in prescribed burning after European contact (Hadley 1999).

Though it is common to consider prescribed burning as a post-hoc explanation of results, to our knowledge no study has been designed explicitly to document the history of prescribed burning on the Northwest Coast. We suspect this is because traditional burning practices may mimic natural disturbances (Mills 1986), and it may be very difficult to distinguish the two kinds of phenomena. In our research program, we are attempting to distinguish the signatures left by "natural" fires (those where lightning is the primary ignition source) versus "cultural" fires (where humans are the ignition source), and to make predictions about the ways in which these two phenomena will be manifested in the paleoecological and archaeological records (e.g., Lepofsky et al. 2003).

In this chapter, we summarize our research on prescribed burning among

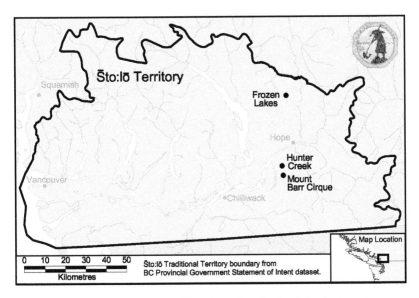

Map 8.1. Stó:lō Nation territory and location of study sites (Map by Steve Young, Sierra Club of British Columbia)

the Stó:lō of the Fraser Valley (Map 8.1). Based on information from Stó:lō elders, we believe that their historically documented practice of prescribed burning extends well back into the precontact period. The goal of our research was to document how prescribed burning was practiced in the historic period, and then to determine how far back into the precontact era this practice extends. Our research was a collaborative effort between many different individuals and institutions, and its success relied on several different lines of evidence and kinds of expertise. In particular, we integrated knowledge of Stó:lō traditional land-use practices, paleoecology, forest ecology, and archaeology.

The project had two main components: interviews with Stó:lō elders and the application of ecological and paleoecological methods. We had two primary goals for the interviews with elders: (1) to document the cultural ecology of prescribed burning; and (2) to identify sites that were subject to prescribed burning in the historic period. Information on prescribed burning practices provided by the Stó:lō elders, as well as our own knowledge of fire ecology, allowed us to formulate a priori predictions about the signatures left by natural versus cultural fires in the paleoecological record. We then used paleoecological methods to evaluate whether our study sites had evidence of prescribed burning.

In the end, we found we could not detect evidence of fires left by prescribed burns—even in study sites where we knew prescribed burning had occurred. We attribute this to the fact that prescribed burning leaves a light ecological footprint that is difficult to detect in the record of our study system. Despite the lack of a clear signature of prescribed burning, we learned a considerable amount about traditional burning practices among the Stó:lō, as well as the potential for detecting evidence of prescribed burning in the paleo-ecological record, and the overall ecology of fire in the ecosystems we studied (e.g., Hallet et al. 2003; Lertzman et al. 2002). We suggest that the multidisciplinary approach used in this study has great potential for investigating the history of traditional land-management practices, especially in systems where the cultural disturbances closely mimic those resulting from nonhuman agents.

The Role of Plant Management in the Precontact Era of the Northwest Coast

Before discussing our approach for documenting prescribed burning, it is important to consider the role of burning and other plant-management practices in the precontact era of Coastal First Nations. To date, previous discussions of precontact resource management, and its role in sociocultural development on the coast, have largely been restricted to examining the importance of salmon (e.g., Burley 1980, 1983; Matson 1983, 1992; Matson and Coupland 1995; Schalk 1977). Other resources, particularly plant resources, have been largely ignored.

The primary role of salmon production in the development of Northwest Coast society has been most strongly argued by R. G. Matson. According to Matson (1983, 1992), the management (intensification) of salmon was a necessary and sufficient cause for the rise of social complexity on the coast. Because salmon are localized, reliable, and abundant, this allowed the development of sedentism and resource ownership and control. Essentially, the argument states that various technologies were developed so salmon could be harvested, processed, and stored more efficiently. As a result of these technological developments, as well as differential access to salmon harvesting locales, social inequalities developed.

Not all researchers agree that there is a causal relationship between salmon and the rise of social complexity (see review in Ames 1994; Ames, this volume), but most recognize that there is a concomitant change in salmon harvesting and storage and social change. Support for the adoption of intensive salmon production comes from evidence of salmon storage (Matson 1992), increased use of salmon weirs (Moss et al. 1990), and the adoption of thinner ground-slate knives (Burley 1980) around 2400 years ago. These technological developments seem to co-occur with other changes that indicate, if not the rise of social inequality, at least important shifts in the social structure on the coast (e.g., Burley and Knusel 1989; Matson and Coupland 1995; Lepofsky et al. 1995).

Like the other contributors to this volume, our purpose is not to argue specifically for the timing or nature of the rise of social inequality on the coast, but to point out that plant resources, in addition to salmon, likely played a critical role in the socioeconomic development of Northwest Coast societies.

Many plant resources are well suited to the process of resource intensification outlined for salmon. In particular, we argue that plants—particularly when concentrated through such human interventions as fire—are at least as reliable, abundant, and localized as has been suggested for salmon. This is especially true for certain species of "root crops" (cf. Deur, this volume; Peacock 1998; Thoms 1989) and for several species of berries (Lepofsky and Peacock, 2004; Turner and Peacock, this volume).

However, the means of intensifying salmon differ from those for intensifying plant resources in several fundamental ways. Whereas salmon harvests were primarily increased through specialized extractive and processing techniques (e.g., weirs, knives), plant production was increased both through technological innovations (e.g., berry rakes, methods of drying) and by managing the resource itself (e.g., pruning, burning, selective). Attempts to increase production by managing salmon stocks are not well documented on the coast. Recorded instances of such management in the literature addressing the early historic period are rare (Jones 2002; Sproat 1868 [1987]: 148). In contrast to their use of salmon, Coastal First Nations employed several means to manage plant resources (Turner and Peacock, this volume). Among them are practices such as the pruning of berry bushes or the cultivation of "roots" to increase productivity and facilitate harvesting, and the prescribed burning of forested areas to encourage the growth of various economically important species. Clearly, plants are every bit as intensifiable as other resources available to Coastal First Nations.

Thus, we hypothesized that if there is a strong correlation for evidence for the intensification of salmon production and changes in social structure on the coast, then there should also be evidence for the concomitant intensification of plant resources. In this study we focused on the documentation of one kind of plant-management practice, prescribed burning, among the Stó:lō of the precontact period. Specifically, we predicted that the onset (or at least intensification) of prescribed burning practices in the Fraser Valley occurred approximately 2400 years ago, at the same time as other significant changes in the socioeconomic systems of the Stó:lō (Burley and Knusel 1989; Lepofsky et al. 2005; Matson and Coupland 1995).

The Cultural Ecology of High Elevation Plant Management

Our interviews with Stó:lō elders yielded a wealth of information on traditional burning practices in the Fraser Valley as they existed in the late 19th and early 20th centuries. This is despite the fact that elders either have not witnessed a burn since they were quite young or did not actually witness prescribed burns themselves and learned about it from their parents. The Stó:lō primarily used fire as a management tool to increase the productivity of blueberries in high elevation areas (1000 to 1500 meters above sea level). These high-elevation areas fall within what is commonly referred to as the Mountain Hemlock biogeoclimatic zone (Meidinger and Pojar 1991). Species that were

8.1. Black huckleberry (Photo by R. D. Turner)

the targets of management include black huckleberry (*Vaccinium mem-branaceum*) (Figure 8.1), oval-leaved blueberry (*V. ovalifolium*) (Figure 8.2), blue-leaved huckleberry (*V. deliciosum*), and Alaska blueberry (*V. alaskense*). Dominant tree species in Mountain Hemlock zone are mountain hemlock (*Tsuga mertensiana*), Pacific silver fir (*Abies amabilis*), and yellow-cedar (*Chamaecyparis nootkatensis*), though yellow-cedar is rarer in the transition between the coast and the drier interior, such as the upper end of the Fraser Valley. The Mountain Hemlock zone has a lower-elevation forested subzone, characterized by continuous forest, and a higher-elevation parkland subzone, where the forest becomes discontinuous as it breaks up into tree islands sep-arated by shrub meadows. There is ethnographic evidence for cultural burn-ing in both the forested and parkland subzones.

Families from up and down the Fraser Valley gathered at the larger berry sites in the late fall to harvest their year's supply of blueberries. During the late 19th and early 20th centuries, Labor Day weekend was the time when fam-ilies commonly traveled to these sites. Family groups camped together any-where from four days to four weeks while they gathered berries and hunted game. Longer stays seemed to be restricted to sites with mule or horse access. At these sites, berries were periodically packed down to the lower elevations as they were harvested. At some sites, berries were dried over racks or on blan-kets in the sun before being transported. Several elders noted that, unlike today, it was easy to fill a large burden basket (with dimensions of 60 cm x 60 cm x 45 cm) at berry-gathering sites that were maintained through prescribed burns.

Traditionally, the practice of prescribed burning was carefully controlled and monitored. According to Lawrence Hope, "special people . . . burned it,

8.2. Oval-leaved blueberry (Photo by R. D. Turner)

they knew the weather" (interview on file with Stó:lō Nation Archives). In more recent years, the practice of burning does not seem to have been restricted to specialists. For instance, Helen Angus reports that her brother used to set fires when he went to the subalpine to hunt. Thus, at least by the middle of the twentieth century, prescribed burning was apparently a more casual and incidental event than traditionally. This shift in burning practices may have been coincident with other shifts away from traditional occupations among the Stó:lō in the late 19th and early 20th centuries (Carlson and Lutz 1997).

We have few details about how fire was actually used to enhance berry growth. There seems to be agreement that berry bushes were more productive in the years immediately following a prescribed burn. Yet contemporary elders have no knowledge of how the fire was lit or controlled, or the size of the patches that were burned. The following statement by Lawrence Hope of Yale, B.C., suggests the nature of the fire itself depended on the species of blueberry:

> [There was] one way of doing it for the blueberry [*V. ovalifolium, V. deliciosum,* and *V. alaskaense*] and one way of doing it for black huckleberry [*V. membranaceum*]. Black huckleberry you pretty near had to burn all the trees down. It's got to be a very hot fire.

Ralph George stated that the Stó:lō tended to set prescribed fires on south-facing slopes, to take advantage of the drier microenvironment. The tendency to burn on south-facing slopes appears to have been a widespread practice among First Nations elsewhere on the coast (Boyd 1999; Deur 1999; Gottesfeld 1994a).

The timing of cultural burning among the Stó:lō is better documented. Burning seems to have occurred after the fall berry-harvest season, when the leaves began to drop from the bushes. As Lawrence Hope indicated, specialists timed the burning "just before it rained . . . otherwise they'd burn the berry patch out." Patches were burned every third year or so, but bushes infested with pests were burned more frequently. Interestingly, modern experiments with another species of blueberry (*V. myrtilloides*) have demonstrated that this species also reached its maximum productivity in the third year of a prune-burn cycle (Vander Kloet 1994).

Interviews suggest that prescribed fires were set for three main reasons: (1) to increase the size of the berry bushes and the number and size of berry fruits, (2) to discourage pests that infest the blueberry bushes, and (3) to encourage herbaceous plants that grow near the blueberries (e.g., "wild potatoes" [probably *Claytonia*, though possibly also *Erythronum grandiflorum* and *Lilium columbianum*]). However, not all high-elevation berry sites were burned. Ralph George notes that the organic layer of soils in the subalpine is often too shallow to withstand intense burns. In these areas, people actually hand-pruned the heather (*Cassiope mertensiana* and *Phyllodoce empetriformis*) to eliminate competition with the blueberry bushes, tree seedlings, and other "bushy" plants.

THE ECOLOGY OF BURNING HIGH ELEVATION AREAS

Based on elders' knowledge and our own observations, prescribed burns increased the productivity of the blueberry patch through several mechanisms. A primary outcome of prescribed burning was to discourage the invasion of less economically useful species, such as conifers and heathers, which compete with the blueberries for sun and nutrients. Periodic fires also seem to have discouraged pests, which presumably inhibit fruit and bush growth. As is well documented in prescribed fires from elsewhere (e.g., Lewis 1973), fires also release into the soil nutrients that were tied up in the woody tissues of plants. This addition of readily available nutrients into the soil likely accounts in part for the elders' observations that blueberry shrubs were productive the years immediately following a prescribed fire. Finally, prescribed fires also served to keep woody regrowth on the berry bushes to a minimum, thereby allowing more of the shrubs' energy to be put into fruit production.

The seasonality of the burns also appears to have been of some importance in the ecology of the cultural fires. In other areas, such as the Boreal forest and the Rockies, fires were set in the spring, when the ground was wet and fires could be controlled more easily (Barrett 1980; Lewis and Ferguson 1988). In our high-elevation study sites, however, snowpack completely covers the ground until midsummer. Thus, the window for setting fires is July though September. Burning at the end of this period, as the Stó:lō did, was ideal for several reasons. The fallen leaves would have provided a readily combustible source of fuel for the fire, yet the autumn rains prevented the fires from burning out of control.

Clearly, the nature of the fire itself was critical to the success of the burn. The fact that different kinds of fires were used for different species of blueberries indicates that factors such as the extent of the patch, shrub size, and root structure were likely taken into account by traditional fire specialists. If a burn was to be successful, it had to kill competing vegetation and burn back unproductive berry bushes, yet not damage the underground structure of the berry bushes. The elders' observation that bushes can be harvested soon after a burn suggests that the blueberry roots don't burn in a prescribed fire and that the above-ground vegetation is able to resprout quickly from the undamaged rootstock (Hamilton and Yearsley 1987).

Investigating the History of Prescribed Burning

In order to make the link between historically-documented burning practices among the Stó:lō and the precontact occurrence of prescribed fires, we developed a series of testable predictions to distinguish "natural" versus "cultural" fires in our study area (Tables 8.1 and 8.2).The predictions are based on three lines of evidence: the identification and dating of charcoal recovered from soil profiles, pollen and sediment analyses from forest hollows and small lakes, and the analysis of forest age structure in forests presumed to have been managed through burning. Below, we discuss in more detail the predictions laid out in Tables 8.1 and 8.2.

The first of our predictions concerns the location of the natural and cultural fires (Table 8.1, attribute 1). In principle, natural fires could occur anywhere, but are most likely to occur on drier, south-facing slopes. In some cases in the Mountain Hemlock zone, fuels are too wet for lightning strikes to propagate as a stand replacing fire and the fires don't spread beyond the tree that was struck. Lightning ignitions are more likely to spread on the south-facing slopes where snow melts earlier in the spring and fuels are typically drier (Agee 1993).

It makes sense that sites for cultural burning would have specific characteristics selected to facilitate access and fire management, and we hoped elders would be able to describe the specific characteristics of sites that were chosen for burning. Although elders have been invaluable in identifying general locales that were burned historically, we were able to gather only limited information on the specific attributes of sites. We have learned that cultural fires may be located more often on south-facing slopes since they are easier to ignite, and that high-elevation sites with poor soil development may have been tended by pruning rather than prescribed fires. We have no information on the size of the areas burned, or what kinds of vegetation were the focus of the burns.

The next sets of attributes (Table 8.1, attributes 2–4; Table 8.2, attributes 1–3) address potential differences in the relative frequency of natural versus cultural fires in the Mountain Hemlock zone. Research on the dynamics of high-elevation forests in coastal British Columbia indicates that very long periods between fires are common. Estimates of fire intervals range from a min-

TABLE 8.1. ATTRIBUTES FOR DISTINGUISHING BETWEEN NATURAL
AND CULTURAL FIRE HISTORIES USING SOIL CHARCOAL
PROFILES AND FOREST STRUCTURE

	Attribute	Natural Fire Histories	Cultural Fire Histories
1	Location	Not in specific kinds of sites; more common on south-facing slopes	Possibly in specific kinds of sites; more common on south facing slopes
2	Identification of charcoal	Primarily tree species	Primarily shrub and herb species
3	Frequency of charcoal lenses	Infrequent	Common
4	Fire return intervals based on C^{14} dates	> 500–800 year intervals depending on aspect and slope position	Decadal intervals based on overlapping or tightly clustered dates
5	Forest age *	Old growth, often > 400 years	Young to mature stands, maximum 120 years

* This attribute applies only to the lower elevation forested study sites. See text for further explanation.

imum of about 500 years up to several thousand years, with intervals of 1000 years or more being common (Hallett et al. 2003; Lertzman 1992; Lertzman et al. 1996; Lertzman et al. 2002). Similar forest types in western Washington State experience natural fires approximately every 800 years, with the incidence of fire concentrated on south-facing slopes (Agee 1993; Fahnestock and Agee 1983). A higher incidence of fire on south-facing slopes is consistent regionally in British Columbia and Washington (Agee and Smith 1984, Gavin et al. 2002). In contrast, the Stó:lō elders state that cultural fires were set every three to five years depending on the site being maintained. The possibility exists that there was some rotation among which patches were burned, but at any given site cultural fires should be close to 100 times more frequent than natural fires.

The large difference between natural versus cultural fire intervals should manifest in several attributes of the paleoecological record. In particular, the distribution, identity, and abundance of charcoal in soils and of charcoal and pollen in lake sediments should both provide information on vegetation growing around a site in the past and the history of fire at that site (e.g., Gavin et al. 2002; Hallett et al. 2003). Different kinds of vegetation should be burned under cultural versus natural fire scenarios. Under a natural fire scenario, fires should be so infrequent that stands of conifers will establish and mature to

	Attribute	Natural Fire Histories	Cultural Fire Histories
1	Dominant vegetation based on pollen	Primarily conifer pollen	Conifer pollen with more abundant shrub and herb pollen
2	Distribution of macroscopic charcoal	Low amounts throughout but with sporadic peaks	Higher amounts throughout
3	Fire intervals based on ^{14}C dates and sedimentation rates	500-800 year intervals	Decadal intervals
4	Sedimentation rates	Increase in sedimentation	No change in sedimentation

old growth between fire events; thus, natural fires will burn mostly woody vegetation. In contrast, areas maintained through prescribed fires should be largely dominated by early successional vegetation such as shrubs and herbs. These differences should be reflected in both the species of charcoal represented in the soil profile (Table 8.1, attribute 2) and the types of pollen and macrofossils in the lake sediments (Table 8.2, attribute 1).

The contrasting frequency of fires between the natural and cultural scenarios should also be reflected in the amount of charcoal that is deposited in the soils and lake sediments. In general, charcoal should be more abundant and more continuously distributed where sedimentary records represent repeated cultural burning. In soil profiles arising from a cultural fire history, there should be many more charcoal lenses than in profiles resulting solely from natural fires (Table 8.1, attribute 3).

In contrast to charcoal in soil profiles, which represent very local fires, charcoal records from lake sediments are a complex mix of charcoal from local and more distant sources. Since airborne pieces of *microscopic* charcoal can be transported long distances, we focused our analysis of lake charcoal on *macroscopic* (>125 (m) charcoal, which is more likely to only represent a local fire history (cf. Clark and Royall 1995, 1996; Millspaugh and Whitlock 1995). Because some amount of charcoal is always present on the landscape, small amounts of background charcoal can enter the lake at any time. Local fire events, however, result in much greater inputs of charcoal into the lake basin. Thus, sporadic peaks of charcoal should indicate infrequent natural fires, while cultural fires should be represented by higher amounts of background charcoal and frequent peaks throughout the record (Table 8.2, attribute 2).

The most direct means for measuring the intervals between fires comes from radiocarbon dating of charcoal from the soil and lake sediment profiles (Table 8.1, attribute 4; Table 8.2, attribute 3). Given the difference in the intervals between natural versus cultural fires, radiocarbon determinations of fire frequency will be vastly different in sites with or without a history of prescribed burning. Since standard deviations of at least 70 years are common for most radiocarbon determinations, they will not identify individual cultural fires occurring on sub-decadal time scales. In an area regularly maintained by prescribed burning, as many as 35 controlled burns may have occurred in the 140-year period surrounding a radiocarbon determination. However, what is important to our study is that a sequence of radiocarbon ages from soil and sediment profiles exhibiting cultural fires will look very different from a natural sequence that has experienced a fire with an order of magnitude two orders lower in frequency. In the latter scenario, radiocarbon determinations should be non-overlapping, and temporally distinct.

Predictions concerning forest age (Table 8.1, attribute 5) differ dramatically for stands that are subject to natural versus cultural fires. Since stand-replacing fires occur at intervals of at least 500 to 800 years, and possibly much more, then forests that have been subject only to such natural fires should tend to be dominated by older age classes than forests experiencing cultural fires. After a fire, natural or cultural, forests proceed through a characteristic sequence of developmental stages, culminating in old growth. In the Mountain Hemlock zone, it typically takes >200 years for forests to reach the old-growth stage (Wells et al. 1998), and most forests that haven't been logged range in age from several hundred years to over 1000 years (e.g., Lertzman and Krebs 1991).

Stands established in areas previously maintained by cultural fires should exhibit a very different distribution of ages. If one of the reasons for burning berry patches was to maintain open shrub meadows by keeping out the encroaching forest, then berry patches under active management should not have had trees growing in them. Trees now growing in former berry patches should thus represent relatively young individuals established since the last cultural burn, and are likely to be found encroaching into shrub meadows from their margins.

The timing of the cessation of cultural burning practices at each berry patch will determine the actual age of such trees. We expect that there was a significant decline in prescribed burning practices beginning at least 120 years ago when Europeans began settling in the Fraser Valley (Carlson and Lutz 1997: 116) and likely began to prohibit traditional burning practices. Aboriginal burning in southwestern British Columbia was restricted by European colonists with passage of the Bush Fire Act in 1874, although this act was not widely enforced until the early 1900s (MacDonald 1929; J. Parminter, personal communication, 2002; MacDonald 1929). The fact that some of the elders today witnessed prescribed burns when they were young indicates that in some areas cultural burning continued until relatively recently. We thus expect that trees found associated with cultural burn patches typically will be less than 120 years in age. Tree invasions of previously open areas dating to 50 to 60 years ago, however, may also be related to the widespread invasion of meadows across western North America during a favorable climatic period in the 1940s to 1950s (Rochefort et al. 1994). Stands older than 120 years either indicate that the forest was not subject to cultural fires, or that cultural burning practices stopped prior to contact, possibly as a result of dramatic depopulation and cultural disruption associated with introduced diseases.

The final attribute of fire regimes we considered reflects our understanding of the differences in extent and severity of natural versus cultural fires (Table 8.2, attribute 4).[1] "Extent" is defined as the area disturbed by a fire, and "severity" is the effect of the fire on plants, typically a measure of fire-induced mortality (Agee 1993). Stand-replacing natural fires in coastal British Columbia are typically both severe and relatively extensive. Cultural fires, however, should be much more limited in their extent, being focused on specific berry patches. We have little information on the severity of cultural burns (but see Turner 1999), except that they may vary in intensity depending on which species are being targeted.

Since we cannot measure extent and severity directly, we use rates of sedimentation in small lakes as a proxy (e.g., Millspaugh and Whitlock 1995). We expect natural fires to be more extensive and severe than cultural fires and thus to result in larger areas of unvegetated terrain. Such exposed areas are more susceptible to erosion and thus should result in increased sedimentation of mineral matter into nearby lake basins. We expect cultural fires, on the other hand, to result in little or no appreciable change in sedimentation into the lakes.

FINDING STUDY SITES

A large component of our initial fieldwork involved locating study sites meeting a variety of criteria relating to cultural history and modern sampling protocols. First, we wished to investigate at least one site with no history of cultural burning and one locality identified by elders where blueberry patches were managed through cultural fires within the historic period. Much to our dismay, we found that several of the sites identified by elders had been significantly altered by recent logging. Logging sufficiently transforms both the vegetation and geomorphology so that such sites could not be included in our study.

Accessibility was another consideration. Although the elders reported regularly hiking up to the harvesting sites, they did so on well-maintained trails (see McDonald, this volume). Today, such trails leading to the Mountain

Hemlock zone are overgrown with young alders (*Alnus* spp.) that colonize after an area has been logged or otherwise cleared. In such cases, we were usually limited to helicopter access.

Another important criterion was finding a study site with suitable lakes for extracting undisturbed sediment cores. Such lakes must be sufficiently small and deep to reflect a local sequence of changes in vegetation and fire frequency (Jacobsen and Bradshaw 1981; Larsen and MacDonald 1994). Further, sediment cores from lakes had to be long enough to yield vegetation histories that pre-date the onset of cultural burning practices.

After numerous conversations with elders, and multiple field reconnaissance trips, we focused on three study sites that met these criteria: Frozen Lakes, East Hunter Creek, and Mount Barr Cirque. Of the three sites, only Frozen Lakes (*xoletsa*, "many lakes" in Halkomelem) was identified by the elders as a berry-collecting site that was maintained by prescribed burning (Figure 8.3). Although elders identified neighboring drainages as having been subject to cultural fires, there is no memory that either East Hunter Creek or Mount Barr Cirque was managed in this way. Thus, since we had no a priori evidence for cultural burning at East Hunter Creek and Mount Barr Cirque these represented control sites.

The three study sites lie within the Mountain Hemlock biogeoclimatic zone. The Frozen Lakes (1180–1450 m), Mount Barr Cirque (1376–1480 m), and East Hunter Creek (1100–1600 m) sites each contain some areas in both the forested and the higher elevation parkland subzone. The Hunter Creek site begins in wet sedge (*Carex nigricans*) meadows in the valley bottom that grade into conifer forest going up the valley wall. This old forest grades into rolling subalpine parkland as the slopes ease to ridges and a high plateau. The Frozen Lake and Mount Barr Cirque sites are small lake basins dominated by open parkland on level areas and with more continuous forest on the slopes. All parkland sites are characterized by rolling terrain with wet sedge meadows surrounded by blueberry patches and tree islands. Both Frozen Lakes and Mount Barr Cirque have lakes suitable for extracting sediment cores, as well as countless small depressions ideal for extracting soil monoliths (Figure 8.4). Because the Stó:lō believe *stl'áleqem*—spiritual beings who occupy natural places— live in the Frozen Lakes, we consulted elders about specific ways we should behave while extracting lake cores.

DATA COLLECTION AND ANALYSES

Our fieldwork focused on collecting two main sources of data: soil monoliths and lake sediment cores. The collection and analysis of soil monoliths followed techniques developed by the authors to document natural fire sequences in old-growth forests of the Northwest Coast (Gavin et al. 2002; Hallett et al. 2003; Lertzman et al. 2002). Standard methods were employed for gathering and analyzing charcoal in sediment cores (Millspaugh and Whitlock 1995). We originally intended to core trees for a detailed evaluation of forest age at

8.3. Elsie Charlie picking berries at Frozen Lakes

8.4. Frozen Lakes. Sediment cores were extracted from both of these lakes.

the lower-elevation forested portions of the study sites. We abandoned this approach early in the study because the forest at East Hunter Creek was logged between our field visits, and because at the other two study sites air photos indicated that the forest was old growth (so that more detailed analysis was not necessary). Refer to Hallett (2001) for further details on our methods of sampling and analyses, especially for sampling of lake sediments at Frozen Lake and Mount Barr Cirque Lake. We provide more information on our strategy for sampling soil charcoal here because the method is novel and information is not generally available (Lertzman et al. 2002).

The nature of the terrain in the two subzones of our study sites required distinct sampling strategies for the soil monoliths. In the forested subzone of Hunter Creek, we dug soil monoliths along a transect from the wet meadow in the valley bottom upslope into the old-growth forest and eventually to the parkland. In the forested and parkland zones of Frozen Lakes and Mount Barr Cirque, we restricted our sampling of soil monoliths to open areas with depositional basins that had the greatest chance of long-term stratigraphic preservation. Small depressions between the base of tree islands and the edges of sedge meadows best met these criteria. The length of record was long enough at these sites (Lertzman et al. 2002) that the early to middle part of the record has the potential to serve as a temporal control for the later periods with greater potential for cultural influence. This temporal control potentially confounds cultural changes in fire incidence with climatically-driven change. We reasoned, however, that such climatically-driven changes in vegetation would be visible in the pollen and macrofossil sequences, and that even substantial increases in fire frequency arising from climate change would not result in frequencies close to those found in culturally managed berry patches (Hallett et al. 2003).

A substantial amount of the laboratory analyses have been devoted to the soil-charcoal profiles from the three study sites. We focused our attention on soil monoliths with distinct layers of soil and charcoal. In the laboratory, we sectioned these monoliths in 1-cm slices and then extracted the charcoal present in these continuous sections. We sent charcoal samples successively from these sections for dating by accelerator mass spectrometry (AMS) to determine the age of fire events represented by charcoal layers in the monoliths. We obtained 23 AMS dates from the monoliths at Hunter Creek, 50 from Frozen Lakes, and 23 from Mount Barr Cirque. We identified charcoal from only a few of the monoliths, because the data were repetitive—most charcoal fragments were mountain hemlock or Pacific silver fir.

Our analysis of the lake cores focused on sediment and charcoal accumulation. We submitted nine AMS dates from the two cores at Frozen Lakes and nine AMS dates from the core at Mount Barr Cirque. From these dates, we were able to estimate the age of the entire core sequence. In addition, we calculated the amount of sedimentation by linearly interpolating between the AMS and volcanic tephra dates. The average deposition time for our 1-cm sampling intervals was 45 years/cm at Frozen Lake and 15 years/cm at Mount Barr Cirque Lake. Our analysis of the pollen in the core is ongoing, but preliminary plant macrofossil results indicate the modern Mountain Hemlock zone forest assemblage persisted from the mid-Holocene to present at our sites (Hallett, unpublished data). At about 6000 B.P. subalpine fir (*Albies lasiocarpa*) and Engelmann spruce (*Picea engelmannii*) needles appear in the sediments, suggesting the early-to-mid-Holocene forests were likely drier and more fire prone than later in the Holocene.

	Attribute	Results	Evaluation
1	Location	Where we could sample sites with different aspects, fires were more frequent on south-facing slopes and open aspects	Consistent with natural and cultural fires
2	Identification of charcoal	Only a few samples were identified, but in all cases the charcoal was from conifer trees	Consistent with natural fires and inconsistent with cultural fires
3	Frequency of charcoal lenses	Relatively frequent in soil monoliths	Consistent with cultural fires
4	Fire intervals based on ^{14}C dates	Variable throughout the Holocene (mean = 1,200 years; range: 300 to 7000 years).	Consistent with natural fires and inconsistent with cultural fires
5	Forest age	Old-growth based on field observation and air photo analysis	Consistent with natural fires

WHAT WE HAVE LEARNED

Since the inception of this project in 1996, we have learned a considerable amount about prescribed burning among the Stó:lō, as well as the fire history of the Mountain Hemlock zone and our ability to detect evidence of past prescribed burning in this zone. What follows is a brief summary of our results as they relate to tracking the history of prescribed burning. Further details of the results of our research are presented in Hallett (2001), Hallet et al. (2003), and Lertzman et al. (2002).

Despite a substantial effort to identify a signature of cultural fires, we could not detect clear evidence of prescribed burning in the soil charcoal or lake sediment records (Tables 8.3 and 8.4). This was true even at Frozen Lakes, where there is a clear cultural record of prescribed fires to increase blueberry production. There are several possible reasons why we were unsuccessful in finding evidence of prescribed fires.

TABLE 8.4. EVALUATION OF PREDICTIONS TO DISTINGUISH
NATURAL VERSUS CULTURAL FIRES FROM LAKE
SEDIMENTS

	Attribute	Results	Evaluation
1	Dominant vegetation based on pollen	Initial results show overwhelming dominance of conifer pollen and few shrubs or herbs	Consistent with natural fires
2	Distribution of macroscopic charcoal	Sporadic peaks throughout the sequence	Consistent with natural fires
3	Fire intervals based on ^{14}C dates and sedimentation rates	Intervals were more frequent than soil charcoal dates; most local fire intervals ranged from 200-500 years*	Consistent with natural fires
4	Sedimentation rates	Peaks in sedimentation occurred during the frequent fire periods	Consistent with natural fires

* See Hallett (2001) and Hallett et al. (2003) for a discussion of the differences between fire intervals based on soil charcoal and lake sediment records.

First, we have come to rethink some of our original predictions about distinguishing natural versus cultural fires. In particular, many of our predictions about cultural fires were based on the assumption that more frequent prescribed fires would produce more charcoal and more charcoal lenses in the record, and that this charcoal would be from shrubs and possibly herbs. However, in making these predictions, we did not consider that, since they are much finer and potentially drier fuels, stems from these less-woody and nonwoody plants are more likely to burn to ash during a fire. As a result, tree charcoal resulting from natural, stand-replacing fires, and from trees at the edge of berry meadows, will dominate the soil charcoal record. This would have an impact on the samples available, both for identification and for dating. Such natural fires may even re-burn charcoal left over from previous, smaller fire events. However, the soil profiles in our study sites are very well stratified, suggesting that soil charcoal preserves a relatively complete record of local fires (Lertzman et al. 2002).

A second, and perhaps more significant reason for our inability to find a record of cultural fires is that prescribed fires may never have made much of an impact on the ecology of high-elevation ecosystems. In the lower-elevation parkland and forested areas, if the burns were small patches that were carefully controlled by burning specialists, it would substantially decrease our chances of finding evidence of them. In the highest-elevation areas, where prescribed fires might have had a significant enough impact on the vegetation that we could detect the burns in the paleoecological record, the Stó:lō used hand-pruning to manage blueberries, just to avoid the kind of impacts we might be able to detect! We suspect that this "light ecological footprint" has also stymied our efforts to find definitive evidence of prescribed fires in low-elevation meadows elsewhere in the region (Lepofsky et al. 2003).

Although the record of cultural fires in the Mountain Hemlock zone remains unknown, this study produced a detailed natural fire sequence that spans much of the Holocene. The soil charcoal and lake sediment records show that the frequency of fires varied throughout this 10,000-year period, with distinct periods when over a 1000 years passed between fires in a study system, and other times when fires were considerably more frequent (e.g., every 100 to 200 years). These shifts in fire frequency throughout the Holocene are likely driven by regional climate shifts such as the cool, moist periods that caused Neoglacial advances from 3500 to 2400 years B.P. (Ryder and Thomson 1986). During this time, we see very few fires in the soil and lake sediment charcoal records at our study sites. A shift to much more frequent fires occurred between 2400 to 1200 years B.P., suggesting prolonged summer drought occurred more frequently during this interval. From 1200 calendar years B.P. until the present, fires were relatively infrequent, with intervals ranging from 500 to 1000 years (Hallett et al. 2003).

Of particular interest here is the active fire period from 2400 to 1200 years B.P., which we call the "Fraser Valley Fire Period" (Hallett 2001, Hallett et al. 2003). This roughly 1100-year period corresponds with a period of significant socioeconomic changes in the Fraser region, as discussed at the beginning of this chapter. During this time, frequent fires, and the climatic anomalies associated with them, would have had significant impacts on a broad range of terrestrial and aquatic resources. We hypothesize that these changes in resource abundance and availability were linked to the socioeconomic dynamism of the period (Lepofsky et al. 2005).

Discussion

Our interviews established that prescribed burning, particularly in high-elevation meadows, was widespread among the Stó:lō of the historic period. Although much knowledge about the details of burning practices has been lost, it is clear that prescribed burning was well integrated into traditional Stó:lō social and economic systems. Burns were carefully timed and were the explicit job of specialists. It would be consistent with other aspects of Northwest Coast society to assume that these specialists held relatively high status in their communities (cf. Ames 1995). The question still remains, however, as to when in the past the Stó:lō began to manage these high-elevation berry patches to increase production.

We initially hypothesized that burning practices may have become established about 2400 years ago during a period of increasing social and economic complexity. Our results suggest an alternative timing for the onset of cultural burning practices. The climatically driven increase in fire frequency from 2400 to 1200 years B.P. would have resulted in the opening up of high-elevation forests and an increase in *Vaccinium* species. During this period, the Stó:lō likely incorporated this increase in berry production into their socioeconomic system, which, as we know from the archaeological record, was becoming increasingly more complex. After 1200 years ago, when the climate again shifted to cooler and wetter conditions, blueberries would have become less readily available. The social and economic pressures to produce blueberries may have been sufficiently great by this time that berry patches had to be maintained through prescribed burning. The position of "burning specialist" may have been created to keep berry production at levels consistent with the preceding hotter and drier climate. In the historic period, the decline of prescribed burning may be associated with reduced populations, shifts in Stó:lō social structure and subsistence strategy, and growing prohibitions on burning by forest managers.

To evaluate these ideas requires continued efforts by researchers to document prescribed burning and other traditional resource-management practices. In the end, it may be that the ecological impact of some practices was of insufficient magnitude to be detected in some ecological systems. However, if we are ever to confidently tease out a cultural signal in the paleoecological record, we must conduct interdisciplinary research that is both geographically broad and covers a long time depth, such as the approach we have taken here.

Notes

Funding for this project is provided by Forest Renewal B.C. (# HQ96239-RE). We are grateful to Elsie Charlie, Ralph George, Les Fraser, and Helen Angus, who graciously shared their knowledge with us about traditional resource-management practices. We also appreciate useful discussions with Hans Roemer (B.C. Parks), Peter Schoonmaker, and John Parminter (British Columbia Forest Service) about vegetation and evidence of Aboriginal burning. Steve Young of the Sierra Club of British Columbia, Victoria, prepared the map of Stó:lō traditional territory; help from Fiona Hamersley Chambers is also gratefully acknowledged.

1. Some researchers have attempted to distinguish natural versus cultural fires in the paleoecological record by the season in which the fire burned. This has been applied in areas where cultural fires tend to be lit in the spring and natural fires tend to occur in the late summer and fall. In our region, where both natural and cultural fires will occur most often in the late summer and fall, seasonality can not be used as a distinguishing criterion.

Chapter 9

Cultivating in the Northwest

Early Accounts of Tsimshian Horticulture

JAMES M^CDONALD

The literature on Northwest Coast peoples prior to European contact provides the general impression that these people were mariners, not oriented to the land. Of course this is not true. The land was a well-utilized resource, full of game to hunt, plants to gather, and many other resources, all carefully managed by titled property holders, the people called "chiefs" in English. This chapter discusses the plant-management practices of the Tsimshian people. The available evidence suggests forms of plant cultivation were integral to the indigenous Tsimshian economy. My intentions are to explore the role plant care had in the aboriginal economy, and to consider the colonial context that distorted and contributed to the interruption of management knowledge and practices.

Very early in my work with the Tsimshian community of Kitsumkalum, I was surprised to encounter historical information about their gardening practices. Some of this information was recorded in published and archival sources, but much of it lived in the oral archives of the Tsimshian Nation. As we spoke about their household economies, people frequently told me about their own gardens or reminisced about the gardens tended by their parents and grandparents. Elders shared their knowledge of gardening local plants and of caring for berry bushes. On hikes, friends taught me to gather and eat roots, berries, and other plant products. Taken as it came, in bits and parts, the pieces of information were not remarkable in themselves and I may not have made much of it but I was trying to systematically cover every sector of the Tsimshian economy for a study on the impact of colonization (McDonald 1985), so I was inclined to record everything I learned. Consequently, over time, a body of information was collected which, when I collated the pieces, revealed the nature of Tsimshian plant management practices, including gardening practices, and how integral plant cultivation was to the indigenous Tsimshian productive economy.

240

This chapter presents that information, using the Tsimshian community of Kitsumkalum as the central reference point. The information does not provide a full account of precontact gardening practices but it establishes two key points: (1) The Tsimshians did manage plant resources in more ways than simply conserving them during foraging. In other words, Tsimshians traditionally took care of certain plant species and assisted their growth. (2) The deeply entrenched stereotype of the Tsimshians as a fishing people without cultivation techniques is wrong, profoundly wrong; but the published record treats the Tsimshian as if they were and are unconcerned with plant management and cultivation. This paper will focus on a sector of the Tsimshian economy that, historically, has been ignored and made invisible by outside observers.

The People

Tsimshian territory encompasses the lower Skeena River and the archipelago of islands spilling out of its mouth, from the Nass, south to the Estevan Group. The people lived in lineage House groups throughout this territory during most of the year to harvest the abundant resources that supported their complex social organization, and then consolidated themselves for the winter season into several towns. Each town was associated with a particular "tribal" population, called a *galts'ap*, and territory. On the Skeena there were at least eleven such tribes. The lower nine formed a loose confederation during the merchant period of Tsimshian history, and became known as the Port Simpson tribes, after the name of the Hudson's Bay Company post where they settled. A tenth tribe that is said to have gone extinct is often included in this group. Kitsumkalum is the next group upriver and has a strong alliance with Kitselas, the most eastern Tsimshian *galts'ap* on the Skeena.

During the nineteenth century, Franz Boas did fieldwork in Port Essington, a cannery village closely associated with the Kitsumkalum people. Boas gathered information from Kitsumkalum fishermen and canners and some of his early descriptions of Tsimshian society are based upon information and narratives collected from unidentified Kitsumkalum sources. Despite this contact, Boas makes little specific reference to that *galts'ap*. He listed "G.its!Emaga'lon" as one of the tribes of the Tsimshian proper, found below the canyon of the Skeena River (Boas 1916: 482), and provided a description of their town as having had three rows of houses, arranged side by side, facing the water, with the street stretching in front of the houses parallel with the river (Boas 1916: 395). (I identify the town Boas describes to be the now abandoned Robin Town located at Kitsumkalum Canyon rather than the contemporary village at the mouth of the Kitsumkalum River.) He also mentions that some of their hunting grounds and berry-picking areas were on the shores of a lake (Kitsumkalum Lake). Hunters had hunting huts in their territories, and one man had a hut in each of the four valleys owned by his House (Boas 1916: 401). Boas's description does not mention the other lands used by the Kitsumkalum in the Zimacord Valley, along the Skeena to its mouth, on certain islands off the Skeena mouth, or at the fishery on the Nass River. Individual Kitsumkalum people could also activate social connections with other towns, both along the coast and on the Nass, thereby extending the territories available to the people.

The bits of information Boas presented about Kitsumkalum were not developed further for nearly a century. Viola Garfield's classic but socio-geographically restricted study of Tsimshian society concentrated upon the Port Simpson people and only refers to Kitsumkalum in a list of several neighbouring Aboriginal groups (Garfield 1939: 176). This neglect of Kitsumkalum in the professional literature has continued to the end of the century. An example is Miller's (1997) overview of the Tsimshian, which ignored Kitsumkalum and Kitselas except to mention the Kitselas archaeological site. My own research during the 1980s and 1990s has addressed this gap by using community-based research with the Kitsumkalum people to explore issues in Tsimshian ethnohistory, colonization, and contemporary conditions (McDonald 1983 to 2003).

Turning to archival sources, the first reference to Kitsumkalum can be found in an entry in the Hudson's Bay Company journal for November 13, 1852, when a canoe of people came to trade at the fort. The traders had recorded other Skeena River canoes, but this was the first specific mention of the community of "Kith lum ki lum" (Hudson's Bay Company Archives, 1832–70, B.201/a/ 7 fo.40d).

Seven years later, the first recorded visit to Kitsumkalum by Europeans was an expedition lead by Major Downie of the Royal Navy, as he surveyed his way inland in 1859. The Major was filled with praise for what he saw:

> a large stream, called the Kitchumsala [sic], comes in from the north; the land on it is good, and well adapted to farming, and that the Indians grow plenty of potatoes. To the south . . . is the Plumbago Mountain[1] . . . [which] runs in veins of quartz." [Downie, in Mayne 1862: 451]

Thus the people of Kitsumkalum, like other Tsimshian groups, entered the written record of history as potato gardeners. Unfortunately the economic practices of the Tsimshians interested Mayne less than the economic potential of their land, especially the surrounding mountains with signs of mineral wealth, and he provided no details of their gardening or other plant management practices.

We must turn to the Tsimshian themselves for a clearer picture of Tsimshian plant management techniques and knowledge.

The Plant Resources

The late Elder, Lucy Hayward, described the *laxyuup* (House lands and estates) as a storage box of food (interview, June 15, 1980). Miriam Temple expressed the same idea with an analogy between managing the wealth of the land and managing your wealth in a bank.

You go up the Nelson Creek or Starr Creek, . . . it belongs to the tribe. And open it up, just like a trunk. [That's what] he do. Now you put the money away. When you need it, you go and open it up and get what you want. Just like the bank now. [Interview September 29, 1980]

These two elders gave witness to the rich biotic diversity of the Tsimshian area. I have references to the use of many plants, but the full range of plant resources that were gathered and used by the Tsimshians before contact with Europeans was probably more extensive than we can expect to reconstruct now. The following list is based on major ethnographies and on my own fieldwork.

Franz Boas. berries, roots (1889: 816, 1916: 182, 251), crabapples (1916: 240) maple wood (1916: 396), fern roots (1916: 337, 404), hemlock sap, lichen (1916: 192, 404), devil's club bark as medicine (1916: 448), skunk cabbage (1916: 405); "what is called rich food . . . [includes] berries; elderberries, currants; and others of similar kind . . . hemlock sap [inner bark or cambium] . . . [was] considered poor food" (1916: 406).

Viola Garfield. inner bark of hemlock, soapberries, berries, low-bush cranberries, crabapples (1939: 199); barks, shoots, crabapples (1966: 13).

Brian Compton.[2] Food plants include the fungal organism *Exobasidium vaccinii*, wood fern, lupine roots, Pacific silverweed roots, springbank clover roots, cambial materials from three confers (the western hemlock, Sitka spruce, and Pacific silver fir), rice-root and probably other members of the lily family, cowparsnip, western dock,[3] salmonberry, labrador tea leaves for a beverage and a medicinal tea, twenty-two fruit-bearing bushes[4]—including wild lily-of-the-valley, red elderberry, highbush cranberry, Western Cordillera and/or Canadian bunchberry, soapberry, black crowberry, salal, Alaska blueberry, black mountain huckleberry [mountain bilberry], oval-leaved blueberry, bog cranberry, red huckleberry, stink currant, saskatoon, strawberries, Pacific crabapple, red raspberry, thimbleberry, trailing wild raspberry, black raspberry—and possibly also Indian plum[5] (1993a: 402), salmonberry (dark red fruit form and golden fruit form identified separately from red fruit form) (1993a: 461); a "real sweet crab apple" (1993a: 461); species used for materials include the western red-cedar, yellow-cedar, and western yew (1993a: 403); nine medicinal plants, including green algae, licorice fern, false Solomon's seal, water parsley, copperbush, labrador tea, and Sylvan goat's beard (1993a: 403); ritual medicinal and purification plants, including devil's club, Indian hellebore, and common juniper, also mosses for insect-repelling smudges (1993a: 404).

Nancy Turner 1995. (pagination is shown in parentheses): kinnikinnick berries (76), salal berries (77), bog blueberry (89), black currant [white-flowered currant] (101), fireweed (106), black hawthorn berries (111), common wild rose (120), cloud berries (121).

Jim McDonald. (archived research interviews and 1987): My own research confirms many of the above references and adds the following additional plants: hazelnuts, grasses, wild onion bulbs, black gooseberries, firewood, mushrooms; potatoes; tobacco.

To these lists of plants that are specifically identified and referenced as used by the Tsimshians can be added three more plants that are used generally along the coast and that, therefore, probably can be listed as also used by the Tsimshian. Some of these species may already be identified above by their generic English name.

Nancy Turner 1995. Sword fern (as a starvation food used generally along the coast in place of spiny wood fern, p. 28), bracken fern (used generally along the coast, p. 30), and cinquefoil were eaten by most groups (p. 115).

The following plants were used by one or more of the immediate neighbors of the Tsimshian and, therefore, may also have been used by the Tsimshian.

Nancy Turner 1995. (pagination in parentheses): The Haida ate calypso corms (52), eelgrass (53), blue lupine rhizomes (92), and sometimes Sitka mountain ash berries (129); Haida children chewed purple hedge nettles (105); the Haida and Nisga'a ate stonecrop (*Sedum divergens*) (72); the Haida and Tlingit used arrow-grass (39); the Haida and Nuxalk ate wild carrots (probably Pacific hemlock-parsley) (Compton 1993b), although the botanical species cannot be confirmed (58); the Nuxalk ate water parsnip (60), swamp gooseberries (101), and black cottonwood cambium (130); and the Gitxsans made a tea from stinging nettle leaves (131).

Harlan Smith 1997, 150–51. (These Gitxsan references are only to plants that have not been listed above; the references are from a summary table that also provides pagination to fuller discussion of each plant in Smith's text): Cambium of lodgepole pine, kneeling angelica, Arctic lupine roots and fruit, sheep sorrel stems and leaves, the nectar from the flowers of red columbine, pin cherries, chokecherries, cambium from the balsam poplar and trembling aspen, the entire nodding onion.

These lists include plants used for medicinal or technological purposes, but I have not tried to be comprehensive in including all plants used for these purposes.

To my list of land plants used by the Tsimshians should be added several water plants: seaweeds, [giant] kelp (Boas 1889: 816, 1916: 44); green algae (Compton 1993a: 403), "broad leaved marine algae" (*op. cit.*: 404); eelgrass and red laver (Turner 1995: 21); and freshwater alga which had a possible use among the Southern Tsimshian (*op. cit.*: 401). I have no information on the management of water plants and have not developed a discussion of that possibility here.

Aboriginal Cultivation

Fundamental to this chapter is the notion that there is a difference between the use of resources as they are found naturally and the cultivation of the resources to enhance their usefulness to a community. Anthropologists use the difference to distinguish between foraging and cultivating, or, as Meillasoux put it (1972: 98–99), between treating the land as a subject of labor and as an instrument of labor. The distinction is of fundamental importance with respect to the way the Tsimshian related to their environment and to the way colonials evaluated the colonized. Land used as a subject of labor involves the simple extraction of resources on a rather opportunistic basis by relatively simple and unstable societies. Land used as an instrument of labor involves more complex social and productive relationships because people are investing their labor in expectation of greater future returns. The evolutionary implications of this distinction may not be palatable to social scientists today but, I suggest, they remain influential in the dominant society's worldview that is too often grounded in the neolithic pride and agrarian imperative of controlling the environment as an instrument of human progress. The stereotype of Aboriginal people as hunters and gatherers who did not invest in their lands serves to diminish Aboriginal resource management, and plants the conceptual seeds that justify the displacement of Aboriginal peoples. The Aboriginal landscape becomes a "wilderness" that needs to be turned into instruments of labor that will drive the new economy. And so the Aboriginal forest becomes a commercial forest under a new regime of resource management.

The evidence presented below shows some ways the Tsimshian people invested in their environment and brought plants into the Tsimshian world.

CARE OF FRUITS AND BERRIES

There are several types of evidence of indigenous management of fruits and berries. One was to bring certain wild plants into a garden environment within the villages and towns. Berries that I specifically know to have been transplanted, in the "old days" as well as more recently, into residential areas and cultivated were highbush cranberries, soapberries, and blueberries. I also was told that highbush cranberries were transplanted to the edges of garden plots.[6]

Other techniques involved modification of the plant's environment. Some mountain berry grounds could be considered a garden environment. They would not have looked wild when they were properly managed. While picking berries on a mountain slope with people trying to push through the tangled bushes of the less-used berry patches, I learned that formerly the old people did not have to put up with such inconveniences. They took care of the bushes. Mildred Roberts told me the brush that grows after logging (and other disturbances) makes it too difficult to pick berries. There used to be trails, which were kept clear so people could use them (interview, July 9, 1980). Favorite berry patches had trails to them that were maintained so that people carrying cedar baskets strung from their necks could easily reach and harvest the berries.

Little wonder that there was a common practice of naming specific berry "grounds" or territories, a convention that effectively inscribes those plants with social meaning and acknowledges Tsimshian plant management. An example is a large territory surrounding the old Kitsumkalum village of Gitxondaxfi, "people of binding." In 1926, a chief, Charles Nelson, said this area was called Txasawdaw Baxia Ik (place where they moved up into the hills.), which was a rich hinterland with important berry bushes. (Beynon and Barbeau field notes, B-F-49, n.d.) and a specific place where those people gathered berries (McDonald 2003). Today, the old names may not be used regularly but people in Kitsumkalum still identify specific places for their cranberries,[7] for example, the cranberry flats at the north end of Kitsumkalum Lake, Lakelse, Lockerby Creek, or the Ecstall River camp.[8] The concept of the territories is still important and is incorporated into Treaty negotiations with the term *laxyuup*.

Another result of my mountain hikes with community members was a recognition that fruits of different berry bushes taste different, even though they are of the same kind. In my mind, I made the comparison to different varieties of wine grapes and all the environmental and genetic factors that contribute to those differences. Elder Winnie Wesley noted that berries were sweeter high "up mountain" and called these "mountain berries." Leslie Johnson Gottesfeld learned that Gitxsan people "refer to traditional berry patches as occurring 'half way up the mountain', that is, in the montane and lower subalpine forest zones dominated by confers . . . at about 3000–4000 feet in elevation" (1993: 96). Knowledge of such differences in quality suggests another reason for managing preferred berry patches to enhance quality and quantity.

Such differences in the quality of local food plants may provide greater understanding of the often-reported specialty foods prepared by the residential communities or tribes. As Eddie Feak, an elder who was close to the people of Kitsumkalum, put it:

> Some places have a lot of dried fish, see, and some places had dried berries, and some had different foods, all along [the Skeena] and down along the coast, they had different [foods]. [interview, 1980]

In the story of "The Water Being Who Married the Princess" (Boas 1916: 274–75), a young chief sent word to each of the tribes to prepare certain items for a supernatural feast. The Lakelse people were to provide "cranberries" [probably highbush cranberries] and crabapples, the Kitwilksebau[9] were to make many hundred score of dried cases of hemlock sap, the Kitselas were to dry many bundles of berries (blueberries, cranberries, and soapberries), and the Kitsumkalum were to dry hundreds of salmon. The berries were provided preserved in eulachon grease. The Kitkaata people were told to bring tobacco and eagle down, while the Kitkatla brought burnt clamshells (presumably the shell was to go with the tobacco). Usually these ceremonial differences are explained in terms of differential environmental availability of resources, but subjective evaluations of quality (i.e., taste) today direct people to environmental zones that are thought to produce better quality. Such local differences may have underlain the specialties that formerly were identified with different communities.

The mountain berry grounds were managed in several ways. Some plants that naturally grew in scattered patches were tended to stimulate a concentration. I was told berries must be picked to come back and, in particular, that blueberry, black currant, and gooseberry bushes have to be cleaned and pruned or they will die. Sarah Wesley, a Nisga'a woman married to a Kitsumkalum man, said berry patches were picked every other year.

Berry bushes were cleared of overgrowth after the fruit season to ensure a healthy supply during the following year, and to allow pickers to move more easily through them. Sometimes pruning was accomplished by setting fire to the bushes (e.g., cranberries).

My notes from a discussion with Alex Bolton (personal communication, August 1982) concerning the management of the cranberry patch at Lockerby Creek camp provide some further points:

> was burned in the Fall—keeps patch clear of trees, etc.—regrows better the next year—bigger berries.

Mark and Rebecca Bolton, two community leaders in the first half of the century, had a lowbush cranberry patch on the Ecstall River (Ben Bolton, interview, March 31, 1980) that was also burned regularly (Art Collins, personal communication, April 30, 1987).

Stephen McNeary, in his study of Nisga'a social and economic life, described Nisga'a plant-collecting schedules without exploring management techniques. However, he did note that "the Niska [sic] used to burn off sections of mountain side to encourage berry growth" (1976: 109) and "yields of hillside berries were increased by burning off clear areas" (1976: 113).

Leslie Gottesfeld emphasized the importance of both pruning and burning for increasing productivity of soapberry patches (Figure 9.1) (1993: 98). Her study of Gitxsan and Witsuwit'en burning of berry patches supports my notes from the Tsimshian. "A light burn stimulates vigorous sprouting and enhances berry patch production" (1993: 97), although it took up to four years for a patch to return to peak productivity after a burn.

Gottesfeld uses her Gitxsan material to explain the dietary importance of plant management:

> Given the low caloric value and small size of individual fresh berries, the location and maintenance of large and productive berry patches with predictable harvests was necessary. . . . Burning was the mechanism to enhance or maintain berry patch quality [Gottesfeld 1993: 94].

9.1. Soapberry (Photo by R. D. Turner)

The name of one or possibly two areas in the Kitsumkalum Valley suggests the use of fire as a plant-management technique. In 1927, the Kitsumkalum leader Arthur Steven told Marius Barbeau a "Story of Gitsemralem," in which he mentioned a territory called *Milgeelde,* and translated as "burnt shrubs"[10] (Beynon and Barbeau field notes B-F-47.1, n.d.). Elsewhere in the Beynon/Barbeau fieldnotes (B-F-50), Sam Kennedy (of the Laxsgiik phratry) applied this name to a piece of Ganhada territory by Goat Creek, which is at the south end of Kitsumkalum Lake. Kennedy translated the name as "burnt mountain tops" (see McDonald 1983).

Another important aspect of Tsimshian berry management is its collective nature. Gathering is a major social effort:

> In traditional times, the collecting of large stores of berries was a late summer activity which involved the congregation of groups of people at productive berry patches, a sustained harvesting effort, and processing of the berries into large dried berry cakes which were then transported back to village sites for winter provisioning. [Gottesfeld 1993: 94]

Stephen McNeary wrote, "parties of [Nisga'a] women picked the berries" (1976: 115). Such work parties are featured in some Tsimshian legends, such as Franz Boas's version of the Kitsumkalum story about Part Summer, the princess who married a bear (Boas 1916: 278–84), which starts with a group of women picking berries together on a mountain. The collective effort is often assumed and not explored, as when Viola Garfield describes how a Tsimshian clan gave away "the products of two days picking from their grounds before they picked for their own use. These berries were gathered for him [the chief] by all the women of the tribe, each from her own territory" (Garfield 1939: 199).

The people responsible for managing these berry grounds were and, where the House structure is still forceful, remain the House titleholders.[11] Leslie Gottesfeld's comment on how the burning of Gitxsan berry patches was managed by the chief of the House groups encapsulates what was involved in the management system.

> When it is the right time he [the Chief] burns the berry patches so the berries are fat and plump. If he didn't do that the berry patches would become old and overgrown and there would be berries but they would just be small. But he knows when to burn so that it cleans up just the berry patch and doesn't spread to the trees . . . (Pat Namox, quoted by A. Mills n.d.: 156). [Gottesfeld 1993: 96]

My information on Aboriginal berry management, with its contrast to contemporary practices, may seem contradictory to the way most people now harvest only wild, noncultivated plants. This contemporary foraging pattern seems to have little to do with plant management of any significant type. But it is important to keep in mind that the use of plants has changed drastically during the colonial period. Physical changes (e.g., clear-cut logging, agriculture) have redefined the human landscape from one that reflected the management practices and knowledge of the indigenous Tsimshian people to one that is colonized and defined by the logging companies, settlers, and other agents of the new social order who have occupied Tsimshian territories. The impact on plant management was extreme.

Elder Lucy Hayward once tried to explain to me where she used to harvest foods with her parents, but quickly became frustrated and exclaimed that the berry bush *lxyuup* she was trying to describe around Kitsumkalum Lake had been ruined by the newcomers and that the old trails had been replaced by roads (interview, 1980). As Miriam Temple put it,

> All I know is [that there were] a lot of people, early days, here. All the little rivers, Kalum, [were used] . . . for hunting. That's all I know. I used to go out to the Kalum myself. But, I don't, I get lost there now. I don't know, there's too much logging back there. I lost the trails. [Interview, August 21, 1980]

ROOT CROPS

The Tsimshian took care of other plants. Jack Darling provides a piece of information on precontact plant management in a brief comment that the early post-contact gardens were situated at sites previously used only for gathering edible roots (1955: 27). The main root crops are bracken fern and rice-root (*Fritillaria*)[12] These are known to have been tended on the coast and it seems

9.2. Labrador tea (Photo by R. D. Turner)

safe to assume that practices similar to those used elsewhere (and described elsewhere) were applied by the Tsimshian. Unfortunately, I found little information on root crops to fill out Jack Darling's comment, although Leslie Gottesfeld postulated that periodic harvesting of roots may have enhanced the productivity of patches of fern rhizomes and lily bulbs (see also Turner et al. 1992, Turner and Peacock, this volume).

Mark Bolton had a rice [rice-root] patch on the Ecstall that was weeded, transplanted, and prepared in the early years of the century (Art Collins, personal communication, April 30, 1987). Sarah Wesley told me wild rice [rice-root] (Figure 9.2) was weeded in the early spring and summer, and (Labrador) tea was also weeded when necessary. People tried to keep the growing areas clean of competing plants (interview, August, 1981). Also, anthropologist Leslie Gottesfeld (1993) postulated that periodic harvesting of roots may have enhanced the productivity of a patch of fern rhizomes or lily bulbs.

FRUIT TREES

Fruit trees were frequently tended and people now consider the presence of hazelnut trees (Figure 9.3) and crabapple trees a sure sign of an old residential community or campsite. I was often told that "wherever the old people camped, there are crabapples."[13] In particular, the old people planted nut tree saplings from Kitselas Canyon around their Kitsumkalum town site (Alex Bolton, interview, June 15, 1979). "Chestnuts"[14] were also harvested at

9.3. Hazelnut (Photo by R. D. Turner)

Kitsumkalum Canyon (Priscilla Nelson, interview, October 23, 1980). I have no information on care or management practices for nut trees but people used to prune the crabapple trees. "Big trees were cut to fall over but leave part so it will regrow as a small tree closer to the ground for pickers" (notes from interview with Alex Bolton August 1982). Anthropologist Margaret Anderson also has heard about "stands of wild crabapple . . . [that] were cared for, including cutting down the crabapple trees when they get too tall to harvest" (personal communication, October 29, 1997).

BARK AND TREE ROOTS

Bark and tree roots are harvested routinely for a variety of purposes. Today, these are important for making many different items, including baskets, masks, and headdresses. The bark is stripped in the old way so as not to girdle and kill the tree. Similarly, roots are taken from living trees where the ground has exposed the roots (for example on a bank or a partial windfall). Conservation principles are applied to protect the living tree.

SEAWEED

When I made inquiries about plant-management practices, several people stated I also should look at how seaweed was tended (see Turner 2003). Although I found no specific references, the conditions were appropriate for some type of management. As Margaret Anderson noted for the Tsimshian coastal regions, sites for this were certainly also owned/controlled by specific people, and on the coast it was probably the biggest volume plant food [for those Tsimshians], or at least equal to the total volume of the various berries [they] harvested. [personal communication, October 29, 1997]

Storage

A piece of evidence on the antiquity of plant management comes in the form of recipes. The very non-European way of preserving crabapples and cranberries in eulachon grease was repeatedly mentioned to me as a favorite technique used by the old people in the early years of the century. Cakes of dried berries also were prepared aboriginally, and traded to the earliest land-based traders, although none of the people with whom I discussed food production mentioned this as a contemporary practice. Nancy Turner indicates crabapples and highbush cranberries were also kept fresh in storage boxes filled with water (1995: 118). Other fruits, apparently, were eaten fresh.

Potatoes

One does not think of horticultural crops in association with the aboriginal Tsimshian so it is surprising to discover the potato making an early and dramatic appearance in the first records of the land-based fur trade. When Chief Factor Ogden of Fort Simpson recorded the arrival of trade canoes on November 14, 1835, he wrote: "Some of the Potato people also arrive" (HBCA, folio 45d). This is the first Fort Simpson Journal and the casual reference "Potato people" stands out as an observation wanting explanation, but who these people were and why they were identified as the Potato people is unknown. Ogden seemed to think the label too obvious to deserve explanation. But now, nearly two centuries later, Ogden's report raises questions. In particular, when did the Tsimshian begin growing a cultivated crop like potatoes, and was this an indigenous cultivar?

I cannot resolve those questions, other than to say that the first land-based traders depended on Aboriginal trade for fresh produce and that at least a part of the "fur trade" was for a cultivated garden crop, the potato. This conclusion means that, with the exception of the brief descriptions by exploring ships, the written record on the Tsimshian has always described a people who grew potatoes and grew them in sufficient quantity for an international trade (see Moss, this volume). The importance of gardens to the seasonal cycle has been noted (McDonald 1987), but the habit of ignoring this importance in ethnographic analysis continues at the expense of a holistic understanding of the Tsimshian productive economy. Here are some of the earliest historical records on Tsimshian potatoes.

The Hudson's Bay Company established Fort Naas (later called Fort Simpson) at Fort Point on the Nass River in 1832, relocating it in the summer of 1834. This was the first trading post on British Columbia's North Coast. Daily trade journals allow us to know, with some accuracy, who was trading what and when. October 17 was the ceremonial opening day, celebrated by raising the British flag and firing a five-gun salute from the ship (folio 3). The first recorded trade occurred the following week, on October 23, when a "canoe with 4 Indians from Nass arrived with a few Beaver skins most of which were traded" (folio 9d).

The entry for November 23, 1834, tabulating the week's trade, listed "2 bush. potatoes." The HBC traders continued to purchase potatoes over the course of the next year, until October 8, 1835 when they "traded ca 20 bush potatoes, [but] stopped since we have enough for winter" (folio 42). At this time, the Fort's own garden had also started to produce and its first crop yielded 204 bushels of potatoes, not enough for the fort but giving the men some independence from the Aboriginal gardens.

The recorded quantity of potatoes obtained in trade from November 1834 through to the end of 1835, the first year of operation, totaled 1655.75 bushels. During this time, some canoes were specifically identified in the HBC records as trading potatoes and of these, some were identified by their community of origin. We know there was potato trade with the four Tsimshian communities of Port Essington, Sebossa (i.e., Tsibassa, a Kitkatla chief), Dundass, and Pearl Harbour; and with the three Haida communities of Cumshawas, Masset, and Skidigate [sic]. The Nisga'a canoes were only described, simply, as being from the Nass River. And on November 14, 1835, "some of the Potato people also arrive," an unfortunately unidentified group of canoes (folio 45d).

The traffic in potatoes accelerated quickly from its start in the Fort's first season. Five noteworthy features of this trade are: First, the trade was engaged very soon after the Fort's construction. Second, the busiest periods during 1835 were May and October/November. Third, there was some apparent hesitation on the part of the Aboriginal traders. On May 18, when 6 canoes passed by the fort, only some stopped and traded 6 bushels potatoes. The rest kept at a distance (folio 30). Fourth, the Fort was only one of the trading stops for potatoes, as suggested in the following references: on May 15, 8 canoes arrived from Cumshewas with potatoes, etc., bound for Nass (folio 30); on August 7, Skidigate Indians started for Pearl Harbour after trading most of their potatoes to Captain Domini of the competing trading ship *Bolivar Liberator* (folio 38); on October 10, 7–8 Masset canoes left for Pearl Harbour loaded with potatoes (folio 42d). These observations suggest a trading route that incorporated other market places and also suggest that the initial hesitation of some canoes to trade with the post may have been associated with a need to evaluate adding the HBC fort to the trading route.

Finally, all three northern Aboriginal nations (Tsimshian, Nisga'a, and Haida) were trading potatoes, so the potato was widely available and plentiful. Canoes were described in HBC records as "loaded with potatoes." An indication of the level of trade is provided by the Factor's comment on October 7 that trade was dull during the past 2 days bringing in only 100 bush potatoes, probably from the 2 large "Tumgass" canoes that arrived that day (folio

42). By comparison, the fort's own garden produced 204 bushels in October 1835 (folio 43d).

Judge J. P. Howay (1929) described the potato trade as extensive and as the source of international tension. The Tsimshian were trying to maintain a monopoly on the potato trade at Port Simpson, with a series of incidents occurring between them and the Haida and British. The British were trying to establish some independence by having their own gardens. These observations seem to be the source of Howay's impression that the Haida grew the potatoes and the Tsimshian only traded them, a theme taken up by Robert Grumet (1975: 308). I do not feel the case against the Tsimshian growing potatoes is convincing, but Judge Howay was mainly concerned about the trade aspects rather than production. This leads to another question.

Given the early availability of potatoes and an apparent indigenous trade in the crop independent of and prior to the Hudson's Bay Company's presence, there must have been gardens. But how old is the practice of gardening?

The usual explanation credits Europeans for the origin of the Haida potato. European traders sometimes planted potatoes for future harvest or gave them to Aboriginal people to plant and harvest, thus introducing potatoes to the coast. John Dunn, an early explorer and visitor to Port Simpson, wrote:

> attached to their houses most of them have large potatoe gardens: this vegetable was first given to them by an American captain; it is now grown in abundance, is traded by them to vessels visiting their harbour, and to the traders at Fort Simpson. I have known for 5–8 hundred bushels being traded in one season, from these indians at Fort Simpson. [Dunn 1844: n.p.]

Dunn's estimate of the trade was published ten years after the HBC's own gardens began production, reducing the post's need to trade for potatoes. Judge Howay names Captain Gray of the American ship the *Columbia* as the one who gave the potato to the northern nations and notes that Gray was like other traders who followed the example of Captain Cook "in leaving with the Indians seeds of various kinds of vegetables" (Howay 1929: 155). Fresh vegetables were a welcome and healthy trade item for sailors halfway around the world away from their familiar tables.

Wayne Suttles (1951b [1987: 140]), in reflecting on Salish potato growing, found the earliest references to Salish potatoes came from the 1840s, and was content to attribute potato cultivation to contact with the fur traders. His comment that "potatoes were raised by the Haida by 1841" (1951b [1987: 140]) seems the basis for the usual understanding of the start of potato cultivation. "By the mid 1800s, it had become a stable food and a valuable trading item for almost all coastal groups" (Turner 1995: 135). The evidence presented here indicates the potato already had achieved its status by the early part of the nineteenth century, and forces a revision of Suttles's dates for the appearance of the potato on the North Coast. Given his comments that a possible eighteenth-century source was the Russian fur-trade fort, this revision would be in keeping with the theme of his 1951b article.

Scott Lawrence, in a popularized account appearing in the BC Coast Historical Society's magazine, turns the question of the origin of the "Haida potato" into a glamorous mystery, but some of his sparsely referenced comments are instructive. My own review of the Hudson's Bay Company archives confirms his statement that

> Among the first native peoples to trade at this post, Fort Simpson, were the Haida who . . . brought very large amounts of potato, which was a considerable surprise to the British. [Lawrence 1972: 44]

I did not, however, note any surprise expressed in the journal's entries.

After providing a description of the circum-global diffusion of potato cultivation as a result of Spanish and Portugese trade, Lawrence goes on to write that Marius Barbeau suggested the source of the Haida potato was a transpacific voyage, possibly to Japan or China. He cites the

> old-time chief H. F. Clifton, of the Kitka'ata tribe [the Tsimshian of Hartley Bay], [who] recalled stories of his tribal elders about the early days of Haida travel. Once, when a band was out fishing, they were blown far to sea and finally discovered an island where the people ate maggots (rice) and were called Kakyoren. These strange people brought the Haida in and taught them new ways, one of which was the cultivation of the potato, which they brought back with them and cultivated when they returned home." [Lawrence 1972: 44].

There is little evidence to support Scott's confident suggestion, presented complete with a line drawing of the "Haida potato" plant, that the plant is not the same as the species used by Europeans. Wayne Suttles examines and rejects the possibility that one of the native plants that get called "Indian potatoes" may have been mistakenly identified as a "white potato" (1951b [1987: 141]).

Generally, ethnobotanists may doubt Lawrence's theory because there is a likely English etymology for the indigenous words for potato in the languages of the Haida, Tsimshian, and Nuxalk. Linguistic evidence suggests the words may be derived from "the English words 'good seed', which were repeated over and over . . . in an effort to show the natives how to plant it" (Turner 1995: 136).

Before leaving the "surprised" Scottish Officers trading potatoes in Fort Simpson, it is worth noting that the Scots themselves had started cultivating potatoes less than a century earlier, in 1739, although the potato spread rapidly in Scotland because it improved the peasant's diet so greatly (see Mackie 1964: 287). Andrew Nikiforuk (1991: 112–14) points out that the potato, a plant taken from the Americas to Europe, was for a long period viewed with suspicion in Europe, where it was originally associated with leprosy, scrofula, or typhus, or thought to be poisonous or the devil's food. At first, only the Irish took to the potato; the plant did not have any widespread acceptance in Europe until the eighteenth century, finally gaining official sanction when the Medical Faculty of Paris declared the potato safe to eat. The potato's rapid rise in popularity was due to its cheap and easy production as a filler of the empty bellies of the poor peasants and workers of Europe.

If the "Haida potato" does have a European origin then potatoes must have caught on rapidly and spread like wildfire on the North Coast, unlike their history in Europe. The explanation for the potato's acceptance on the coast cannot be the same as for Europe. Unlike the Europeans who were malnourished and exposed to the risk of starvation, the Tsimshian had a variety of other, rich foods available. Nonetheless, Marian Smith suggested in 1949 that the Salish adopted the potato so quickly because of a previously felt starch deficiency (Suttles 1951b [1987: 145]). She does not report why the Salish felt that way. The coastal people had other carbohydrate foods available, although these were generally less productive or spatially concentrated. Nor can we assume the potato's taste was a significant factor. Potatoes, by themselves, are not a delicious food. Good, but rather plain tasting. Wayne Suttles suggests the Russians had a hard time persuading Kaniagmiut to eat introduced, planted vegetables (1951b [1987: 138, fn 3]). Most people prefer to eat potatoes with an accompaniment. The Northwest Coast tradition of dipping potatoes into fish oil is similar to the current North American habit of smothering them in butter, sour cream, or some other rich food. If potatoes do not, on their own, cause diners to salivate, why would a basically bland, starchy food become so popular so fast? Suttles (1951b [1987; 145]) suggests potatoes were accepted readily because they had a cash value. On the North Coast, this explanation may not account for the presence of potatoes when Fort Simpson opened its trade room. At that time, the maritime fur trade had long been in doldrums, and a doubtful stimulant to indigenous potato production. There is evidence, however, that indigenous trading networks could have provided incentives, in which case the academic problem of ignoring the role of potato gardens in aboriginal society surfaces again.

What other explanations could there be for why the potato was adopted so quickly on the Northwest Coast? Possibly because the people identified it as a dietary alternative to an indigenous plant, similarly starchy and bland but already incorporated into culinary traditions and recipes. Suttles (1951b [1987]) considered several native plants that have been called Indian potato, but the plants he mentions are not in the Skeena area.

Perhaps the potato was accepted quickly because it could be grown according to existing plant-management techniques. Tobacco may have provided the model, and this will be discussed in the next section. A relevant citation is John Darling, who attributed Tsimshian gardening practices to the traders. The relevant reference concerns edible roots:

During the early years of contact a new use was made of land. Potatoes began
to be cultivated in small patches by the household groups. . . . Although agricultural land was scarce around the fort, the various households made use
of productive soil at sites previously used only for gathering edible roots.
Since these sites were already claimed, the new practice of cultivation did
not disturb traditional rights inland. [Darling 1955: 26]

To this can be added Nancy Turner's comments that the Haida and other groups
grew and harvested springbank clover in a way that "had characteristics akin
to agriculture":

They divided extensive patches of Clover growing along river flats into rectangular beds that were owned by families or individuals in a village group
and passed from generation to generation . . . they carefully removed larger
rocks and sticks and only dug up the largest rhizomes, leaving the smaller
ones to grow. [1995: 94]

Similarly, patches of cinquefoil were owned and managed by Haida chiefs.
Both springbank clover and silverweed were generally used on the coast.

As for technology, the digging sticks women used to gather other root foods
were equally useful for planting and harvesting potatoes.

The Northwest Coast also had certain social prerequisites for the acceptance of potatoes. Wayne Suttles specifies a root-gathering tradition, a sedentary life, concepts of ownership, and techniques of plant management (1951b
[1987: 146–47]). Douglas Deur, elsewhere in this volume, comes to similar conclusions, and John Darling's comment above about potatoes replacing other
root crops all support the conclusion. Thus, potatoes should be seen to have
intensified the Tsimshian use of the edible plant environment without acquiring the role of a dietary staple as they did in Europe, where the social system
limited or precluded other sources of nutrition.

Tobacco

There is evidence that tobacco was cultivated and used as a narcotic by the
Haida and their northern trade partners before the merchant traders appeared,
although the native plant is thought to now be extinct in northern British
Columbia and Alaska (see Turner and Taylor 1972).

Captain Vancouver provided the following description of the growing of
tobacco on the Queen Charlotte Islands in July 1794:

On each side of the entrance some new habitations were constructing, and
for the first time during our intercourse with the North West American
Indians, in the vicinity of these habitations were found some square patches
of ground in a state of cultivation, producing a plant that appeared to be a

species of tobacco; and which, we understood, is by no means uncommon amongst the inhabitants of Queen Charlotte's islands, who cultivate much of this plant. [Vancouver 1984: 1359]

The village, however, showed significant signs of contact with Europeans. Vancouver commented on the European dress of the messenger from the Haida chief. "He sent a young man dressed in a scarlet coat and blue trowsers to invite our party on shore" (Vancouver 1984: 1359). Archibald Menzies, Vancouver's surgeon and botanist, observed there was trade in European clothing near Cape Edgecombe (Vancouver 1984: 1360 fn 2), which is another gardening locale Menzies mentions in his own comment on the gardens Vancouver was describing:

> Our curiosity [at the sight of the gardens] was much excited & it was natural to regret our not being able to have a nearer view of them. I happened to be the only one of the party who had formerly seen similar gardens near Cape Edgecombe about eighteen leagues in a south West Direction from this place; others we have been informed have seen them about the Queen Charlotte's Isles & northward of Cross Sound under mount Fairweather on the exterior edge of the Coast. . . . Here we see the first dawns of Agriculture excited among these savages, not in rearing any article of real utility to their comfort or support, as might readily be expected, but in cultivating a mere drug to satisfy the cravings of a fanciful appetite that can be no ways necessary to their existence. [Vancouver 1984: 1359 fn 2]

Although tobacco is often said to be limited to the Haida and Tlingit, Franz Boas (1916) thought the Tsimshian cultivated it in the residential gardens along the Skeena.

> On Queen Charlotte Islands, and perhaps also among the Tsimshian, tobacco was raised in olden times gardens cleared near the villages. The tobacco was not smoked but chewed mixed with calcined shells. [1916: 52]

The case for Tsimshian cultivation of tobacco is weak. Nancy Turner and Roy Taylor assert the Tsimshian did not grow it themselves but do not provide an authority for this statement (Turner and Taylor 1972: 251). Viola Garfield simply stated that "tobacco was obtained from the Hudson's Bay Company in return for crabs (crabapples), wild celery,[15] and other food articles" for trade with the Nisga'a (Garfield 1939: 199). This does not illuminate the question, because she was describing twentieth-century preparations for potlatch and not trying to establish the antiquity of tobacco use on the coast.

The use and cultivation of native tobacco (*Nicotiana quadrivalvis* variety) had been discontinued on the Queen Charlottes by the 1880s, replaced by commercial tobacco (Turner and Taylor 1972: 250). This date may provide the explanation as to why Tsimshian tobacco gardens are not described in the ethnographic literature: The Tsimshian had converted to commercial tobacco and had abandoned their tobacco gardens before the first ethnographers arrived.

Garfield did, however, specify a ceremonial use for tobacco:

Tobacco was in demand among the Nisqa [sic] and Tlingit because of its use in death ceremonies, but was not so used by the Tsimshian. [Garfield 1939: 199]

Today, at Kitsumkalum, sacrificial burnings for the dead may include tobacco, either in the form of pouch tobacco or cigarettes. These are offered to satisfy the need of the dead for tobacco and/or to show respect. Of course modern, commercially available tobacco is now used, but the practice itself is not a European one.

Whether or not they cultivated tobacco themselves, or traded with neighbors who did, a clear case can be made for the Aboriginal use of tobacco by the Tsimshian. Taken together, the following references are suggestive of the precontact antiquity of tobacco use and cultivation. Boas records tobacco used, with other things, for sacrifices made to obtain success. The offerings are burned. His sets of offerings include:

Food, fat, tobacco, bird's-down, and red ocher are sacrificed for success; . . . offerings consist of fat, eagle down, red ocher, tobacco, food, blue paint, and lime of burnt clamshells; . . . Coppers, fat of mountain goat, tobacco, fish oil, crabapples, cranberries, red ocher, and eagle down are mentioned as presents to a supernatural being; . . . and tobacco, fat, and other good things are thrown into the water as presents to a supernatural being; . . . the slave of the Killer Whales is given tobacco; . . . and the same personage . . . the slave of the Stars, is given tobacco, red paint, and sling stones. [Boas 1916: 451–52]

. . . other kinds of property used for sacrifices. . . . These are particularly tobacco, red paint, and sling stones . . . ; tobacco, tallow, and coppers . . . , which are taken along on canoe journeys. . . . In order to propitiate a killer whale, tobacco, red paint, and sling stones are thrown backward from the canoe. . . . [Boas 1916: 436]

In the story of "The Water Being Who Married the Princess" (Boas 1916: 274–75), the Kitkatla people bring tobacco to a feast and the Kitkaata people bring burnt clamshells.

In the Story of the Town of Chief Peace (Boas 1916, story #28), a boy

decided to leave [his father's] house while the people were asleep. He arose from his bed, took mountain-goat fat and some tobacco to chew, and some small coppers. Then, before going out, he went to one of his father's slaves, and said to him that he was leaving his father's house because he was angry.

Then the boy encountered Mouse Woman in the town of Chief Peace. This chief had a beautiful daughter and Mouse Woman instructed the boy:

> "Have you a little fat, tobacco, or a small piece of copper?" The prince said, "Yes I have fat, and tobacco, and copper." Then the Mouse Woman said, "Ask the chief's attendants to spread a mat in front of the chief and the chieftainess and the three uncles of his daughter; then throw the fat on the mat, and also the tobacco. Then the small amount of fat will enlarge on the mats." . . . The attendants did as they were told. . . . He also throw tobacco on the other mat, and the tobacco became a great pile . . . he threw a piece of copper in front of the chief, and it became a large costly copper. [Boas 1916: 208–9]

These actions created great gifts of wealth for the boy and won him the daughter of Chief Peace as his wife. Significantly, tobacco and fat are associated here with the great symbol of wealth, copper.

Tobacco was also used orally, although it was not smoked in precontact times but chewed in a mixture with lime (from burnt abalone or clam shells) to extract the alkaloids and prevent the tobacco from burning the mouth. When European smoking tobacco became popular, the aboriginal practice died out (in Turner and Taylor 1972: 251). John Fowler told Harlan Smith in 1925 that Kinnikinnick

> leaves were mixed with tobacco and smoked by some old [Gitxsan] men. . . . Luke Fowler said the leaves of this plant [Kinnikinnick] were not smoked in the early days and that the Hudson Bay introduced this idea to the Gitksan. . . . John Fowler on August 9th said the . . . leaves were smoked. [Smith 1997: 87]

Nancy Turner states that after commercial tobacco was introduced, "the practice of mixing dried kinnikinnick leaves with it as a flavoring and to make it last longer was soon learned even in groups previously unfamiliar with kinnikinnick as a smoking substance" (Turner 1995: 137). The Tlingit sometimes substituted the inner bark of pine for lime (Dixon, in Turner and Taylor 1972: 251), which may be significant for the Gunhut group of the Kitsumkalum laxskiik phratry descended from Tlingits who migrated from Alaska around 1740 (Boas 1916: 486, 270).

There is ample evidence for Tsimshian use of tobacco, but this does not prove the Tsimshian cultivated it, and there is no information on how the Tsimshian might have grown the plant. However, given the environmental similarity between the Tsimshian and their tobacco-growing neighbors, Boas's opinion that the Tsimshian cultivated the plant seems reasonable. Lacking a direct description of Tsimshian gardening methods, the practices of their Haida neighbors provide some insight to broaden my speculations:

the seeds were planted at the end of April at the same time as potatoes. Each pod was placed in a small mound of earth, as in planting potatoes. The gardens were deeply cultivated and kept clear of weeds. The plants were harvested and tied in bundles at the beginning of September, and rotten wood was mixed into the soil to enrich it for the next year, a common practice in tobacco cultivation throughout North America. [Setchell, in Turner and Taylor 1972: 250–51]

Again, here is the association, made by both Boas and Setchell, connecting tobacco gardens and potatoes. Nancy Turner speculated that the cultivation of the European potato caught on quickly because the potato and tobacco require similar management (personal communication, Nancy Turner and Dana Lepofsky 1982). Interestingly, tobacco and potato are both species in the potato or nightshade family, Solanaceae.

Celery is also associated with tobacco and potatoes. Viola Garfield associated tobacco, celery [cow-parsnip] and other food items as trade items (1939: 199); and the explorer J. Hoskin associated tobacco with wild celery on the Queen Charlotte Islands (Turner and Taylor 1972: 250).

Nineteenth-Century Gardens

This section turns from the pre- and early-contact period to the gardens tended by the Tsimshian of the nineteenth century. These gardens, it must be remembered, were and remain more important to Tsimshian society than is usually accepted in the literature. The people of Kitsumkalum, like all Tsimshians, were gardeners when they first appeared in the written historical record. Subsequently, their indigenous approaches to gardening were subjected to outside prejudices belittling their plant-management knowledge, and to various efforts to teach them how to garden in European ways. This assault began immediately during the colonial period.

In the Hudson's Bay Company Archives, there are numerous references to the company's efforts to introduce European farming practices, familiar to them and to their culture, into the region for their own purposes. The experimental plots around Fort Simpson were tended by hired Tsimshian and frequently raided by irate, mischievous, hungry, or curious Tsimshian, both adults and youths. The experience gained from both of these activities (farm laborers and garden raiders) provided the Tsimshian with firsthand knowledge of the contemporary European agricultural practices.

After Confederation, the Government of Canada continued efforts to induce European-style farming, with Indian policies and programs that saw fruit trees, vegetables, and other crops planted on reserves. Residential schools also had agricultural training, forcing students to work in fields, growing crops for their own food and for sale. In 1916, when the authors of the McKenna-McBride Royal Commission wrote their report on Indian Reserve allocations in British Columbia, they described the economic potential of the Kitsumkalum reserves in a way that nicely sums up the governmental perspective underlying the government's assimilationist policies. Indian Reserve 1 was portrayed as a "potential farming area little developed and partially timbered, containing tribal village and graveyard." Indian Reserve 2 was a "good timbered area virtually unused—fishing station and old village site." Indian Reserve 3 was a "potential farming area and fishing station—timbered. Old Village." Port Essington was a "village site and fishing base." Killutsal I.R. 1 and I.R. 1a—the Lakelse Reserves—had "good land, timbered and chiefly used by Indians as a berry patch" (Canada and British Columbia 1916: 552).[16] All except Indian Reserve 2 were being used for gardens and produced good crops. All produced fish. All were expected to provide merchantable timber, which they eventually did (Canada and British Columbia 1916: 557). By Euro-Canadian standards, the reserve lands were viewed in terms of their horticultural potential to support the rural, indigenous people.

The appearance of settlers at the end of the nineteenth century and the development of an agricultural region in the Terrace area provided the nearby Tsimshian with further opportunity to adopt European-style produce and methods of cultivation.[17] Miriam Temple (interview August 8, 1980) remembered the training and encouragement the older people received from Henry Frank, one of their new farming neighbors, as well as the small orchards and gardens that they set up with the benefit of direct interaction with the settlers. Community interest in gardens was strong, even causing Kitsumkalum people to participate in local agricultural exhibitions. The Ganhada chief Charles Nelson won first prize for his turnips at a Prince Rupert agricultural fair during the Great Depression. During the 1930s some had started small-scale farming at Copper River, upriver from Kitsumkalum. Given the indigenous Tsimshian system of land tenure, plant tending, and related activities, Kitsumkalum people would have been, in many ways, predisposed to incorporate European agricultural practices they encountered.

Gardens were kept at Kitsumkalum reserves and campsites. For example, Benjamin Bennett had a house and garden at Kitsumkaylum[18] Reserve, and a camp at Zimagord Reserve from which he hunted and trapped. Another leader, Mark Bolton, had a garden at Kitsumkaylum Reserve that his family planted every spring. His grandmother gave the garden land to him (E. Spalding, interview 1979). I have references to other gardens at Port Essington, Lakelse, the abandoned town of Dalk ka gilequoex on I.R. 2, the so-called and abandoned Rosswood village that was situated somewhere at the north end of Kitsumkalum Lake, Old Kitselas, and the camp sites at Kwinitsa, Feak Creek, and Salvus. There are also gardens in the backyards of urban Kitsumkalum people living in various towns and cities. Although I do not have specific references to gardens, I assume there were also nineteenth-century gardens at the old village of Gitxandakhl[19] located south of Kitsumkalum Lake.

These historic gardens were economically and nutritionally important to the community, producing lettuce, potatoes, turnips, cabbages, carrots, onions, radishes, and several types of berries for home consumption and trade.

The gardens were managed in conjunction with crabapple trees, plum trees, nut trees, and small livestock such as chickens, geese, ducks, turkeys, pigs, or a solitary cow.

An indication of the importance of introduced agriculture comes from the traumatic experience of railway construction that destroyed the gardens at Kitsumkalum. The year railroad construction began at Copper River (1908), at least nine family groups were living on the reserve at Kitsumkalum. Some of these residents had houses and other buildings, while the archival papers indicate others only had gardens. Stephen Woods, for example, had a log house on his cleared lot, a work shed, a smokehouse, and a doghouse. His family tended a small garden, probably growing crops like potatoes, carrots, turnips, and cabbages. This was destroyed, like the other gardens, by construction work that covered the gardens with rubble that left them useless.

> Really good ground, you know, [for gardens]. And CN,[20] when they dyna-
> mite the rocks on both sides, you know, they push the rocks right on my
> garden. It's just like cement, you know. And it's a big part, you know. My
> Dad used to plant all over. Good ground there, but I can't work it. [E. Spald-
> ing, interview, June 20, 1979]

Seven decades later, the community still remembered what happened and regretted the loss of the family plots (McDonald 1990: 22). Their attempts to reestablish their gardens at Kitsumkalum after the railroad was completed suffered a second disaster during the severe flooding of 1936 that washed away all their work and as well as part of the reserve land.

European-style gardening peaked in regional economic importance during the period before the Second World War. Deteriorating environmental and market conditions after the war caused small-scale farming in the area to decline. Since the decline of camp life, gardening for Kitsumkalum people has been restricted to backyard plots in residential lots on reserve land and in urban areas.

Technology and the Transformation of Tsimshian Horticulture

The previous section discussed information on Tsimshian plant resources and the ways the Tsimshian utilized these resources. In the process of recording that information, I also gathered details on the Aboriginal tools and on the knowledge the Tsimshian used for plant resource management. While less extensive than the information on plant use, some of this material on Tsimshian technology has not previously been published.

GATHERING TECHNOLOGY

Gathering technology is, in certain aspects, quite simple: the basic tools usually described for gathering plants are the resource sites, hands, carrying con-

tainers, and elementary tools such as a digging stick if roots or clams are being sought. Other aspects of the technology are more complex and sophisticated, specifically their knowledge of the resources, resource locations, seasonal availability, management, and processing. These can be stylized as the most important component of Tsimshian gathering technology.

Much of this knowledge was the monopoly of women, and remains so today, although colonization deeply and adversely affected the transmission of that knowledge and the concomitant experience of working with that knowledge.

COLONIAL INSTITUTIONS AND PLANT MANAGEMENT KNOWLEDGE

Colonization seriously disrupted the technological knowledge of plant management and harvesting. Colonizing governments, industries, and settlers all assumed ownership and control over resources, preventing effective management by the indigenous title-holders and Tsimshian communities. Wage labor patterns and compulsory education in Canadian schools, not only in the residential schools, alienated a large portion of the younger generation from significant blocks of cultural learning and experience. Since then, the attractions of urban centers for the young who are tied to wages, television, and downtown entertainment provide appealing substitutes to the work of the mysterious bush, a place of perceived and real hardship that is not considered necessary so long as there are supermarkets nearby with commercially imported foods. In the 1980s there were some individuals trying to salvage the ancient knowledge, but it seemed to be disappearing as quickly as the forest cover itself.

Aboriginal people are changing this situation. Today, there is a renewed commitment to the land and to the restoration of plant-management techniques, for many reasons: the politics of land claims and Treaty negotiations, the benefits of economic alternatives to supermarkets and malls, and the social and cultural recovery that has captivated the consciousness of so many Aboriginal people across North America and Canada in particular.

TENDING

Jack Darling stated the Tsimshian gardens were cultivated by household groups at the camps (1955: 26). Another pattern during the early part of the twentieth century involved the adult women doing the principal work while the men were away fishing. The women were assisted by young children living at home and by adult children when they came home to visit or to live between periods of wage labor. A different pattern was for the men to plant the garden and then leave it to go fishing. If there was no family living nearby, which was often the case when the women worked in the canneries, the plants grew without tending or with occasional weeding if the women took time off ("laid off") for a long enough period that allowed them to return by train.

They do the garden like other people, when they got time. . . . She come back. She left, she lay off for a penny a week. Come back and hoe the potato patch. . . . And turnips, hoe them around. A week and she left us down Port Essington. And she came back and my sister lived here, beside the bridge. She gave her a hand. . . . She go back. She use train. . . . [Miriam Temple, interview, September 29, 1980]

These crops were harvested in the fall after fishing closed.

FERTILIZERS

In the 1970s and 1980s people told me a little about how the Tsimshian took care of gardens. Winnie Wesley said the old people knew how to take care of plants and not ruin them (Winnie Wesley, interview, June 19, 1979). One woman told me the old people dug holes for fish guts near berry bushes "to feed the plant" but not too close (Winnie Wesley 1979?). Another woman said blood would be spilled over the plants as a fertilizer. Sometimes slices of seal meat were placed in the dirt for a good garden. "Moss from salt water" (seaweed) was used as well. It was suggested that this might have been done for berry bushes as well as the gardens. All these practices correspond to and therefore seem imitative of the gardening of the Hudson's Bay Company. Wayne Suttles suggests a Russian origin for the practice of using seaweed (1951b [1987: 138, fn 3]). Nancy Turner at one time assigned the use of "sea-weeds such as boa kelp [*Egregia menziesii*]" to fertilize potatoes and gardens to the period "following contact with Europeans" (1998: 46). Yet, if the Tsimshian knew how to maintain the soil for tobacco plants by mixing in rotting wood chips, they would have been predisposed, at least, to accept seaweed as a way to maintain the ground. At present, it would be futile to attempt to evaluate the antiquity of the patterns.

CONSUMPTION AND TRADE

Plant materials were consumed domestically and also were part of the Aboriginal trade and the commercial trade. Berry cakes, tobacco, seaweed, possibly potatoes, and many other food items were traded among the various nations in the north (Turner and Loewen 1998). The arrival of European traders added more nations to this trade. The Europeans also opened a market for timber, fish, and meat, which rapidly evolved into the industrialized markets of forestry and commercial fishing.

More recently (twentieth century), domestic garden produce generally was consumed by the family, although in some cases surplus was sold or traded. One Kitsumkalum resident who operated a small store on the reserve in the 1930s sold some of the vegetables to local Aboriginal people and settlers. The types of garden crops grown throughout the twentieth century suggests their dual potential for both subsistence gardening and market gardening: potatoes, turnips, carrots, cabbage, lettuce, onions, radish, nut trees, crabapples, prune plum trees, berry bushes, chickens, turkeys, geese, ducks, pigs, cows.

Ownership and Colonization

I have mentioned the theory that Tsimshian social structure and, in particular, property relationships, made it easy for them to accept European gardening methods and crops. Among the Kitsumkalum people, the ownership and use of garden sites rested with the family. A garden "belonged" to the head of the unit, either male or female. Sometimes people remembered that the location was part of the family's (lineage's) original *laxyuup*. This is all consistent with Darling's statement that the gardens were on land already claimed by householders and that "the new practice of cultivation did not disturb traditional rights in land" (Darling 1955: 27).

The imposition of Canadian property laws and the reserve system usurped legal control over these garden areas. Ownership was universally assumed by Canada but the actual impact this had on use rights was mixed. For the Kitsumkalum people, their reserves protected their main gardening sites and several, but not all, other locations. Camp sites, such as at Feak Creek, isolated from the colonial interests, continued to be used as long as the economy of the camp was viable and people continued to live on the trap line. On the other hand, the site at the mouth of the Nelson River was in a commercial forestry area and not allowed as a Reserve. It was eventually logged by a commercial company and abandoned by its Aboriginal residents, and has now become part of the British Columbia Forest Service Red Sand Lake Demonstration Forest site providing an interpretative wilderness experience. Many Aboriginal communities experienced interference with their off-reserve gardens when conflicting land-use patterns (e.g., settlement, logging) took precedence over the gardens. Even on the protected reserve lands there were problems, for example with rights-of-way that cut through the Reserves, destroying cultivated areas (see McDonald 1990).

The impact of colonization on Tsimshian use of their resources provides a context necessary to interpret what we know about indigenous plant management and plant cultivation. In particular, the establishment of the three reserves for Kitsumkalum in 1891 serves as a benchmark for the Canadian usurpation of Tsimshian ownership of resource sites. The reserve system must be viewed as only one colonizing tool used to appropriate Tsimshian resources (e.g., see McDonald 1985, 1987, 1994) but it is highly symbolic of the enormous damage caused by the loss of control over resources and of the colonial and racist caricature of Aboriginal societies.

In allotting the reserves, the Indian Reserve Commissioner summarized his interpretation of Tsimshian economic relationships:

> You will not be confined to your reserves. You can go on the mountain to hunt and gather berries as you have always done. Some think it would be

a hardship that the hunting grounds should not be defined, but the govt. does not see how that could be done, for an Indian goes where he will to hunt, or gather berries. No survey could be made of them. [Canada, n.d., Public Archives of Canada RG 10 v.1022; handscript of minutes]

This seems a simple statement, but it is encapsulates the colonial mentality and presents much misinformation about the Aboriginal situation. Contrary to the Commissioner's opinion, the Tsimshians gathered much more than berries and did so in areas other than on the mountainsides. Describing their activities simply as "gathering" trivialized and dismissed their plant-management strategies. Further, the Commissioner was grossly ignorant about the Tsimshian, for they could not go wherever they wished but were required to observe strict laws governing resource property. These laws defined property arrangements associated with the Aboriginal Houses and towns (McDonald 2003). Consequently, Tsimshian people easily could have surveyed their properties and, if they had been allowed to do so, a century of land disputes might have been avoided.

The details of the original legal relationships are part of the current Treaty negotiations but are not well documented in the published literature. John Darling's review of the ethnography found that lineages owned patches of edible roots, cedar stands, certain generalized territories, and kelp beds (1955: 10–12). Viola Garfield added that lineages owned berry patches (1966: 23) and stated that "there were no unclaimed land or sea food resources of a kind important to the Indian's economy" (1966: 14), although she did not give detailed data. The Sessional Papers on the "Indian troubles" indicated widespread ownership of "fruit-gathering preserves" and of timber lands (cedar stands?) (British Columbia, Sessional Papers 1885: 289; 1887: 260 ff). Margaret Anderson reports that "there were owned berry patches and stands of wild crabapple, etc., and these were cared for" (personal communication, 1997). I have recently summarized information on the family estates (McDonald, in press) and the Treaty Offices are collating the detailed information they require for post-treaty governance and resource management (e.g., McDonald 2002).

Furthermore, such resource properties generally had names. Today, resource properties owned by a House group are called *laxyuup* (McDonald 2003). Linguistically, there are specific terms:

an owned location for berry picking is often "na–x–NAME OF PLANT", where na– is a prefix indicating possession (alienable); x– is a prefix indicating "to experience or partake" (as in *x–piyaan*, "to eat smoke; to take part in smoke" ... to smoke (as in a cigarette). Hence, people speak of a *naxmoolks*-owned crabapple patch. If there is a specific name for the site, then that replaces this generic term of course. [Anderson, personal communication, 1998]

Specific territorial names of the Kitsumkalum *laxyuup* were identified in McDonald 1983 and McDonald 2003.

For the present paper, I have already discussed names associated with berry grounds that were managed by burning techniques (the two territories identified as "burnt shrubs" and "burnt mountain tops"). Another name I have encountered describing plant resource property is *sqaw'a ms,* "where grows devil's club" (McDonald 1983). Territorial names of this type are associated with specific properties owned by House (lineage) groups. We can speculate that the women who actually harvested the resources may have had other names for more specifically identified productive locales within the territories.

These references provide an indication of what was lost with the imposition of the reserve system. To these losses can be added other legislated restrictions on Tsimshian management practices, especially those in the Forestry Act, which made it illegal to remove any products of the forest on Crown Lands (British Columbia, Government 1924). Some old people still remember and complain about how they were prevented from taking bark from the hemlock or cedar trees. One person told me "I used to eat hemlock sap, now get pinched if you make marks on the trees" (Miriam Temple, interview, August 21, 1980). In specific cases, women were prevented from stripping trees as they had been accustomed, and men could no longer cut timber for boxes or canoes without a timber license or lease.

Leslie Gottesfeld documents the suppression of berry-patch burning in the Prince Rupert Forest District from the 1930s:

> Indian-caused fires have decreased during the past two years. As early as possible in the Spring all Indian settlements were visited and our policy explained in plain words. Notices were written out and posted at Indian trading posts which seemed to get results. [Anonymous 1932: 2, quoted in Gottesfeld 1993: 96]

> Anyone suspected of deliberately setting fires was subject to criminal prosecution, and several convictions were obtained. [Gottesfeld 1993: 99]

In 1887 a delegation of chiefs was told by the province's Premier that timber on crown lands was protected by timber lease and that people cutting trees for house construction or storage boxes could be stopped legally by the owners of the lands if they held a lease (British Columbia 1887: 253ff.). The complicated requirements for tenure inhibited Aboriginals from registering timber lands and there also is documentation of prejudice against Aboriginal registry (Pritchard 1977: 115 ff).

Also important to mention is the destruction of the resources by agriculture and settlement, which started immediately after the reserves were established in the 1890s. By 1910 a community existed at Eby's Landing (Terrace), with homesteaders spreading across the Kitsumkalum plateau, establishing farms and settlements according to Provincial, not Tsimshian, laws and tenures. This extensive homesteading and the settlements associated with it, caused considerable losses to berry patches. In 1980, a woman who was nearly a hundred years old tried to explain to me where her family picked berries but was sure that I would not be able to locate the sites because they were over-grown with flowers (i.e., agricultural use) and destroyed by roadways. She said they had been destroyed at the turn of the century.

The result was the redefinition of the Tsimshian seasonal cycle and plant-management strategies as described in McDonald (1987). Turner and Peacock (this volume) note a number of constraints on the scheduling of plant harvests. These include biological factors and cultural preferences, but colonization also became a strong influence.

To summarize that discussion:

> This is a sentiment I heard expressed many times and it should be empha-sized: agriculture, settlement, and more recently forestry, obliterated large sections of the Tsimshian environment. Hunters, trappers, gatherers, all felt like strangers on their own territory because they could no longer find the old familiar natural landmarks, or the ancient resource sites that their lin-eages had groomed over the centuries. [McDonald 1985: 210]

All these changes are part of the colonial assumption of ownership. They are also part of the creation of a "new wilderness" that obliterated the anthro-pogenic mark of the Tsimshian presence, transforming the Tsimshian land-scape into one that would serve the new regime. The stereotype of the great Pacific wilderness is a colonial concept that blinds a new nation (Canada) to the sovereign control of an older nation (Tsimshian) and allows governments in Ottawa and Victoria to disregard the rights of the Tsimshian, as if the lands were truly *terra nullius*, an empty land.

Today, the effect of colonization on berry production is manifested in the way Kitsumkalum people wander around the countryside, foraging along log-ging roads and side roads, seeking out good places to gather berries. Clear-cut areas are considered to be good locations for berries, as are rights-of-way for roads, hydro lines, and other areas subject to regular clearing of the for-est canopy. Gathering plant materials in the Kitsumkalum Valley is now a hit-and-miss affair, conducted without significant or sustained management. The management of berry patches is not practical in a world where clear-cut logging may suddenly eradicate a berry patch without notice and where techniques such as burning may result in legal action. The situation for Kitsumkalum seems similar to the Gitxsan case where Leslie Gottesfeld pictured traditional berry patches that are overgrown and no longer productive (Gottesfeld 1993: 70). Stephen McNeary's reference to soapberry gathering also suggests this change:

> I found in mid-summer that there was a great interest in the location of soapberry patches, which today are exploited on a first-come, first-serve basis. It was said that in former times . . . [a chief] owned a particularly large patch. [1976: 115]

Recent liberalization of provincial Forestry practices and court cases (e.g.,

Delgamuukw v. British Columbia) may alleviate some of the pressure against Aboriginal management techniques but the new regulations do not restore the ownership rights necessary for long-term management strategies and investment. Those rights are still to be negotiated through the Treaty process or reestablished through the courts.

Conclusion

A number of different indigenous plant-management strategies can be gleaned from information on the Tsimshian. Elsewhere in this volume, Turner and Peacock provide a picture of a continuum of increasing human effort and impact on the landscapes. This continuum has three major nodes: foraging activities that represent the least effort and impact, a variety of cultivation activities, and domestication activities that require the greatest effort and cause the greatest impact. The entire range of the practices described by Turner and Peacock can be documented for the Tsimshian. Given my focus on the Kitsumkalum materials, further study of the other Tsimshian communities may expand and deepen our understanding of these practices.

Much of the plant gathering I have done with community members is best described as foraging— collecting materials while visiting a remote site or while gathering berries with a family on a clear-cut area they spotted earlier. In light of the range of contemporary foraging activities, I would add to Turner and Peacock's discussion a differentiation that recognizes both the opportunistic harvesting of plant foods and materials, and a more deliberate seeking of plant materials that lacks any intention of cultivating or otherwise caring for the plants, beyond basic conservation principles. Plant materials (e.g., berry bushes, trees suitable for bark, mushroom grounds, roots exposed in cut banks, good-looking Christmas trees) are spotted while driving along provincial roadways or logging roads or while people are hiking through the woods, perhaps while hunting or surveying the land on Band Council administrative projects. The materials may be taken immediately, or the knowledge of a good harvesting location may be retained for future use.

Foraging is a very significant harvesting method today. A fundamental reason for the contemporary importance of foraging lies in the colonial context of Tsimshian land ownership and management. Contemporary land tenure and Forestry regulations protecting the forest for commercial use reduce the effectiveness and utility of indigenous plant management, leaving foraging the most sensible harvest method, at least from a legal perspective. Even basic conservation principles suffer from the "tragedy of the commons," as people evaluate the potential for someone else to take what is left behind or the potential for a logging company to clear-cut the forest.

Cultivation activities involve the encouragement of plant populations so as to ensure continued or enhanced productivity. The evidence presented here indicates a range of aboriginal cultivation activities, from the low-intensity techniques of tending (but not tilling), to transplanting, burning, and gardening. Fruit and nut trees were transplanted, as were different types of berry bushes. Burning and pruning were techniques used to maintain berry patches. There is evidence of preferential use of particular berry patches, with special care taken of these plants and of the trails leading to them. The special care included pruning, burning, and harvesting according to growth cycles. Weeding, clearing, and fertilizing were also recorded as management practices.

There is also evidence suggesting the Tsimshian should be added to the list of Aboriginal people planting and caring for tobacco, the one crop generally accepted as having been domesticated on the northern coast. The record on the cultivation of the so-called "Haida potato" is less clear. Although the potato trade may have been possible because of the early adoption of the potato from Europeans or Americans, the origin of this crop and the reasons potato gardens were so quickly popular remains enigmatic.

Thus, the overall conclusion is that the Tsimshian have a history of cultivating plants, including gardening, that is rooted in Tsimshian mythology and recorded in early archival materials. The wilderness of the Tsimshian landscape is a wilderness created by colonial forces that obliterated the anthropogenic mark of the Tsimshian presence to establish a new regime serving a new people. Further, whatever the antiquity of the practice, for the past two centuries, at least, the Tsimshian plant-management techniques have included gardens. This is a considerable period of time and predates the earliest ethnographic description of their way of life. Perhaps the most important point I can make is that, whether they independently invented or borrowed the practice of gardening, the Tsimshian have been horticulturalists for as long as Europeans, ethnologists and historians in particular, have known them. Only our stereotypes of the Tsimshian productive economy, society, or culture describe a people without gardens.

Notes

I wish to acknowledge the support and encouragement the Tsimshian people, especially those of Kitsumkalum, have provided for my research. I thank all those who contributed this paper and the two community members who reviewed and commented on the manuscript: Dianne Collins and Alex Bolton of the Kitsumkalum Treaty Office.

1. There is no Plumbago Mountain but plumbago is a graphite sought by nineteenth-century explorers. The reference is probably to Kitselas Mountain, which did produce some gold.

2. I am not employing the otherwise useful distinction Compton makes between his Coastal and Southern Tsimshian sources.

3. Compton (1993a: 460) notes that wild rhubarb (rhubarb is an introduced plant) is the South Tsimshian term for western dock and that the characteristics of both plants are perceived as similar. Nancy Turner notes this is a coastal usage. In the Interior, she says, wild rhubarb refers to cow-parsnip, which is often called wild celery on the coast (personal communication, 1997).

4. Compton (1993a: 454) indicates the Southern Tsimshian use a term for berry that is a labeled category containing several important named utilitarian taxa.

5. The identification of this species is doubtful as its range is not known to be close to Tsimshian territory. See Brayshaw (1966), *Trees and Shrubs of British Columbia* (N. Turner, personal communication, 1997)

6. The late *sm'oogyit* Walter Wright, in his Wars of Medeek, mentioned a different use of such knowledge: the planting of bushes to camouflage armories "so the land looked as if it had not been touched" by the building of a cache of munitions for Kitselas (n.d., Chapter 6).

7. These may be bog cranberries (N. Turner, personal communication, 1997), but throughout this paper I avoid reporting my references as more specific or precise than the original source material.

8. Some of these locations are associated with the residential arrangements made during the Port Essington period and reflect indigenous social ties but not necessarily territorial claims.

9. Allaire (1984: 92) identifies this group as assimilated to the Gilludzar at Lakelse Lake.

10. This may be a territory of the Ganhada phratry, close to the Clear River north of Kitsumkalum Lake.

11. Certainly, contemporary political efforts support, reenforce, and recover the legal system associated with such managerial authority.

12. Among contemporary Tsimshian, the English term "rice" is applied to the rice-root (*Fritillaria camschatcensis*) and to true rice (*Oryza sativa*, a grass). Compton (1993a: 460) notes that rice (an introduced plant) is called by the South Tsimshian term for rice-root and that the characteristics of both plants are comparable. (A similar application of endemic names to related, introduced cultivars is found in the case of indigenous and commercial strawberries, raspberries, and blackberries.) The interchangeable use of the term "rice" for rice-root is due to the white bulblets that form around the bulb of *Fritillaria* and that resemble rice grains. Since true rice is not native to the coast, I assume all references I have to wild rice are actually to the bulb. The terms lily, kamchatka lily, and chocolate lily also are applied to the *Fritillaria camschatcensis* plant, although the chocolate lily, *Fritillaria lanceolata*, is not found in the Tsimshian area.

13. The correlation between concentrated crabapples and settlement has been noted elsewhere on the coast, for example Bella Bella (N. Turner, personal communication, 1998) and Kwakwaka'wakaw and Nuu-chah-nulth territory (D. Deur, personal communication 1998).

14. This and the previous use of the term "nut trees" probably refers to hazelnuts, since the northern environment is not conducive to chestnut trees (N. Turner, personal communication, December 11, 1997).

15. Nancy Turner suggests this refers to cow-parsnip (personal communication, 1997).

16. These Reserves were transferred from Kitsumkalum to Port Simpson in 1959 and subsequently sold.

17. Residential schools also provided training to Tsimshians on the coast. The Terrace area did not have a residential school and, since I have no references to Kitsumkalum children attending any residential school in this time period, I do not consider this factor here.

18. The name of the reserve is spelled with a "y."

19. Barbeau recorded this as a village occupied by a Lagybuu clan related to the Wild Rice [rice-root] clan of the Gitanyow Lagybuu phratry and the Wild Rice Clan of the Witsuwit'en (Barbeau 1929: 156 ff). Oddly, Barbeau locates the village at the headwaters of the Kitsumkalum River, near the Nass River, which is out of the ter-

ritory of the Lagybuu phratry. The confusion may be related to the older custom of naming the portion of the Kitsumkalum River above Kitsumkalum Lake as the Wiigwenks (the Aboriginal name) or the Beaver River (the local English name). In this case, the headwaters of the Kitsumkalum River would be south of the lake, which is Lagybuu territory.

20. "CN" refers to the railway company, which originally was the Grand Trunk Pacific and became the Canadian National. For the story of the impact of the railway see McDonald (1990).

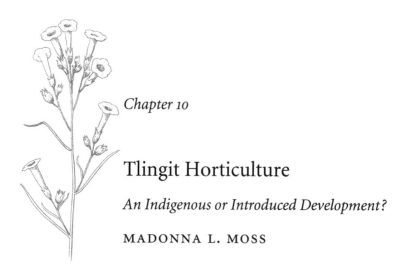

Chapter 10

Tlingit Horticulture

An Indigenous or Introduced Development?

MADONNA L. MOSS

Due to its geographic position at the extreme northernmost end of the Northwest Coast, Tlingit territory has the lowest diversity of terrestrial plant resources of any subregion of the culture area. While ethnographers have listed culturally important plants and recorded some gathering methods, the economic and dietary importance of plants in Tlingit subsistence has been neglected. As far as I know, there has been no previous focused effort to systematically examine the evidence of cultivation of indigenous plants among the Tlingit. Similarly, Tlingit cultivation of potatoes and other introduced plants has not been seriously investigated. In this paper, I examine the evidence for pre-contact and post-contact horticulture, with reference to developments occurring elsewhere on the Northwest Coast. I will employ ethnographic and ethnohistorical data to document the intensification of gardening for production of the potato and other introduced crops in the nineteenth century. I will also evaluate the archaeological record of gardens in an effort to assess the importance of horticulture and its effects on land use and settlement patterns during the nineteenth and twentieth centuries. Because the Tlingit had access to fewer species of culturally significant plants than any other Northwest Coast group, this analysis will assess whether the Tlingit relied less on plants than other Northwest Coast groups, or intensified their use of available plants in response to fewer opportunities.

Environmental Background

Extending from Icy Cape, Alaska, in the north to the international boundary between the United States and Canada in the south, Tlingit territory is most frequently referred to as "Southeast Alaska." It encompasses the islands of the Alexander Archipelago and the adjacent narrow strip of coastal mainland. The northern boundary marks the northern limit of the ethnographic territory of the Yakutat Tlingit in addition to an important physiographic discontinuity. The coastline northwest of this point to Cape Suckling is straight and highly exposed, unlike the highly convoluted shorelines to the southeast. Dixon Entrance and the Canadian boundary provide a convenient southern border for the area to be discussed here. Distinctive named groups (*kwaans*) of Tlingits occupied this region at European contact, and Tlingit people make up the largest component of the region's contemporary Native American population. At some time prior to European contact, the Kaigani Haida settled the southern half of Prince of Wales Island, Dall, Sukkwan and Long islands, displacing Tlingit groups (Blackman 1990: 240; Langdon 1979). Annette Island is home to the descendants of Christianized Tsimshian Indians who moved from the Prince Rupert region of British Columbia in 1887 to found "New Metlakatla" (Dunn and Booth 1990). For the purposes of this paper, however, I will draw upon Tlingit archaeological, ethnographic, and ethnohistorical sources (which are strongest for the central and northern areas of Tlingit territory) to address questions related to horticulture in aboriginal Southeast Alaska.

Due to its northernmost position, Tlingit territory has a lower terrestrial biodiversity than the Northwest Coast subregions to the south. Latitudes in Southeast Alaska range from 54° 30' N to 60° 8' N, limiting the range and, in some cases, the abundance of a number of economically important plants found to the south. Most notable among these is western red-cedar (*Thuja plicata*), which is restricted to areas south of Frederick Sound at an approximate latitude of 57° N. Also absent are plants associated with the comparatively dry Douglas-fir (*Pseudotsuga menziesii*) forests, a vegetation zone entirely absent from Southeast Alaska. Plants with edible bulbs or roots used in the south that do not occur in Southeast Alaska include camas (*Camassia* spp.), two species of onions (*Allium cernuum, A. acuminatum*), two fawn lilies (*Erythronium revolutum, E. grandiflorum*), harvest lily (*Brodiaea coronaria*), a chocolate lily (*Fritillaria lanceolata*), wild hyacinth (*Brodiaea hyacinthina*), Tolmie's mariposa (*Calochortus tolmiei*), seashore lupine (*Lupinus littoralis*), wild caraway (*Perideridia gairdneri*), and spring gold (*Lomatium utriculatum*) (Pojar and MacKinnon 1994; Suttles 1990). In addition, the economically important tuber, wapato (*Sagittaria latifolia*) is restricted to southern British Columbia and areas south. Edible roots used by Tlingit peoples that have been documented in the anthropological literature are listed in Table 10.1. While Southeast Alaska does have its share of berries and fruits, the following Northwest Coast species are absent or extremely rare: common gooseberry (*Ribes divaricatum*), sticky gooseberry (*R. lobbii*), red-flowering currant (*R. sanguineum*), trailing blackberry (*Rubus ursinus*), blackcap (*R. leucodermis*), woodland strawberry (*Fragaria vesca*), two huckleberry species (*Vaccinium membranaceum, V. ovatum*), two species of wild rose (*Rosa gymnocarpa, R. pisocarpa*), Indian plum (*Oemleria cerasiformis*), and Solomon's seal (*Smilacina racemosa, S. stellata*) (Pojar and MacKinnon 1994; Suttles 1990). Nut-bearing trees and shrubs absent from Southeast Alaska include hazelnut (*Corylus cornuta*) and Garry oak (*Quercus garryana*). Clearly, Southeast Alaska contains a more restricted range of options for indigenous plant use and management than was found on the south and central portions of the Northwest Coast.

Tlingit Use of Indigenous Plants

The main plants used for food and medicine, which do occur in Southeast Alaska and are documented in Tlingit ethnography, are listed alphabetically by species in Table 10.1. As readers of the other chapters in this volume will note, nearly all of these plants were used by other Northwest Coast peoples. However, while several other contributors to this volume (see chapters by Deur, M[c]Donald, and Turner and Peacock, this volume) have found substantial evidence of horticultural practices involving many of these plants, I have encountered little comparable evidence in my study of Tlingit materials. Let me first qualify this statement by defining my usage of the term "horticulture." Following Ford (1985a,b), Turner and Peacock (this volume) employ a model of people–plant interactions as a continuum between foraging and food production, the latter of which can be typed as cultivation and domestication. The available ethnographic evidence suggests that Tlingit use of indigenous plants is best categorized as "foraging," with selective harvesting as a common practice. There is almost no documentation of practices I define as horticulture directed at indigenous plants: no intentional cultivation, clearing, tilling, tending, weeding, pruning, transplanting, sowing, burning, or fertilizing, save a limited number of practices used to maintain the productivity of berry patches (Thornton 1999). Since the horticulture practices have been described for other Northwest Coast groups, I am surprised by their apparent absence in the Tlingit record. However, Turner and Taylor (1972) have documented strong evidence for pre-European cultivation of tobacco by the Tlingit, including seeding, weeding, and fertilization of tobacco plots; similarly, Thornton (1999) has encountered evidence of berry plot enhancement. As Suttles (1951b) suggested for the Coast Salish, I will argue that Tlingit horticultural expertise with tobacco set the stage for Tlingit success in raising potatoes and other root crops after their eighteenth-century introduction into the region.

Before discussing Tlingit horticulture of tobacco and introduced root crops, however, let us more carefully evaluate Tlingit use of indigenous plants. Table 10.1 presents the main uses of Tlingit food and medicinal plants, as documented primarily in de Laguna (1972), Newton and Moss (1984), and Emmons (1991).[1] There is little ethnographic evidence that any of these plants underwent horticultural treatment by the Tlingit, with the one exception to be discussed below. The 20 plants marked with an asterisk were subject to horticultural practices by other Northwest Coast groups as documented by Turner and Peacock (this volume). A detailed synopsis of plant uses and management by individual Northwest Coast groups can be found in their chapter. The purpose of Table 10.1 is simply to identify Tlingit food and medicinal plants and to illustrate the fact that many of these plants had horticultural potential.[2]

(text continues on page 282)

TABLE 10.1. KEY INDIGENOUS PLANTS USED BY THE TLINGIT
(Major Sources: de Laguna 1972; Emmons 1991; Newton and Moss 1984)

Scientific name	Common name	Part used as food indicated; medicine specified	Sources
Achillea millefolium	yarrow	stem, leaves as medicine compress	de Laguna; Emmons
Adiantum pedatum	maidenhair fern	medicine	Emmons
Amelanchier florida (A. alnifolia)	serviceberry, Saskatoon	berries	Newton and Moss
Arabis hirsuta	hairy rock cress	greens? leaves as medicine	de Laguna; Emmons
Arabis lyrata	Kamchatka rock cress	greens?	de Laguna
Arctostaphylos uva-ursi	kinnikinnick	berries	Emmons; Newton and Moss
Arnica cordifolia	heart-leaved arnica	medicine	de Laguna; Emmons
Aruncus dioicus	goatsbeard	root as medicine	de Laguna
Athyrium filix-femina	lady fern	roots	de Laguna; Emmons
Chamaecyparis nootkatensis	yellow-cedar	bark: medicine	Emmons
Claytonia sibirica	Siberian spring beauty	food?, "money dope"	de Laguna
*Conioselinum pacificum	hemlock parsley Indian carrot	roots	Newton and Moss[1]
Coptis trifolia	goldthread	roots as tonic	de Laguna
*Dryopteris expansa	spiny wood fern	roots	de Laguna

TABLE 10.1. (continued)

Scientific name	Common name	Part used as food indicated; medicine specified	Sources
Dryopteris dilatata	wood fern	leaves for medicine	Emmons
*Epilobium angustifolium	fireweed	leaves, stems, medicine tea	de Laguna, Newton and Moss
*Fragaria chiloensis	strawberry	berries	de Laguna; Emmons; Newton and Moss
*Fritillaria camschatcensis	northern rice-root	rhizomes	de Laguna; Emmons; Newton and Moss
*Gaultheria shallon	salal	berries	No published reference for Tlingit use
Gentiana douglasiana	land otter medicine	medicine	Emmons
Gentiana platypetala	mountain gentian	medicine	Emmons
Geum macrophyllum	large-leaved avens	leaves as medicine	Emmons
Gymnocarpium dryopteris	oak fern	leaf bud for medicine	Emmons
Hedysarum alpinum	Indian potatoes, bear root	roots	Emmons
*Heracleum lanatum	cow-parsnip, wild celery	peeled stem, root: poultice	de Laguna; Emmons; Newton and Moss
Iris setosa	Arctic iris	rootstalk: medicinal charm	de Laguna
Lathyrus spp.	wild pea	peas, roots	de Laguna
Ledum groenlandicum	Hudson's Bay tea	leaves as tea, medicine	de Laguna; Emmons
Ledum palustre	Hudson's Bay tea	leaves as tea, medicine	de Laguna; Emmons
Loiseleuria procumbens	alpine-azalea	leaves as medicine	Emmons
*Lupinus nootkatensis	Nootka lupine	roots eaten, intoxicant	de Laguna; Emmons

Scientific name	Common name	Use	Reference
Lysichiton americanum	skunk cabbage	root as medicine for cuts, taken internally	de Laguna; Emmons
Maianthemum dilatum	false lily-of-the-valley	leaves as poultice	de Laguna
Malus fusca	crabapple	fruit	Newton and Moss
Menyanthes trifoliata	buckbean	roots: medicine	de Laguna
Nymphaea tetragona	yellow pond lily	roots: medicine	de Laguna
Oplopanax horridus	devil's club	bark and stem as medicine internally, and poultice	de Laguna; Emmons
			Newton and Moss
Osmorhiza chilensis	mountain sweet-cicely	medicine	Emmons
Palmaria palmata	ribbon seaweed, dulce	fronds	Newton and Moss
Peltigera aphthosa	freckle pelt	medicine	Emmons
Picea sitchensis	Sitka spruce	pitch, inner bark, branch tips, medicine	Emmons; Newton and Moss
			de Laguna
Plantago maritima	goose-tongue	leaves	No published reference for Tlingit use[2]
Polygonum alaskanum	wild rhubarb	leaves	Newton and Moss
Polypodium glycyrrhiza	licorice fern	rhizomes	de Laguna in Emmons
Porphyra spp.	black seaweed, red laver	fronds, medicine	Emmons; Newton and Moss
Potentilla anserina (spp. *pacifica*)	wild sweet potatoes, silverweed	roots	Emmons; Newton and Moss
Pteridium aquilinum	bracken fern	rhizomes	No published references for Tlingit use
Ribes bracteosum	stink currant	berries, stems and leaves: medicine	de Laguna; Emmons
			Newton and Moss
Ribes lacustre	swamp currant, swamp gooseberry	berries, leaves as poultice	Emmons; Newton and Moss

TABLE 10.1. (*continued*)

Scientific name	Common name	Part used as food indicated; medicine specified	Sources
Ribes laxiflorum	trailing black currant	berries	de Laguna; Emmons
Rosa nutkana	Nootka rose	rose hips	Newton and Moss
Rubus arcticus	nagoonberry	berries	Emmons; Newton and Moss
Rubus chamaemorus	cloudberry	berries	de Laguna; Emmons; Newton and Moss
Rubus parviflorus	thimbleberry	berries, leaves as medicine	de Laguna; Emmons; Newton and Moss
Rubus pedatus	creeping raspberry	berries	de Laguna; Emmons; Newton and Moss
Rubus spectabilis	salmonberry	berries, "furry buds"	Emmons; Newton and Moss
Rumex occidentalis	Indian rhubarb, dock	berries, leaves, stems; roots: medicine for cuts	de Laguna; Emmons
Salicornica pacifica	beach asparagus	shoots	No published reference for Tlingit use[2]
Sambucus racemosa	red elderberry	berries	Emmons; Newton and Moss
Shepherdia canadensis	soapberry	berries	de Laguna; Emmons; Newton and Moss
Smilacina racemosa	false Solomon's seal	root as medicine	Emmons
Sorbus sitchensis	Sitka mountain ash	medicine	de Laguna
Streptopus amplexifolius	clasping twisted stalk	roots: intoxicant	Emmons
Trientalis arctica	northern starflower	roots as love medicine	de Laguna
Trifolium wormskjoldii	springbank clover	rhizomes	Oberg 1973
Tsuga heterophylla	western hemlock	gum, cambium, needles, medicine	Emmons; Newton and Moss; de Laguna
Urtica dioica	nettles	leaves	Newton and Moss

Scientific name	Common name	Use	Sources
Vaccinium alaskaense	Alaskan blueberry	berries	de Laguna; Newton and Moss
Vaccinium caespitosum	dwarf blueberry	berries	Emmons
Vaccinium ovalifolium	oval-leaved huckleberry	berries	de Laguna; Emmons; Newton and Moss
Vaccinium oxycoccos	bog cranberry	berries	Emmons
Vaccinium parvifolium	red huckleberry	berries, leaves as medicine	de Laguna; Emmons; Newton and Moss
Vaccinium uliginosum	mountain blueberry	berries	Emmons; Newton and Moss
Vaccinium vitis-idaea	lowbush cranberry	berries	Emmons; Newton and Moss
Valeriana sitchensis	Sitka valerian	root: external medicine	de Laguna; Emmons
Veratrum viride	Indian hellebore	root, leaves: medicine for wounds, hair; intoxicant	de Laguna; Emmons
Viburnum edule	highbush cranberry	berries, bark in lotion bark as medicine	de Laguna; Emmons; Newton and Moss
Viola glabella	yellow violet	syphilis medicine	Emmons

*Plants marked with an asterisk were cultivated by Northwest Coast groups other than the Tlingit as documented by Turner and Peacock (this volume). These practices included clearing, tilling, tending, weeding, and pruning, transplanting, sowing, burning, and fertilizing. I have not considered digging or selective harvesting to be intentional horticultural practices.

1 "Indian carrot" was misidentified as *Daucus carota* in Newton and Moss (1984: 43).

2 Both beach asparagus and goose-tongue are still gathered by Tlingit today (Norton 1981; Turner 2004; Mudie et al., in press).

Please note: Several medicinal plants listed here are considered poisonous.

From a nutritional standpoint, the most important Tlingit food plants were "wild sweet potatoes" (*tset*), "wild rice" or rice-root (*kuh*), fern roots, what I will term "spring greens," a wide variety of berries, and seaweeds. *Tset* are most likely Pacific silverweed (*Potentilla anserina* ssp. *pacifica*) and Indian carrot (*Conioselinum pacificum*). While very important to groups south of Tlingit territory, springbank clover (*Trifolium wormskjoldii*) occurs only in the extreme southern portion of the region on Prince of Wales Island and the adjacent mainland. To my knowledge, the only author who mentions clover is Oberg (1973: 59, 68–69). Springbank clover does not appear to have been of widespread importance to the Tlingit, although it may have been glossed as *tset* in this southern area. *Kuh* is the widely used northern rice-root or chocolate lily, *Fritillaria camschatcensis*. The common fern and berry species are listed in Table 10.1. *Palmaria palmata* and *Porphyra* spp. were the primary seaweed taxa. Primary "spring greens" include the young leaves and shoots of cowparsnip, or Indian celery (*Heracleum lanatum*), salmonberry (*Rubus spectabilis*), wild rhubarb, or western dock (*Rumex occidentalis*), and stinging nettles (*Urtica dioica*).[3]

The gathering, processing, cooking, consumption, and storage of these foods have been described elsewhere (de Laguna 1972; Emmons 1991; Newton and Moss 1984; Oberg 1973) and will not be reiterated here. Although Tlingit food plants were selectively harvested as defined by Turner and Peacock (this volume), there is no published record of their management using the myriad other cultivation methods outlined in earlier chapters. Specific localities were noted for the quality of fern roots, wild sweet potatoes, and Indian celery they yielded (Newton and Moss 1984: 20–22), but this kind of selective harvest and returning to customary areas was not explicitly recognized by informants as having affected the plants themselves.[4]

This does not diminish the economic and nutritional importance of these plants. For example, Oberg (1973: 68–69) notes that during May in Tlingit country, "[t]here is a tremendous surge in plant life. In the pleasant weather people go out in search of green plant foods, like the tender stems of the salmonberry, wild rhubarb, and wild clover. . . . Most of the green plants are consumed at once, for the body has been starving for them." According to Oberg (1973: 75), herb and root gathering occurred from May through September, comprising no more that 20 percent of the time spent collecting food resources. Berry picking took place from June through September, comprising about 30 percent of the time spent food collecting during the month of August.[5] Emmons (1991: 151) mentions that berry fields were hereditary property, but most authors do not specify which species of berries were within these owned patches. An exception is de Laguna (1972: 407), who reported that strawberry patches were owned at Yakutat. Oberg (1973: 59) notes that "berry, root, and clover patches were small and often possessed by single houses," but does not provide additional detail.

In de Laguna's (1972: 655–64) description of Tlingit medicinal plants, there is also a lack of evidence for cultivation. Of the medicinal plants described

10.1. Wild lily-of-the-valley (Photo by R. D. Turner)

(skunk cabbage, cow-parsnip, yarrow, wild rhubarb, devil's club, unspecified lichens, blue currants, thimbleberry, hemlock and spruce trees, mountain ash, Hudson's Bay tea (Labrador tea), goatsbeard, white hellebore, yellow pond lily, buckbean, Arctic iris, wild heliotrope, Siberian spring beauty), only one plant, either "deerberry" or "bunchberry," has a record of a horticultural practice. Internal references cited by de Laguna (1972: 656, 32,) referring back to de Laguna 1972: 32) indicate that this was *Maianthemum dilatatum,* also known as false lily-of-the-valley (Figure 10.1). In 1952, de Laguna (1972: 656) witnessed a party of women transplant this plant (*qet kayani*) from the "woods between the Ankau lagoons and the ocean beach" to areas near their homes in Yakutat. This is the single case I have been able to locate documenting Tlingit cultivation of an indigenous plant. Considering the diversity of management techniques used on many plant species by other Northwest Coast groups, lack of published evidence for Tlingit cultivation is surprising. One possibility is that nineteenth- and early twentieth-century ethnographers did not expend much effort documenting women's activities (Moss 1993), and plant foraging has been portrayed as the nearly exclusive province of Tlingit women.

Tlingit Tobacco Horticulture During the Pre-Contact Period

Initial European contact with the Tlingit occurred with the 1741 Bering-Chirikof expedition, but little is known of the exact provenience or nature of this encounter. More is known of the Spanish expeditions of the 1770s in which both southern Tlingit and Haida were contacted. There are several eighteenth-century accounts of Tlingit cultivation of tobacco, extending from Yakutat

Bay to regions to the south. In 1788, near Cross Sound, Captain Colnett observed "a house & garden neatly fenced in, & European plants growing [!]," a statement de Laguna (1972: 131–32) interpreted as a reference to tobacco. In 1791, Malaspina saw some cultivated fields on the shores of Yakutat Bay that he termed "tobacco patches" (de Laguna 1972: 149, 410). Beresford recorded the use of tobacco at Port Mulgrave in 1787. Vancouver (1967 [1798]: 255–57) was the first to leave a written account of the Angoon Tlingit (*Xutsnoowuwedi*), and at what was probably the entrance to Kootznahoo Inlet, he observed:

> On each side of the entrance some new habitations were constructing, and for the first time during our intercourse with the North West Indians in the vicinity of these habitations, were found some square patches of ground in a state of cultivation, producing a plant that appeared to be a species of tobacco and which, we understood, is by no means uncommon amongst the inhabitants of Queen Charlotte's Islands, who cultivate this plant. . . . [Vol. III, p. 256]

Turner and Taylor (1972) have shown that the tobacco grown by the Tlingit and Haida (Figure 10.2) was probably *Nicotiana quadrivalvis*, an annual species of arid lands, indigenous to the American Southwest, but cultivated by the Indians of the Columbia River rapids, Oregon, Idaho, Montana, and Wyoming. Turner and Taylor state:

> Even in cultivation its natural range extends north only as far as the Columbia River rapids. The means by which it was transported along with the knowledge of cultivation procedures to the Queen Charlotte Islands and Alaska could be highly significant in tracing prehistoric inter-group contacts between the Indians of western North America. [Turner and Taylor 1972: 250]

Even though 30 years have passed since this statement was written, we still have not solved the "mystery" (Turner and Taylor 1972) of tobacco's introduction to the northern Northwest Coast. Based on their analysis of oral literature, Turner and Taylor (1972: 254) tentatively suggested that the Tlingit acquired tobacco and knowledge of its cultivation from the Haida, and that the Haida acquired it from an interior location. Meilleur (1979) hypothesized that trade between Haida or Tlingit and interior Athapaskans may have been the route of introduction. More recently, Nancy Turner (marginalia to author, May 15, 1998) has suggested that genetic fingerprinting of museum specimens may yield new clues.

There is general agreement that, during the contact period, the Tlingit and Haida maintained tobacco in their territories through horticultural practices of this nonlocal plant. These practices included clearing, tilling, sowing, tending, weeding, fertilizing, transplanting, and maintaining garden plots. Tobacco was obviously a highly valued plant to warrant this level of intensification. As

10.2. "Haida tobacco" (*Nicotiana quadrivalvis* Pursh., variety). Photograph of last remaining specimen of Haida tobacco, labeled "Nicotiana," in the collection of the British Museum, London [BM000815987]. Notation on the back reads: "Queen Charlotte Islands on the Northwest Coast of America. Capt. Dixon. Used by the inhabitants as Tobacco." (Photo by Robin Smith)

described by others (de Laguna 1972; Emmons 1991; Taylor and Turner 1972), tobacco was not smoked Aboriginally; it was chewed or held in the mouth, often with crushed lime from burnt clam shells, to help release the nicotine. Tobacco was an important wealth and trade item, also distributed at potlatches.

I strongly agree with Turner and Taylor (1972: 252) that pre-European tobacco horticultural practices contributed to the subsequent success of potato horticulture, particularly among the Tlingit. Turner and Taylor (1972) explain that the Haida grew quantities of potatoes, turnips, beans, and peas for their own use, but also for trade with the Tlingit, the Tsimshian, Euro-American mariners, and the Hudson's Bay Company. They also state that:

> No other Northwest Coast group adapted to agriculture so readily. Undoubtedly, the previous experience of the Haida in growing tobacco and their ability to apply tobacco cultivation techniques to vegetable cultivation were major factors in their success. [Turner and Taylor 1972: 252–53]

I would offer that the Haida were not alone in this success. As I hope to demonstrate, some Tlingit groups, perhaps most notably the Angoon Tlingit, became very successful at potato production, drawing on their expertise with tobacco horticulture.

Tlingit Potato Horticulture During the Historic Period

According to Howay (1920: 7) and implied by Niblack (1888: 277), in 1789, the American trader Robert Gray is said to have introduced potatoes to the Queen Charlotte Islands Haida. Emmons (1991: 152) believed that it was the Russians who introduced the potato to the Tlingit. Robinson (1983: 261–63) doubts that Gray introduced potatoes to the Haida, but suggests that shortly after the reestablishment of the Russian colony at Sitka in 1805, the Russians introduced potatoes to the Tlingit. Although I have been unable to locate any definitive information on the initial introduction of the potato to the Tlingit, quite possibly some southern Tlingit obtained knowledge of potatoes from the Haida, while other Tlingit, most likely the Yakutat and Sitka, acquired comparable knowledge from the Russians in their midst. It is also possible that the Tlingit and Haida received potatoes via intertribal trade networks from other indigenous groups prior to direct European contact (McDonald, this volume; Suttles 1951b, this volume). Certainly, by the turn of the eighteenth century trade contacts between both Euro-Americans and the Tlingit as well as between the Tlingit and their neighbors increased in frequency, and knowledge of potato cultivation appears to have been widespread.

During the maritime fur trade, tobacco cultivation was abandoned by those groups who had easy access to the large quantities of newly available trade tobacco (Robinson 1983: 245). Some of these abandoned plots may have been turned into potato fields. Tlingit potato production seems to have intensified after 1825, coincident with the decline in sea otters (Robinson 1983: 265). By the mid-nineteenth century, at least in some areas of Southeast Alaska, the Tlingit were growing large quantities of surplus potatoes to provision European colonists. Although the Russians and British considered the Tlingit dependent on them for trade goods, Gibson (1978) persuasively argues that it was the Europeans who had become dependent on the Tlingit, especially for potatoes, between 1841 and 1861. In 1845, the Russians in Sitka purchased over 1,000 barrels of potatoes from Indians who arrived in over 160 boats (Gibson 1978: 370). The naval officer Golovin (1979: 38; 1983: 85) found that the Sitka Tlingit provided most of the Russian colonists' fresh food in 1861–62. The Russians themselves apparently did little hunting and had few gardens or domestic livestock. The Tlingit brought potatoes, turnips, greens (as well as deer, game birds, and halibut) to the Russian American Company warehouse where they received blankets, rice, or flour in exchange.

One area that was clearly outstanding in terms of its fame as a potato-producing region was the southwest coast of Admiralty Island, home of the Angoon Tlingit (de Laguna 1960: 50). James Swan (1879), visiting the area in 1875, wrote,

> The houses were surrounded with garden patches planted in rows, well heaped up to admit drainage. Each garden was fenced in, and each had narrow strips of bark stretched across from fence to fence over each bed to keep off the crows, which are exceedingly numerous and great pests. These wary birds, however, are always on the alert for a trap or a snare, and the strips of bark make them think the fowler has spread his net for them, and they keep away. This delusion is kept up by the Indians, who hang up the carcasses of several dead crows in each garden patch, tying their legs to the bark lines as if they had been caught in that position. It is a simple and very effectual contrivance. The Indians raise most excellent potatoes at this place. [Swan 1879: 146–47]

As described earlier, the Angoon area has been noted for contact-period tobacco cultivation. The southwest shore of Admiralty Island is particularly well-suited to horticulture because it falls within the rain shadow of the Baranof Mountains and the heavy precipitation characteristic of the larger Southeast Alaskan region is here a relatively moderate 100 cm per year. Many of the potato garden sites in this area have southerly and southwesterly exposures favorable for plant growth. The reputation of the area for horticulture is widely recognized by Tlingit elders, as illustrated by Ruby Jackson, a Tlingit woman who lived in Juneau:

> On Admiralty Island, across the bay from Killisnoo, is a very large sandy beach—a crescent shaped inlet—that was all a garden. People planted their potatoes, turnips, rutabagas, carrots, whatever. The sandy soil produced the mealiest potatoes in the world—delicious. There were also small red potatoes, the skin was red and orange, on the inside, it was very sweet. I guess

it was known as Indian potatoes, but we knew it as Tlingit *k'onse*, which is potatoes of the Tlingit people . . . they even grew tobacco there. [Newton and Moss 1984: 25]

Additional detail on the practices of Tlingit potato cultivation and use are provided by Emmons (1991), who spent much of the 1880s in Southeast Alaska:

> Potatoes . . . are now generally grown on old village sites where the soil has been enriched through generations by the refuse from the fireplace, and this is further fertilized by covering with seaweed and burning it. The soil is light, and both easy to work and fertile. The gardens are fenced with brush, the beds are made long and narrow and high, being built up from eight to twelve inches above the trenches, and at the top are not more than two feet wide. Potatoes are planted in the early spring in April, and are visited and weeded several times during the summer. The potatoes are dug up by the middle or end of September. The cultivation is done by the women, their only implement being a stake about four feet long, the end sharpened and hardened in the fire. They also use paddle-like sticks and an imitation of our spade. After being gathered, the potatoes are stored in cellars under the floor, or in root houses. They were generally cooked by steaming in the ground oven, or on a hearth of heated stones, and covered with seaweed which not only confined the steam but contributed salt, and were eaten with grease. [Emmons (1991: 152]

Turner and Taylor (1972: 250) note that tobacco "seeds" were planted at the same time as potatoes at the end of April. Some of the same cultivation techniques were employed in both tobacco and potato horticulture, as indicated above. The technique of storing root crops in "cellars under the floor" is similar to that used to store the indigenous springbank clover and silverweed by other Northwest Coast groups (Deur, this volume). However, such practices were also used by Europeans, Americans, and Canadians to store potatoes and other crops, so the origins of the Tlingit practice remain ambiguous. The cooking method described is more clearly an indigenous practice.

Large-scale success in Tlingit potato horticulture was apparently not universal for all Tlingit groups, however. In some areas, there appears to have been relatively little interest in potato horticulture. Writing of the Yakutat Tlingit, de Laguna (1972: 410) states, "although the missionaries tried to induce the natives to raise garden vegetables, their efforts were not very successful. A few men and women did make gardens at their summer fish camps, especially on the Ankau lagoons, but vegetables were not much relished." Quoting an informant, de Laguna continued, "my mother and grandma never cooked no carrots or turnips or other vegetables. They eat it raw with seal grease. . . . They don't care for no potatoes. They got native sweet potatoes" (de Laguna 1972: 410).

However, many Tlingit highly valued potatoes and other introduced root vegetables and these were regularly grown throughout most of the twentieth century, until relatively recent times. For example, in the 1980s, Walter Williams stated,

> Keeping a garden is a thing of the past because you can buy potatoes and other vegetables very cheap. But at one time, every available spot was taken for gardens. They grew carrots, rutabagas; later they added strawberry patches and raspberry bushes. It was fun to grow your own garden and it kept you busy and out of mischief so to speak. In the white man's world they plant with fertilizer, something to make the garden rich. The Tlingit people used seaweed, preferably the old rotten type above the tide line. You pick them up and fill sacks and bury them in your garden. Another thing is *geeshaxwoo*, literally "bull kelp hair." Starfish, and sometimes even left-over dry fish were used. Just bury it in the ground and forget about it over the winter. When they would go back to the garden the next year, it was fertilized. [Newton and Moss 1984: 25–26; italics added]

Before turning to the archaeological record of gardens in Tlingit territory, one often-overlooked by-product of potato horticulture should be considered. A significant use of potatoes was in the production of alcohol. As illustrated in Table 10.1, a few indigenous plants were used as intoxicants, including clasping twisted stalk (*Streptopus amplexifolius*), the very toxic Indian hellebore (*Veratrum viride*), and lupine (*Lupinus nootkatensis*), but these appeared to have been used infrequently (de Laguna 1972: 411; Krause 1956: 109). However, when maritime traders introduced alcoholic beverages to the Tlingit, drinking rum and other liquors became a routine part of trading encounters as well as a popular trade item. De Laguna (1972: 411) wrote that "liquor (*nau*) was relished by the Yakutat people when it was introduced," and there is a story of when the Russians gave "whisky" to Raven (de Laguna 1972: 873; italics added). The Angoon Tlingit reportedly were the first Tlingit group to learn how to distill alcohol (de Laguna 1960: 159). An American ex-soldier is said to have taught the Tlingit how to make rum at Killisnoo, and this would eventually become known as "hoochenoo" or "hooch," a corruption of several Angoon Tlingit names. The extensive gardens near Angoon likely produced potatoes used for this purpose, although hops, molasses, and brown sugar were also used to make alcohol (see Emmons 1991: 157–58).

Archaeological Evidence of Gardens

An electronic search of the 1154 archaeological sites in Southeast Alaska listed on the Alaska Heritage Resources Survey (AHRS) as of December 1997 found that garden features are recorded for 74 sites. Table 10.2 shows the distribution of sites by 1: 250,000 scale map names. Unfortunately, not too much can be made of these data because the presence of gardens has not always been systematically recorded on site records, and information on gardens may not

TABLE 10.2. SOUTHEAST ALASKA ARCHAEOLOGICAL SITES
WITH RECORDED GARDENS (% OF TOTAL NUMBER
OF SITES)

Alaska 1:250,000 Maps	Total No. of Sites	Sites with Gardens	% of Total
Craig	327	14	4%
Juneau	40	5	13%
Ketchikan	85	2	2%
Petersburg	198	13	7%
Sitka	244	25	10%
Sumdum	27	2	7%
Port Alexander	233	13	6%
Sites with Gardens (% of Total Number of Sites)	1154	74	6%

(Sources: Alaska Heritage Resources Survey [AHRS], State of Alaska, Anchorage)

have been picked up in the electronic search. Garden sites as a percentage of site total vary between 2 and 13 percent. I suspect that all these estimates are low, and that this variation probably reflects the different research interests and habits of archaeologists more than it does the geographical distribution of site types. Small sample sizes also restrict interpretation of these data.

In an effort to evaluate the reliability of these data, I have reviewed a portion of the Sitka area and evaluated the sites on two 1: 63,360 quadrangle maps, Sitka B-2 and Sitka C-2. I am personally familiar with this area since it was the subject of my Ph.D. dissertation (Moss 1989; Moss et al. 1989). As Table 10.3 demonstrates, of the 23 garden sites known to me, for some reason, only seven of these turned up in the electronic search. Recognizing that southwest Admiralty Island may have an especially high density of garden sites, it still seems safe to assume that the number of garden sites now listed on the AHRS for all of Southeast Alaska is substantially lower than the actual number of sites with evidence of gardens.

Unfortunately, archaeologists, including myself, have not systematically recorded much data about garden features at archaeological sites. In general, the surface remains of gardens consist of raised rows of mounded earth, separated by furrows and arranged in parallel lines within square or rectangular plots, well above the high tide line. The rows are typically about 40–60 cm wide, and raised approximately 15–30 cm above their associated furrows. The rows are aligned in one of two directions: parallel to the shoreline or perpendicular to the shoreline. The most extensive garden area I have observed occurs at 49-SIT-183, Village Point Village, where garden rows occur along a

TABLE 10.3. POTATO GARDEN SITES ON SOUTHWEST
ADMIRALTY ISLAND

Site number	Site name	Recorded as garden on ahrs?
49-SIT-034	Favorite Bay Midden Garden	yes
49-SIT-124	Killisnoo Picnicground	no
49-SIT-130	Windy Smokehouse	no
49-SIT-132	Yaay Shaanoow	no
49-SIT-159	Ci xani	no
49-SIT-160	Channel Point Village	no
49-SIT-162	Danger Point Fort	no
49-SIT-169	Killisnoo Harbor Village	no
49-SIT-177	South Killisnoo Village	no
49-SIT-182	Turn Point	no
49-SIT-183	Village Point Village	no
49-SIT-244	Daax Haat Kanadaa	no
49-SIT-257	Neltushkin	no
49-SIT-286	Kanalku Bay Garden	yes
49-SIT-295	Taukaan Neeshoo/Sullivan Point	no
49-SIT-298	Dasaxuq Gardens	yes
49-SIT-299	Anteyuq	no
49-SIT-302	Favorite Bay Garden	yes
49-SIT-305	Kootznahoo Roads Garden	yes
49-SIT-306	Scott's Ranch and Midden	no
49-SIT-307	Kenasnow Camp and Midden	no
49-SIT-309	Stillwater Garden	yes
49-SIT-312	Maple Rock Midden	yes

(Sources: Moss 1989 and Moss field notes)

300 meter stretch of shoreline. For the few sites for which I do have mapped data, the estimated surface area covered by garden rows is presented in Table 10.4.

The great range in size of areas cultivated most probably relates to a variety of factors. Individual sites vary in their suitability for cultivation, based on the amount of cleared area, soil type, the quality of drainage, and solar exposure. The scale of site use also depends on the number of people using

TABLE 10.4. SURFACE AREA OF GARDEN ROWS AT SELECTED SITES

Site number	Site name	Square meters in garden rows
49-SIT-124	Killisnoo Picnicground	350
49-SIT-130	Windy Smokehouse	800
49-SIT-132	Yaay Shaanoow	410
49-SIT-244	Daax Haat Kanadaa	65
49-SIT-299	Anteyuq	4825

(Source: Moss 1989)

the site, and the duration of their gardening activity. This, in turn, relates to proximity of permanent residence, patterns of property ownership, and the affiliation of certain sites with specific matrilineages and clans.

Precontact archaeological sites appear to have been the preferred locations for gardens. Village sites, house sites, fish camps, and fort sites, and especially those containing shell middens, were attractive for many reasons. Most of these sites had areas cleared of trees, providing openings in the forest suitable for gardens. Many sites contained culturally enriched soils, as documented by high levels of phosphate that contrast markedly with the low levels of phosphate in the naturally occurring acidic soils (Moss 1985). The shell, bone, and other large particles in such archaeological deposits also appear to have improved drainage in areas that were probably naturally too wet for gardens.

During the nineteenth and twentieth centuries, settlement and resource use patterns changed on Admiralty Island as well as elsewhere in Southeast Alaska. For example, Krause (1956) lists two villages on southwest Admiralty Island in 1882, Angoon and Neltushkin. Over time, use of Neltushkin (49-SIT-257) as a permanent village shifted to seasonal use, as its residents and those of Angoon moved to Killisnoo for cannery work. While the exact timing of this shift in residence is not known, a number of clans maintained rights to Neltushkin and used the area during the summer (de Laguna 1960: 55–56). Similarly, while forts probably continued to serve a defensive function during the early historic period, by the 1880s such use was apparently obsolete (Moss and Erlandson 1992). A number of the fort sites listed in Table 10.3, including 49-SIT-132 (Yaay Shaanoow), 49-SIT-162 (Danger Point Fort), 49-SIT-177 (South Killisnoo Village), and 49-SIT-244 (Daax Haat Kanadaa), were used as gardens. Due to the relatively exposed position of these sites on prominent headlands, these clearings may have received more sunlight and have been particularly well-suited to gardens. Fish camps associated with salmon streams also were used for gardening. Harvesting of potatoes and other introduced root crops seems to have co-occurred to a large extent with salmon fishing and processing as well as fall hunting and trapping (Moss 1989). While I acknowledge that more information is necessary to demonstrate this claim, it appears to me that during the nineteenth century and continuing on into the twentieth century gardening became a key component of the seasonal round of the Angoon Tlingit. Through root-crop horticulture, household, matrilineage, and clan groups retained and maintained their historic ties to parts of their owned landscape.

Perhaps a related phenomenon is that berry bushes, most commonly salmonberries and thimbleberries, grow in large, dense stands on many archaeological sites on southwest Admiralty Island, not only those with garden areas. This association of berries with archaeological sites of various types has usually been considered a natural occurrence. However, my earlier work on the distribution of phosphate (Moss 1985) indicates that such berry patches are associated with culturally enriched soils, suggesting that these may be part of an anthropogenic landscape. While I cannot be sure these berry patches are the result of deliberate activity, more study should be directed to examining this possibility. I have also noticed that crabapple trees are associated with fish camps, and Tlingit experts have suggested to me that this is common. J. McDonald (this volume) has documented this association for the Tsimshian, as have Turner for the Haida, and Deur for the Kwakwaka'wakw and Nuu-chah-nulth (marginalia to author, May 15, 1998). In the Tlingit case, this may also be evidence of horticultural practices, whether in the precontact or post-contact period remains to be determined.

As indicated above, the subject of horticulture and gardening has not attracted much archaeological attention. Future research should include a systematic survey of garden sites, detailed mapping of surface features, and, particularly, effort directed at dating these sites. It will be almost impossible to understand the variation inherent in gardening practices and the historic development of gardening during the eighteenth, nineteenth, and twentieth centuries without chronological control. At a single site, it will be necessary to obtain multiple dates, as many sites undoubtedly contain evidence of many episodes of gardening. One common-sense approach would be to core the trees now growing on old garden rows. The age of the tree growing on a specific plot should provide a minimal date for the abandonment of that plot. While I have been unable to pursue such research yet, I hope to do so at some point in the future. I think that the Angoon area is an excellent locality to examine how horticulture was incorporated into Tlingit land and resource use patterns during the historic period.

An Interim Assessment of Tlingit Horticulture

Tlingit use of indigenous plants was not abandoned with the introduction of potatoes. As Martha James has explained, "Right along with planting the garden, we dug k'wulx, [wood] fern roots" (Newton and Moss 1984: 25). However, there is no published record of pre-European horticulture, except in the case of tobacco, a topic of great importance but with little new information since the pioneering work of Turner and Taylor (1972). There is also no evidence that horticultural practices used in growing tobacco, potatoes, or other introduced vegetable root crops were used to enhance the production of indigenous plants.

The Tlingit situation appears to have differed markedly from that of other Northwest Coast groups who did employ a range of horticultural practices with a variety of indigenous plants, as described elsewhere in this volume. Like others, Tlingit social groups owned and managed many food resource territories, but unlike other groups the only plant resources reported to date that the Tlingit routinely owned were berry patches, and, less commonly, root and clover grounds. Surprisingly little has been documented about which berry species were owned or in which habitats. Even during the historic period, the Tlingit apparently did not intensify their use of indigenous plants as did some of the other Northwest Coast groups discussed in this volume. This may have been due, at least partially, to the absence of key species such as camas, onions, wild caraway, wapato, hazelnut, and oak (and limited springbank clover) in Tlingit territory.

While some Tlingit, like those at Angoon, readily adopted horticultural practices, others, like the Yakutat, were apparently not much involved in potato horticulture. The Angoon Tlingit intensified potato production, taking advantage of the favorable ecological setting of southwest Admiralty Island. The twentieth-century Angoon Tlingit became regionally famous for their potatoes, growing these and other introduced root crops for their own consumption, as well as to trade to neighboring Tlingit groups, Europeans, and Americans. They also grew potatoes for use in distilling liquor. In the twentieth century, gardening decreased in scale with the increasing availability of commercial produce, but it continued to provide strong ties between some groups and their ancestral resource territories. The scale of potato horticulture and its economic and social importance to other Tlingit *kwaans* has not yet been documented.

I believe that Tlingit horticulture is a sorely neglected topic of research, and that the potential of the archaeological and oral historical records has barely been tapped. Since gardening persisted in importance until relatively recent times, ownership and use of these sites is apparent, although often not explicit, in the ethnographic record (e.g., de Laguna 1960; Goldschmidt and Haas 1946). In fact, there is probably a strong correlation between garden sites and sites for which there are relatively abundant oral historical data. Finally, the historical archaeological potential of Tlingit garden sites holds great promise for revealing nineteenth- and twentieth-century changes in land and resource use patterns. These topics are not only of scholarly interest, but of great interest to the descendants of Tlingit horticulturalists.

Notes

I am most grateful to Douglas Deur and Nancy Turner for their extraordinary patience as well as their interest in this work. Their detailed feedback on this paper is most appreciated. I also would like to thank Joan Dale, Alaska State Office of History and Archaeology, Anchorage, for kindly providing information on garden sites from the Alaska Heritage Resource Survey. I also thank Jon Erlandson, Gabriel George, K. J. Metcalf, Ken Mitchell, and John Neary for scraping through the brush with me in search of garden rows. Finally, I would like to dedicate this paper to Tlingit scholar Richard G. Newton, an expert on plant use.

1. The dates of the de Laguna and Emmons publications should not be misinterpreted in terms of temporal priority. George Thornton Emmons's ethnographic work with the Tlingit dates to the 1880s and 1890s. In ca. 1888, he began working on a monograph of the Tlingit that he was unable to finish during his lifetime. For years, his notes and writings have been accessible to researchers at the American Museum of Natural History, and Frederica de Laguna has consulted these writings since she started working with the Tlingit. Her publications on the Tlingit have always drawn upon Emmons, and in 1991 she brought Emmons's monograph to publication. Nevertheless this work, cited as Emmons (1991) is filled with additions and editions by de Laguna. These sources are clearly interdependent.

2. In Table 10.1, I have tried to include all the food and medicinal plants Turner and Peacock (this volume) listed for which plant management techniques were documented if these plants occur within Tlingit territory. For two of these, salal (*Gaultheria shallon*) and bracken fern (*Pteridium aquilinum*), there is no published record of Tlingit use. Nancy Turner (email to D. Deur, April 1, 1998) has suggested that a number of other indigenous plants were probably used by the Tlingit, including *Allium schoenoprasum, Amelanchier alnifolia, Cornus canadensis, C. unalaskchensis, Empetrum nigrum, Polystichum munitum, Sedum roseum, Triglochin maritimum,* and *Zostera marina.* I have not included these plants in Table 10.1 because of the lack of published or unpublished information documenting Tlingit use. Archaeological research may eventually result in the identification of some of these plants in contexts of uniquely favorable preservation.

3. De Laguna (1972: 33) noted that at old sites the nettles "grow thick." D. Deur (marginalia to author, May 15, 1998) notes that nettles do not readily grow in the Spruce-Hemlock Zone and may have required some degree of human intervention to become established or thrive. He suggests that the clearing of settlement areas would facilitate their growth or that they might grow atop refuse heaps, along with other culturally preferred species that favor relatively intense sunlight exposure or neutralized soil pH. In my archaeological survey work in Southeast Alaska, the most common plants I associate with site areas cleared or disturbed by human occupation are salmonberry, thimbleberry, crabapple, and young spruce and hemlock trees. I have observed one or two patches of nettles, but not in association with archaeological indicators. Obviously, the association of certain plants and archaeological features is a topic worthy of systematic study.

4. N. Turner (email to D. Deur, 1998) noted that the Tlingit also selectively harvested spruce roots and ornamental grasses used in basketry. I have not addressed the issues of management of plants used as materials here.

5. As pointed out elsewhere (Moss 1993: 638), Oberg's quantitative data are less precise than they appear.

Chapter 11

Tending the Garden, Making the Soil

Northwest Coast Estuarine Gardens as Engineered Environments

DOUGLAS DEUR

Late in life, Franz Boas labored to compile the geographical data he had accumulated during his lifetime's study of the Kwakwaka'wakw, or "Kwakiutl" peoples of coastal British Columbia. In the resulting volume, amidst the descriptions of village sites and places of religious significance, among the maps of berry-harvesting sites and hunting territories, Boas provided maps and descriptions of elaborate gardens that the Kwakwaka'wakw people had once constructed near their villages. His maps showed complexly subdivided plots on the British Columbia tidal flats, containing two native, estuarine plants with edible starchy roots: springbank clover (*Trifolium wormskjoldii*) and Pacific silverweed (*Potentilla anserina* ssp. *pacifica*). The one garden that Boas mapped in particular detail, as an example of a traditional Kwakwaka'wakw garden, sat at the mouth of the Nimpkish River on northeastern Vancouver Island. Figure 11.1 provides a schematic reconstruction of the Nimpkish garden as it existed in the nineteenth century, compiled on the basis of both Boas's map and the author's field observations of the site.

As many indigenous people reported to Boas and his academic peers, *Potentilla* and *Trifolium* gardens, such as those on the tidal flats of the Nimpkish, commonly occupied estuarine salt marshes where the mouths of rivers and streams met salt water. These two plants were ordinarily grown together in closely planted plots, producing a dense concentration of thin, long, starchy roots and rhizomes.[1] Often, these plants were grown alongside other estuarine plants with edible roots and bulb segments, including northern rice-root lily (*Fritillaria camschatcensis*) and Nootka lupine (*Lupinus nootkatensis*).[2] Specialized digging tools were constructed for root cultivation, and specialized "digging houses" or other domestic structures were built to shelter harvesters alongside some garden sites. At the Nimpkish River site and many others, large gardens were divided into numerous family-owned subplots, which were sometimes encircled by low rock walls.

Boas (1934; n.d.: 166) described the traditional construction of the Kwakwa̱-ka̱'wakw gardens as follows:

> The women clear the ground of pebbles, which are thrown up in large piles or in walls which surround a bed. . . . The garden-beds are separated by stone walls, but often also by blocks [or "planks"] which are put up on edge right into the ground, being held between the pairs of short posts [or "pegs"]. [Boas n.d.: 166]

However, Boas was not alone in his documentation of these gardens. Accounts of similar gardens appeared in ethnographic accounts of several different indigenous populations within this geographically and linguistically diverse region, including:

Randall Bouchard and Dorothy Kennedy (1990: 306) on the Nuu-chah-nulth: People cultivated estuarine plants "by clearing the ground and placing rocks around" their root plots, where rootlets were "placed back in the ground so they would grow the following year."

Darryl Forde (1934: 80) on the Kwakwa̱ka̱'wakw: "Patches of the wild clover root were enclosed in stone fences by Kwakiutl women, each of whom had her individual plot."

George Gibbs (1877: 223) on the Coast Salish: "Inclosures for garden patches were sometimes made by banking up around them with refuse thrown out from clearing the ground, which, after a long while, came to resemble a low wall."

Nancy Turner (1975; personal communication 1998) and Edwards (1979) on the Nuxalk: During the colonial period, women "enclosed their clover plots with fences" and "transplanted rootlets into these plots."

Charles Newcombe (n.d.: 35/4) on the Haida: At estuarine sites, "stones [were] cleared off" of clover gardens and traditionally "people even separated [their plots] with fences," probably of stone.

And, while references to gardens in the earliest explorers' accounts of this coast seem, at best, oblique, these early accounts nonetheless provide hints of the presence and significance of such root plots at the moment of first contact. Archibald Menzies (1923: 116), the botanist for Captain George Vancouver's expeditions, for example, noted a number of Nuu-chah-nulth women working estuarine root plots on the west coast of Vancouver Island in September of 1792:

> In the evening our curiosity was excited in observing a number of Females busily occupied in digging up a part of the Meadow close to us with Sticks, with as much care and assiduity as if it had been a Potato field, in search of a small creeping root about the size of a pack thread. This I found to be the

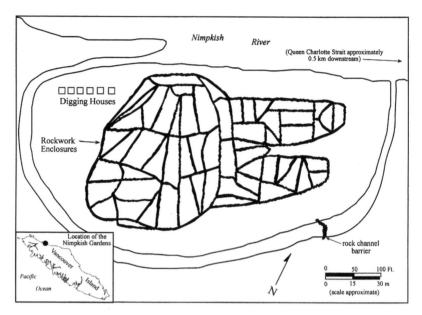

11.1. The Kwakwa̱ka'wakw *Potentilla* and *Trifolium* gardens on the Nimpkish River tidal flats, based on the map produced by Boas (1934) and subsequent field recon-naissance by the author. The map illustrtates the placement of rock walls surrounding individually owned plots within the larger garden site as they existed in the 19th century, when the gardens were still being maintained. Today, only fragments of these walls remain intact. The site is roughly 0.5 km upstream from Queen Charlotte Strait, and is partially inundated with brackish water during peak tides and streamflows. (Drawing by D. Deur)

> Roots of a new species of Trifolium [*T. wormskjoldii*, Figures 11.2 and 11.3] which they always dig up at this time of year for food. . . . Wherever this Trifolium abounds the ground is regularly turned over in quest of its Roots every year. [Menzies 1923: 116]

References to the traditional importance of these estuarine root foods, if not their management, were also to be found in the journals of other early explor-ers of the region, such as Cook (Newcombe n.d. 24/6: 1536), Jewitt (1807), and Moziño (1970). Menzies and his European peers seemed to interpret these root grounds as being entirely natural features of the landscape. Yet, as the other chapters of this volume have demonstrated, such assumptions were often quite incorrect. To be sure, over the last century and a half indigenous consultants have repeatedly asserted that their ancestors maintained, enhanced, and expanded these root gardens through a number of methods: weeding, trans-planting, selective harvesting, the enhancement of soils, and the construc-tion of stone or wood impoundments. Archival accounts and the testimony of twentieth-century elders, taken together, provide a view of estuarine root cultivation that is remarkably consistent along the British Columbia coast, and

11.2. E. S. Curtis, *The Root Digger,* about 1915. The woman is
Virginia Tom of Hesquiaht. Virginia Tom made all of the clothing
she is wearing especially for this photograph. Her digging stick is
of yew wood. (British Columbia Archives, D-08314)

11.3. Springbank clover with edible rhizomes (Photo by N. J. Turner)

arguably manifest many aspects of plant management practices as they existed prior to European contact.

Explicit references to gardens and gardening practices are to be found in some of the region's earliest anthropological writings, dating from the nineteenth century, including the early works of Boas (1895: 133), Newcombe (n.d. 24/6: 1551–52), and Dawson (1887a: 65; 1887b). An abundance of passing references to estuarine plant use and cultivation may be found within ethnographic and linguistic materials recorded since these pioneering works, yet most of these references remain unpublished to this day.[3] Cumulatively, this evidence makes it clear: the written record of estuarine plant use and management predates much of the literature from which the region's "noncultivating" reputation has been derived (Deur 2000).

These anthropological references to the use and management of estuarine roots are revealing on many counts. Not only did these root vegetables serve as an important food source and the primary source of dietary carbohydrates for most precontact Northwest Coast peoples. Estuarine root foods were also intricately and intimately associated with indigenous cosmology and ceremony, and trade in these roots served numerous social and economic ends (Turner and Kuhnlein 1982). In addition, like many other wetland cultivation methods found in the Americas, Oceania, and elsewhere, the methods of root cultivation employed by Northwest Coast peoples reflect a detailed appreciation of the interplay of cause and effect within environmental systems (Deur 2000). The peoples of this coast repeatedly modified estuarine soils, plants, and hydrology in anticipation of a predictable outcome: the qualitatively and quantitatively enhanced output of root foods.[4] Although the scientific community has given little recognition to these practices, it is apparent that the peoples of the Northwest Coast, at some point in their history, learned to engineer the tidal flats in order to harness the productivity of one of the world's most biologically productive environments: the temperate-zone salt marsh. Both the social significance of estuarine root foods, and the methods employed to acquire them, shall receive further consideration in the pages that follow.

Estuarine Root Vegetables in Economic, Ceremonial, and Cosmological Life

According to available ethnographic accounts, the demand for *Potentilla*, *Trifolium*, and other estuarine root vegetables was quite high along much of the Northwest Coast. The importance of these roots may be attributed not only to their dietary importance, but also to the related fact that they held significant roles within Northwest Coast economic and ceremonial life. If material wealth was "convertible" into prestige, and prestige was convertible into material wealth within Northwest Coast societies, these root foods were convertible into both. As such, root production was a means to *many* ends— dietary, economic, political, and ceremonial—a fact that appears to have encouraged their intensified production and use (Hayden 1995, 1996; Wagner 1977, 1996).

Root vegetable production was vital among Northwest Coast elites as a means of meeting ceremonial obligations, which was, in turn, vital for their maintenance of social and economic standing. Along the entire coast, cedarwood boxes of estuarine roots were standard items of exchange at potlatches (McIlwraith 1948, 1: 537), and estuarine roots were consumed as a regular part of winter ceremonial dances (McIlwraith 1948, 1: 194). Boas (1921: 527–31, 535–42) notes the use of these estuarine roots as an important part of the Kwakwa̲ka̲'wakw ceremonial barter economy, with, for example, several cedarwood boxes of estuarine roots being used as a regular part of the bride price. Elsewhere, oral traditions mention several cedarwood boxes of estuarine roots being exchanged for high-prestige ceremonial items, such as elaborately decorated copper shields, which served as a mnemonic of chiefly prestige, at once asserting and insuring the owner's social, economic, and political clout (Boas 1910: 93).

Northwest Coast ethnographers, most notably Boas (1921), also discuss entire ceremonialized feasts, regularly held among the Kwakwa̲ka̲'wakw, alternatively devoted entirely to the consumption of either *Potentilla* roots (Figure 11.4) or *Trifolium* rhizomes. Drucker (1951: 62) noted similar feasts among the Nuu-chah-nulth. With the Nuu-chah-nulth, these feasts were commonly held in the early winter, following two to three months of post-harvest storage; the roots reportedly grew sweeter with this duration of storage (Drucker n.d. Box 2, 23/1: 81). Each feast had its own intricate etiquette sur-

11.4. *Potentilla*, Pacific silverweed (Photo by R. D. Turner)

rounding the act of root preparation and consumption. Drucker's elderly consultants expressed pride in great root feasts of their ancestors: "Tales of famous feasts speak of young men having to go up on the roof of the house to pour in water to make steam, so high were the piles of clover roots" (Drucker 1951: 62). In ceremonial speeches at these events, chiefs identified the estuarine root plots in their ownership that had supplied the feast, and recounted the lineages through which they had inherited these plots (Drucker n.d. Box 2, 23/1: 64). So great was the demand for these roots for social, economic, ceremonial, and dietary purposes that the peoples of this coast, such as the Kwakwa̲ka̲'wakw, reportedly traveled hundreds of kilometers by canoe in order to trade these estuarine root foods in conjunction with other plant foods used for ceremonial and subsistence purposes (Turner and Bell 1971).[5]

Kwakwa̲ka̲'wakw elder Daisy Sewid-Smith (personal communication 1998) notes that her ancestors from the village of Haada reported that they "used to plant enough [*Trifolium, Potentilla*, and *Fritillaria*] to trade with the northern people. . . . The northern people would come and trade with them . . . [the Heiltsuk and] the Nuxalk." While some of the northern peoples had productive gardens of their own, many lived in places with very restricted salt-marsh environments, which gave the Kwakwa̲ka̲'wakw—with their large deltaic tidal flats—a comparative advantage in root production. She notes that the people of Haada dried their surplus output of estuarine root foods for shipment and that trade in these articles was a major part of the region's intervillage economy, both prior to and during the time of European contact. This trade effectively ceased during well-documented intertribal warfare of the mid-1850s, which accompanied the demographic and economic disruption of the colonial period and effectively brought an end to these trade alliances (see Galois 1994).

References to the importance of these estuarine roots as a trade good and feast food also may be found in the very earliest explorer's accounts from the Northwest Coast. While held captive by Chief Maquinna of the Mowachaht Nuu-chah-nulth, John Jewitt purchased bundles of *Potentilla* roots for a Christmas meal in 1803. Jewitt noted that bundles of these estuarine roots were coveted trade items, but—as a captive stationed at the mouth of Nootka Sound—he did not have the opportunity to observe garden sites: "the plant that produces it I have never seen" (Jewitt 1807: 120).[6] Likewise, during his 1792 stay at Nootka Sound, Moziño (1970: 21) encountered bundled estuarine roots in the course of bartering with the Nuu-chah-nulth, and reported that their most important plant foods appeared to include "the silver weed . . . the roots of the trailing clover, and the scaly, onionlike bulb of the Kamchatka [rice-root] lily."

A number of ethnographic consultants have reported that the longest, thickest roots of both plants were associated with high status. As Boas's primary field assistant George Hunt (1922) reported, the largest clover roots were called by a special name, "*laxabales*" (or "*lhaxabális*"), and "is a food specially for the chief and his family. No one else may use it." These large roots were actively sought by harvesters in estuarine gardens, and were given to the Kwakwa̱-ka̱'wakw clan chiefs as a form of tribute, validating the chief's ownership of the garden site (Boas 1921: 1333–39). Boas (n.d.: 67–68) reported that "In digging cinquefoil-roots, the smaller upper roots are distinguished from the long lower ones. . . . The baskets with the long roots are kept in the right corner of the house, [i.e., the side associated with the chief's immediate family] those with the short roots in the left corner of the house [the side containing less-prestigious individuals and items]." Kwakwa̱ka̱'wakw plank houses commonly contained rhizome storage areas beneath the sitting area of the clan chief (Daisy Sewid-Smith, personal communication; Boas 1889b; 1909). In addition, chiefs were referred to with a number of laudatory, metaphorical terms in ceremonial contexts, including "*lhagwa'nawe*," which Boas (1929) translates as "the thick root of the tribe," apparently a term etymologically related to the Kwa-k'wala term for large clover and silverweed roots (Boas 1929: 235, 1947). This term appears to have been used frequently in ceremonial contexts in which one spoke of chiefs or ancestors (Boas 1930).

The social, economic, and ceremonial significance of estuarine roots no doubt enhanced, and was enhanced by, their importance in Northwest Coast indigenous cosmology. As Haskins (1934: 184) suggested, *Potentilla* roots held a "prominent . . . place in the tales and myths of the Coast Indians from Oregon to British Columbia." According to Kwakwa̱ka̱'wakw consultants, in the distant time before the arrival of the transformers, when the world was still dark, chaotic, and devoid of human mortals, there was no water for the ancestral beings to drink; these ancestral beings survived on the moisture inside of starchy estuarine roots. When discussing later periods in the oral historical corpus, the period after the arrival of the transformers, *Potentilla* and *Trifolium* roots were commonly depicted by Kwakwa̱ka̱'wakw consultants as one of the primary foods eaten in "heaven" by these ancestral and supernatural beings (Boas 1908: 167, 1935: 85). In addition to being of importance to supernatural beings, Northwest Coast oral traditions commonly assert the importance of *Potentilla* and *Trifolium* rhizomes in the diet of human mortals. In diverse folkloric contexts, it is suggested that estuarine roots are among the most important foods in the human diet (e.g., Boas 1905: 178; Boas 1908: 45). The socioeconomic importance of estuarine root vegetables also receives frequent mention in these oral traditions, with human characters bolstering their status by giving away cedarwood boxes of estuarine roots in special root feasts, potlatches, marriages, and in trade (Boas 1905: 82; Boas 1910: 95, 357).

There is also some evidence in Northwest Coast oral traditions that estuarine roots served as a more dependable food source than animal foods, over time and space. This lends some support to the contention that estuarine root gardening played a "risk-reducing" role in an overall subsistence strategy oriented toward the harvest of generally less predictable marine animal resources. Many tales speak of "roots" and animal foods alternately eclipsing one another in importance, depending specifically upon the abundance of prized fish and game. A Nuxalk account tells of a famine in the winter (probably the result of a poor salmon harvest) before the arrival of Europeans, during which the people survived only by rafting from marsh to marsh, digging clover roots (Edwards 1979: 11).[7] Likewise, Northwest Coast oral traditions contain several references to the storage of estuarine roots as provisions for the lean winter months (Boas 1908: 54; Boas 1910: 457).

Methods of Estuarine Plant Cultivation

In order to acquire their coveted estuarine root foods, Northwest Coast peoples expended considerable labor to enhance root production and control access to root plots. Ethnographic accounts and contemporary biophysical data, when taken together, make it clear that the levels of root productivity found on contemporary unmanaged tidal flats would not have met the rates of root vegetable procurement and consumption indicated in many of these ethnographic sources. To achieve such rates of production, human intervention was required. Accordingly, Northwest Coast peoples appear to have developed methods of root enhancement and reproduction well-suited to the distinctive conditions of the region's salt marshes. In the process, they became competent cultivators.[8] Returning to the contemporary definitions discussed in this volume's introduction, the diagnostic characteristics of "plant cultivation" include such actions as the seeding or transplanting of propagules, the intentional fertilization or modification of soils, improvements of irrigation or drainage, and the clearing or "weeding" of competing plants. All of these practices were integral components of the estuarine gardening tradition of the Northwest Coast, as reported by indigenous consultants, early ethnographers, and explorers.[9]

OWNERSHIP

According to numerous ethnographic accounts, cultivated patches of estuarine roots were subject to some form of ownership and land tenure along much of the Northwest Coast. Indeed, Boas (1921) documented patterns of chiefly ownership and tenure of plots of each of the plants traditionally cultivated in estuarine gardens: silverweed, clover, sea milkwort, "wild carrot," Nootka lupine, and rice-root lily (see Turner and Peacock, this volume). Among the Kwakwaka'wakw, Boas (1934) noted that clans or households owned each large garden site, while family sub-units of the clan or household owned and maintained individual plots within this garden. Among the Kwakwaka'wakw, each garden, as well as each garden subplot, was given a name; some, but not all, of these names were descriptive of biophysical properties of the site (Boas 1934; Adam Dick, personal communication 1998; Daisy Sewid-Smith, personal communication 1998). Occasionally, these garden subplots appear to have been partitioned into smaller plots, reflecting changes in the composition of families within the village or clan that owned the larger garden. Thus, as Boas (n.d.:

166) noted, "generally the various garden-beds belonging to the women of one [clan] are close together." This seems to reflect both family groups working together in garden construction and maintenance, as well as the cumulative effects of long-term plot apportionment and subdivision within larger plots owned by kinship groups. This pattern of garden plot tenure is reflected in the appearance of large garden sites in archaeological contexts, often consisting of large oval or crescentic sections, which appear to have been subsequently subdivided by walls lining individual plots, or added on to by fragmentary extensions (Figure 11.1). This Kwakwaka'wakw pattern of garden plot tenure is similar to the pattern found elsewhere along the central and northern Northwest Coast, wherein chiefs controlled large garden sites, and families owned and maintained subdivided garden plots. Similar patterns of estuarine root garden or subplot ownership have been reported among almost every ethnolinguistic group in the region, including the Nuu-chah-nulth (Drucker n.d.; Sapir 1913–14; Sapir and Swadesh 1939), the Kwakwaka'wakw (Boas 1934, 1921), the Tsimshian (Compton 1993a; Darling 1955:10–12), the Haida (Blackman 1990: 249; Newcombe n.d.: 46/18; Turner, personal communication 1998), the Tlingit (Oberg 1973: 59), and others.

Traditionally, digging on a chief's estuarine garden or a family's sub-plot without their permission was a grave offense, even among members of the same clan or household, and was considered grounds for compensatory demands or violent reprisals (Boas 1910: 187, 383; 1921: 1345–48;). "There were gardens on all sides of [your plot] and there's no crossing that boundary; no, you never take [roots] from someone else's garden . . . you don't do that!" (Adam Dick, personal communication 1999). Kwakwaka'wakw elder Charles James Nowell Owadi (in Ford 1941: 51) reported that "in the olden days . . . if one woman gets in another's [clover] patch, they fight over it." In some cases, garden sites even appear to have been guarded against root theft. Turner et al. (1983: 120), for example, describe stories among the Ditidaht Nuu-chah-nulth of a Pacheedaht Nuu-chah-nulth chief who had between six and ten slaves guard a particularly productive garden to insure that no unauthorized person would dig there; these slaves would also dig these roots when they were ready. This rigid system of land tenure provides some insights into the importance of garden site ownership and, by extension, the high demand for estuarine plant resources; it also tentatively suggests that the demand for these resources had, at some times, and for whatever reasons, exceeded the readily foraged supply (Millon 1955). Though the demand for these root foods declined during the contact period, patterns of land tenure persisted. These large fields of herbaceous plants, sitting between impenetrable forest and saltwater, represented the only grazing lands available along much of the coast and became prime targets for agricultural reoccupation by arriving settlers. In the 1890s, according to multiple accounts, white agricultural settlers' violation of estuarine land title and damage of individual families' estuarine root gardens— such as the multiple Kwakwaka'wakw rhizome gardens at the head of Kingcome

Inlet—were among the primary reasons for inter-ethnic hostility at this time (Cotton 1894: 801; Adam Dick, personal communications 1998, 1999; Galois 1994; Kwawkewlth [Kwagiulth] Agency n.d., vol. 1648: 407–10, 572).[10]

The scale of these garden plots appears to have varied considerably between sites, and may have also varied considerably over time. The size of the entire garden in Figure 11.1 is estimated to be slightly over 2 acres or 0.81 hectares in size. Adam Dick (personal communications 1998, 1999) indicated that the gardens at Kingcome and Knight Inlets were considerably larger, possibly due to larger deltas and the higher level of surplus root production at these sites. Pointing to the tidal flats on the western bank of Kingcome Inlet, he indicated that at one time "There wasn't a square inch of these lands that was not part of someone's gardens." The salt marsh area he alluded to in this case was over 10 acres or roughly 4 hectares in extent. The individual family subplots he witnessed (in largely relict form) as a boy were roughly five to seven meters wide and at least twice as long; these measurements are similar to those of the subplots mapped by Boas at Nimpkish River. Both of these Kwakwa̱ka'wakw dimensions are somewhat smaller, however, than the Nuu-chah-nulth subplots of up to one acre, as reported by Sapir (1913–14) at the Somass River flats. In some peripheral locations, far from the prime tidal flats, gardens sometimes appear to have been much smaller, responding to limits on suitable estuarine habitat.

WEEDING AND GARDEN HUNTING

Numerous ethnographic accounts note that both weeds and pests were eliminated from these garden plots. By most accounts, competing plants, such as grasses, rushes, and sedges, were weeded out of estuarine root plots, leaving only *Potentilla, Trifolium,* and other preferred plants, and allowing the expansion of beds of edible roots into areas formerly containing competing plants. Among the Kwakwa̱ka'wakw, Boas (n.d.) and Forde (1934: 80) reported that estuarine gardens were "kept weeded to insure good growth." Similar practices, involving the weeding of all unwanted plants from garden plots, have been reported by Nuu-chah-nulth, Nuxalk, and Haida ethnographic consultants (Edwards 1979; Newcombe n.d.: 2/4; Nancy Turner, personal communication 1998; Turner et al. 1983). Kwakwa̱ka'wakw elder Adam Dick's (personal communication 1998) testimony provides additional insights:

> They worked on it all day . . . when [they] used to go down there and clean them up. *All* the weeds that grow: they wouldn't let anything . . . one little grass on there. They'd go there and clean them up. . . . *Just* to weed . . . They'd go down there, I don't know how many times a year, just to keep that *t'aki'lakw* clean. [The elders I knew in my childhood] told me that. I remember they used to talk about . . . when you keep working on the garden, [you] keep it alive! You keep working on the garden. You don't . . . walk away from it!

On other occasions, both Adam Dick and Daisy Sewid-Smith shared the term used for weeding estuarine gardens: "We *siixa* it, when they go down to the flats. *Siixa*—that's what they called it . . . when they pulled all of the weeds out of the *t'aki'lakw.*" Both insisted that weeding was a separate activity from harvesting, sometimes carried out in conjunction with soil-churning activities performed with digging sticks several times a year. By removing a diverse assortment of wetland plant species, which repeatedly invaded garden plots both by seed and by rhizomatous shoots, plot owners intentionally fostered the profusion of a small number of culturally preferred species. These practices, therefore, resulted in the considerable expansion of the portion of the estuary in which culturally preferred species could grow, and resulted in localized, biotic simplification of the salt-marsh flora (cf Harris 1997: 219).[11] After gardens were abandoned, the invasion of other native estuarine species into these gardens pushed *Potentilla, Trifolium*, and other culturally preferred species back to their much smaller natural ranges; as more than one ethnographic consultant has lamented, "you can't see where our [ancestors'] gardens used to be any more; they're all overgrown with grasses."

Similarly, some ethnographic references refer to these gardens being guarded by hunters, because waterfowl eat *Potentilla* and *Trifolium* roots and rhizomes (Turner and Kuhnlein 1982; Edwards 1979). (Indeed, considerable ethnographic and folkloric evidence suggests that the peoples of the Northwest Coast recognized a close general association between waterfowl and rhizome gardens, and geese and ducks are sometimes depicted as the estuarine gardeners of the time before the transformer's arrival.) Hunting both allowed for the protection of estuarine gardens' vegetable output and provided a secondary dietary benefit from the gardening process. Edwards (1979) reports that the Nuxalk constructed waterfowl traps adjacent to their gardens prior to European contact, consisting of nets that were dropped from poles alongside garden plots. Turner et al. (1983) have reported similar waterfowl traps around tidal flats among the Ditidaht Nuu-chah-nulth. Traps were reportedly abandoned for guns once these weapons became available through trade with Europeans. Numerous contemporary consultants mention that men continued to hunt at garden sites through the end of the twentieth century, even long after these sites have been abandoned for root production purposes; the author has found many recent shotgun shells at long-abandoned, ethnographically documented estuarine garden sites during archaeological reconnaissance fieldwork, providing corroboration of their testimony.

VEGETATIVE TRANSPLANTING, REPLANTING, AND SELECTIVE HARVESTING

Though ethnographic references to plant propagation techniques are less numerous, much evidence supports the contention that vegetative planting and selective harvesting were commonly used to enhance the output of garden plots. References to transplanting and selective harvesting of *Potentilla*,

Trifolium, and other estuarine root foods appear frequently in those few ethnographic accounts in which the question of estuarine plant use has received thorough attention (e.g., Edwards 1979; Turner et al. 1983). Further, these accounts appear among geographically distant indigenous groups, and consultants from each of these groups have depicted estuarine root transplanting and selective harvesting as ancient practices, long predating European contact.[12]

Numerous ethnographic consultants have reported that root fragments and small individual plants of desirable specimens of *Potentilla, Trifolium, Fritillaria, Lupinus*, and *Conioselinum* were replanted in situ from specimens within, or in the immediate vicinity of, traditional estuarine root garden plots. This appears to have been done regularly, ordinarily at the time of the harvest. Hesquiat Nuu-chah-nulth consultants told Turner and Efrat (1982: 68, 73) that the ends of *Trifolium* rhizomes, and entire *Potentilla* roots, were traditionally "placed back in the ground so they would grow the following year." Nuu-chah-nulth consultants told Bouchard and Kennedy (1990: 23) about the same vegetative replanting practices, and identified a number of locations that were traditionally managed in this way; during the harvest, this "had to be done 'just right' in order to ensure there would always be more plants" (see also Craig and Smith 1997: 73). Nuu-chah-nulth consultants suggested that vegetative replanting of garden sites demonstrated individual or collective proprietorship, and partially explained why owners of these root beds were so possessive of them. Turner (personal communication 1999), Compton (1993a), and Edwards (1979) documented similar testimony when working with elderly Nuxalk and Tsimshian consultants: during the fall, "In consideration for the next year's clover harvest, immature white roots were returned to the earth" from within and immediately around garden sites (Edwards 1979: 5–6). Similar accounts of vegetative replanting have been provided by Kwakwa̱-ka̱'wakw elders (e.g., Daisy Sewid-Smith personal communication 1998). Traditionally, Adam Dick (personal communication 1999) asserted,

> You don't take those little pieces [of root]. You leave them here. They come back. You put them back in the ground because that's going to be your *texw-sus* [*Trifolium*] and *tliksem* [*Potentilla*] next year.

Adam Dick reports that children were given the job of gathering *Fritillaria* bulbs exhumed by adults, removing the edible bulblets and a segment of the main bulb from the plant, and replanting the remaining portion of the main bulb within the garden plot.[13] During the early twentieth century, and perhaps earlier, children were provided with sticks of particular lengths, in order to determine how far apart the bulbs were to be replanted. Though it appears that all of these estuarine plants were commonly grown together in polycultural plots, this was not always the case. Reportedly, over time, the combined practice of weeding and in situ transplanting sometimes resulted in the creation of contiguous monocultural plots containing only *Trifolium, Potentilla, Fritillaria*, or *Lupinus*.

There is also evidence, albeit less widespread, of long-distance transplanting of propagules. According to Edwards's Nuxalk consultants (Edwards 1979a,b: 6), they traditionally enhanced their gardens in the springtime by "transplanting clover roots from elsewhere." Compton (1993a,b: 251) also encountered northern Wakashan consultants who described long-distance transplanting of estuarine roots between indigenous territories. Nancy Turner (personal communication 2000) heard similar reports from Nuxalk consultants, and learned that that a *Trifolium* patch in the tidal flats near Kitlope Lake had been transplanted there from the Kimsquit Flats early in the twentieth century by Nuxalk elders, Margaret and Stephen Siwallace; prior to this time, no *Trifolium* had grown there. Early in the twentieth century, Chief Humseet of the Knight Inlet Kwakwaka'wakw testified to the McKenna-McBride Commission (1913–16: 188) that he still visited a portion of Knight Inlet far from the large, natural tidal flats at the head of the inlet, in order to harvest the "roots there which my forefathers planted there." When asked to identify what sorts of roots had been planted there, he provided a list of Kwak'wala names; translated, they consisted of springbank clover, silverweed, lupine, rice-root lily, and "wild carrot." While Adam Dick only recalled the in situ vegetative replanting of propagules from his childhood, he nonetheless said that explanations that he heard as a child indicate that the elders before his time "must have planted [*Potentilla, Trifolium, Fritillaria*, and other] roots from other places," to their gardens. This was true, he felt, if only because the gardens described to him as a child were grown in relatively geometric plantings, and in places outside of the plants' normal habitats.

Remnant linguistic evidence hints at the role of vegetative propagation in garden contexts. When discussing terms for "planting," turn-of-the-century Nuu-chah-nulth consultants reported two different terms: "*c'opqa*," to "stick something into" the soil and "*tokwa*" or "to cover something with soil" (Sapir and Swadesh 1939). Both terms appear to denote vegetative planting methods, as employed within estuarine gardens. Adam Dick (personal communication 1998) provided a Kwak'wala term that was used for the replanted propagules of *Fritillaria* (and possibly other plants) that the Kwakwaka'wakw formerly transplanted, "*gagemp*," which translates literally as "grandfather." This metaphorical term alludes to its production of subsequent generations of offspring, and is consistent with the rich vocabulary of metaphorical terms employed in Kwak'wala (Boas 1929, 1947). A small amount of placename evidence lends additional insight into the role of transplanting in the maintenance of estuarine gardens. Bouchard and Kennedy (1990: 43) document a garden site mentioned by Nuu-chah-nulth elders called "*shishp'iḵa*" because of the traditional maintenance of *Potentilla* plots there. Bouchard and Kennedy (1990) translate this term literally as "cultivated"—the term appears to be etymologically related to terms for vegetative transplanting.[14]

According to several ethnographic sources, the peoples of the Northwest Coast created their gardens by rearranging large amounts of rock and soil on the tidal flats. As Boas and others suggested, the relocation of rocks, soil, and debris within gardens served to create a level planting surface. However, the ways in which this was accomplished appear to have varied from site to site, depending in part on local physiographic conditions and the extent of human demand for estuarine roots. The same basic technologies were applied in different ways, and with different intensities, to achieve the same basic ends of enhanced root production.

In large, low-gradient tidal flats, such as at the Kingcome and Somass estuaries, soil was churned and sometimes slightly mounded within individual plots. The sod of these low-gradient sites was regularly turned and broken up to keep the soil porous, an act temporally and functionally separate from root harvests. In the sheer glacial topography of the Northwest Coast, well-churned soil, itself, was a rare resource, and these actions appear to have enhanced both the size of individual roots and the spatial extent of root plots. Within these low-gradient gardens, paths of compressed soil marked the edges of garden plots containing churned soil, and plot boundaries were sometimes demarcated with horizontal logs or vertical corner markers made of rock piles or posts. As Adam Dick (personal communication 1998) recalled of Kingcome Inlet,

> Every family's got their own little garden. They've got pegs in the four corners, and there's no overlapping. . . . And they go continuously, because you're moving the soil around [inside the entire garden area].

Adam Dick reported that, at Kingcome, these corner boundary markers traditionally consisted of cedar poles, one to two meters high, with "flags" of pounded red cedar bark tied at their tops. Charles James Nowell *Owadi* (Ford 1941: 51) described traditional ownership of clover gardens on deltaic tidal flats similarly: "In the olden days, the women had their own clover patches marked with sticks on the four corners." This matches accounts from other ethnographic contexts, such as the Nuu-chah-nulth notes of Sapir (1913–14: 23), whose consultants described root plots at the low-gradient Somass estuary that

> had four cedar stakes marking the boundaries of [each plot] which were about one acre in extent. The stakes were six feet high and called *tlh'ayaqiyak-tlhama*. These posts were changed about every 10 years to prevent rotting.[15]

In some cases, these Nuu-chah-nulth marker posts also appear to have been marked or decorated with cedar bark "flags." On rare occasions, decayed remnants of such marker posts can still be found today at garden sites that were maintained well into the colonial period.

In small, rocky, or medium- to high-gradient estuaries, however, soil was reportedly mounded. In sites of lower gradient, gardens appear to have been managed in the manner outlined above, except that estuarine soil was often mounded within individual plots. Daisy Sewid-Smith (personal communication 1998) notes that people built mounds of soil in which to plant at some estuarine locations, such as near the village of Haada, where her ancestors lived. People used to push the marsh soil together with their digging sticks, adding height to existing plots or pushing mounded soil out laterally from the most productive parts of the flats: "[they] mound . . . to extend it . . . to make their garden bigger." On some tidal flats, poles appear to have been laid out around garden plots; while serving to mark off boundaries, this also appears to have sometimes aided in retaining mounded soil (Boas 1934, 1921; Bouchard and Kennedy 1990; Daisy Sewid-Smith, personal communication 1998).

Elsewhere, there appear to have been gardens, such as the Nimpkish garden in Figure 11.1, in which all patches were encircled by rock enclosures, as described in the documentary sources mentioned above. Again, Boas (1921: 186–94, 1934: 37, n.d.) suggested that these gardens were the product of a labor-intensive process: garden beds were constructed by the removal of rocks and boulders down to a level rock-and-soil surface. These rocks were placed in "large piles or in walls which surround a bed," and appear to have served in part to retain mounded and churned marsh soils (Boas n.d.: 166). Such rock-work gardens appear to have been quite widespread in high-relief shorelines or in places with comparatively small natural salt marshes, such as the more diminutive deltas of the Kwakwaka'wakw territories, the Nuu-chah-nulth territories in Clayoquot Sound (Bouchard and Kennedy 1990: 306), Haida Gwaii (Newcombe n.d.: 35/4), and possibly within the territories of the Coast Salish (Gibbs 1877: 223).[16]

Ethnographic accounts suggest that the extent of human modifications, such as soil mounds or retaining walls, was a function of the tidal-flat gradient, the demand for roots, and sometimes the availability of rock or wood construction materials. Clearly, high-gradient marsh gardens required much greater inputs of labor to maintain than low-gradient sites lacking rock or wood abutments. Asked why their ancestors opted to build rockworks in some places and not others, Kwakwaka'wakw elder Daisy Sewid-Smith (personal communication 1998) noted that "it depends on the land, how much land there is." She suggested that soil mounding and rockwork construction expanded the amount of area in which these plants might grow. Adam Dick noted of rockworks that "you have to have something like that to make flat ground" in places where it isn't already available. Conceivably, in the absence of available *natural* estuarine root grounds, demands on the resource created incentives to engage in such labor-intensive methods of plant enhancement. These constructions allowed the indigenous peoples of this region to modify site hydrology, pushing back the salt water, if only slightly, to open new land to the production of a crop with limited tolerance for prolonged saltwater inundation. Elders' testimony suggests that, whether or not retaining features were used, the practice of mounding soil expanded the area where these important root foods could grow within the estuary; durable retaining features, however, assured that boundary markers persisted, and the labor invested in mounding would not have to be repeated annually due to the erosive forces of waves, peak tides, floods, and freshets that frequently affect many of the region's smaller or more exposed tidal areas.

In the course of reconnaissance archaeological surveys, Deur (2000) has located remnants of such rock features at certain ethnographically documented estuarine garden sites. The fragmentary garden rockworks that remain at these sites exhibit designs ranging from complex, subdivided plots of the sort mapped by Boas, to singular rectangular or oval enclosures, to crescent-shaped rock terraces. All are situated in the high marsh, meters above the elevation of the stone fish-traps (for which they have sometimes been mistaken in previous archaeological surveys). When well preserved, these features often are abutted on at least one side by organically rich soils; they are very different in design than stone fish-trap features, reflecting the different functions and engineering constraints characteristic of each.

In these locations, the practice of mounding soils and reinforcing them with rock or wood abutments appears to have served to elevate lower portions of the salt marsh on a backfill surface, a fact that seems to corroborate ethnographic testimony that these features "made the garden bigger." Importantly, mounding and the construction of rock or wood retaining walls allowed the seaward expansion of the very narrow band of the high salt marsh in which *Potentilla*, *Trifolium*, and other culturally preferred species can grow (Figure 11.5). Among the region's intertidal vascular plants, these species have particularly narrow ranges of distribution within the tidal column in undisturbed estuarine sites, owing to their solar requirements and their osmotic intolerance for regular saltwater inundation (Jefferson 1973). By raising the position of the planting surface relative to the tidal column, rockworks, log alignments, and mounded soils appear to have served to alter local hydrology, dramatically expanding this otherwise narrow portion of the intertidal zone.

Moreover, the estuarine soil of garden sites, itself, was recognized as an important resource by Northwest Coast peoples. Ethnographic accounts from the early twentieth century hint that Northwest Coast peoples recognized a strong correlation between the atypical soil of the estuarine tidal flats and the growth of these edible plants (Harrington n.d. b: 373). Accordingly, Edwards (1979: 6) reports that the Nuxalk stored living *Potentilla* and *Trifolium* rhizomes in boxes of soil taken from salt-marsh gardens and placed within recessed areas in the floors of longhouses. Boas (1934) reports a sediment dam made of hemlock boughs in the estuarine channel adjacent to the Nimpkish River gardens, though he does not explain its function or significance. The soils that accumulate in the upper salt marsh, in the vicinity of the garden plots, are rich in fresh sediments and organic detritus from riverine, estuarine, and marine sources, carried to the high tide line by peak tides and floods.

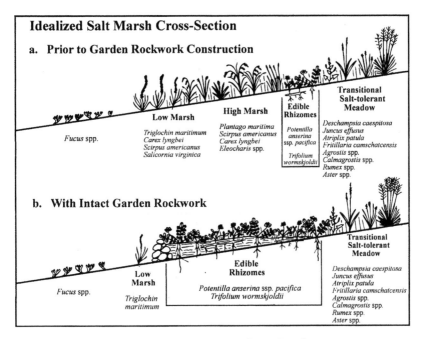

Idealized Salt Marsh Cross-Section

a. Prior to Garden Rockwork Construction

Transitional
Salt-tolerant
Meadow

Low Marsh   High Marsh   Edible
                                   Rhizomes

Fucus spp.

Triglochin maritimum   *Plantago maritima*   *Potentilla*   *Deschampsia caespitosa*
Carex lyngbei   *Scirpus americanus*   *anserina*   *Juncus effusus*
Scirpus americanus   *Carex lyngbei*   ssp. *pacifica* *Atriplix patula*
Salicornia virginica   *Eleocharis* spp.   *Fritillaria camschatcensis*
                                                 Trifolium   *Agrostis* spp.
                                                 wormskjoldii *Calmagrostis* spp.
                                                                Rumex spp.
                                                                Aster spp.

b. With Intact Garden Rockwork

Transitional
Salt-tolerant
Meadow

Low
Marsh   Edible
                   Rhizomes
Fucus spp.

Triglochin   *Potentilla anserina* ssp. *pacifica*   *Deschampsia caespitosa*
maritimum   *Trifolium wormskjoldii*   *Juncus effusus*
                                                           Atriplix patula
                                                           Fritillaria camschatcensis
                                                           Agrostis spp.
                                                           Calmagrostis spp.
                                                           Rumex spp.
                                                           Aster spp.

11.5. Idealized salt marsh cross-section (Drawing by D. Deur)

A significant portion of each estuary's total organic output is redeposited to the upper salt marsh each year, making that portion of the Northwest Coast estuary among the most productive environments in the world, if measured in terms of carbon produced per unit of area (Alaback and Pojar 1997; Thom 1987). Such soils are typically much higher in nutrient composition than the majority of the region's rain-leached soils, and are characterized by pronounced seasonal plant growth (Deur 2000).[17] Northwest Coast peoples appear to have learned how to subtly modify the salt marsh in order to maximize the productivity of these unique areas.

Adam Dick, Daisy Sewid-Smith, and other Kwakwa̱ka'wakw elders (personal communication 1998) insist that the soil at all garden sites, regardless of the presence of mounds or rockworks, was traditionally churned up as an intentional act, sometimes done separately from the actions of garden plot weeding or harvesting. It appears that this act both increased the porosity of the soil, and mixed in the most recent deposits of estuarine or alluvial debris that builds up each year, which, as Adam Dick put it, "feeds the garden" with fresh soil.[18] Adam Dick (personal communication 1998) explains:

> You've got to keep that earth soft . . . soften it up! You'd . . . "plow" it, I guess . . . And then you continue on that, digging the soft ground so it will grow better every year . . . "fertilizing?" I guess that's . . . the word for it.

Moreover, such Northwest Coast peoples as the Nuu-chah-nulth had an elaborate traditional taxonomy of soil properties, with distinct terms for "soil that is easily broken up," "soft, yielding soil in which one sinks easily," "level soil," "muddy soil," or "sandy soil," for example (Sapir and Swadesh 1939: 281 ff).[19] This taxonomy parallels patterns of traditional soil knowledge found among wetland cultivators elsewhere in the world (Wilken 1987). Adam Dick (personal communication 1999) recalled that the soil at his grandmother's former garden contained diverse materials, in a wide range of sizes. The best gardens, he learned from the elders of his childhood, had a little sand in the soil but it was very important that they "have a little bit of everything" including soil materials of diverse size and composition. Importantly, soils of such diverse texture are rare natural features on the tidal flats of the Northwest Coast, where waves and fluvial processes commonly produce a sedimentary structure with distinctly banded strata of uniformly textured soils. To "have a little bit of everything," natural estuarine soils would commonly require plowing through these distinct strata, or augmentation from external sources. And significantly, texturally diverse and structurally amorphous soils appear to be diagnostic of managed root gardens in archaeological contexts, a fact that appears to corroborate this testimony.

The removal of rocks and the creation of a porous, texturally diverse soil had palpable benefits for the size and quality of estuarine roots. Large, straight roots were clearly sought and fostered, for reasons both dietary and ceremonial, as described above. Indeed, gnarled roots, as are typically encountered in unmodified or rocky soils, are implicated as a source of supernatural misfortune in some Northwest Coast oral traditions, and rituals were prescribed for those who encountered them (see McIlwraith 1948, 1: 537).[20] Further, it is almost impossible to remove unbroken roots from the dense sod that develops on unmanaged estuarine soil, or on gardens that were abandoned long ago (Lepofsky et al. 1985). Yet the Kwakwaka'wakw considered it essential that estuarine roots and rhizomes not be broken while being removed from the soil; both Boas (n.d.) and contemporary consultants, including Adam Dick, mention that it was considered inappropriate, even shameful, for a person to break these estuarine roots as they were extracted from the soil. In addition to facilitating the bundling of estuarine roots, the harvesting of unbroken root segments appears to have aided in the procurement of the long roots associated with high status.

To underscore the point that indigenous peoples were actively "engineering" the tidal flats to enhance root production, it is important to note the grammatical categories employed within terms for plant-resource sites, as recorded by Boas (1921, 1934, 1947) and others. What becomes apparent in the grammatical analysis of resource-site terms and placenames is that the Kwakwaka'wakw, at least, viewed estuarine gardens and their soils as the "manufactured" products of human agency, and quite distinct from naturally occurring plant resource sites. The Kwak'wala terms for "gardens" and "natural root grounds" are grammatically distinct, and allude to different qualities of

Naturally occurring root grounds		
ts'o'yadi'	-adi'	"having" roots
	-is	"beach" with roots
Berry grounds		
go'ladi'	-adi'	"having" salmonberries
	-ats'ᵉ	"receptacle" of berries
	-as	"place" of berries
	-nukw	"having" berries
Rhizome gardens		
t̲aki'lakw	-i'lakw	"[human-]manufactured" soil
k'agekw		-"[human-]placed" logs, crosswise
and other constructions, such as:		
	k'agis	-"[human-]placed" logs, crosswise
	k'aqewilkw	-"[human-]placed" logs, crosswise
bi's		[Koskimo Kwakwa̲ka̲'wakw dialect: etymology uncertain]

(Sources include: Boas 1934, 1947; Adam Dick personal communication 1999;
Daisy Sewid-Smith personal communication 1998)

each site (see Table 11.1). Naturally occurring plant-resource sites are referred
to with semantic constructions that suggest that sites naturally "have" or "contain" these resources, such as roots or berries, or that they are, for example,
rhizome "places" or "beaches." Gardens, by contrast, are referred to by terms
that semantically denote that they are products of human agency. Most
commonly, the Kwakwa̲ka̲'wakw called their gardens "t'aki'lakw" or a "[place
of] human-manufactured soil."[21] Some Kwakwa̲ka̲'wakw elders still know the
former term. Adam Dick (personal communication 1998) explains that they
used the term "t'aki'lakw" "because you *made* that soil. . . . It's *yours*!" Daisy
Sewid-Smith (personal communication to Nancy Turner 1998) explained that
the term, with its suffix, "–*lakw*" grammatically indicated that "it's not natural . . . [it is] made by human hands."[22] Similarly, there is at least one Nuu-
chah-nulth garden site, with remnant garden rockworks, that bears the
placename "*ts'isakis*," or "place with soil," a name that is etymologically distinct from terms referring to naturally occurring root grounds (Bouchard and
Kennedy 1990: 464).[23]

Harvest

Traditionally, the Kwakwa̱ka̱'wakw selectively harvested estuarine gardens during the fall, when the foliage died back following the first hard frost. However, the season for digging appears to have varied among different peoples within the region (Boas 1921; Turner and Kuhnlein 1982). Some oral traditions depict the maintenance of rhizome plots as the long-standing domain of women or slaves—within much Northwest Coast oral history addressing the precontact period, it is a humbling experience for a free man to harvest estuarine roots (Boas 1908: 45). This does not appear to have been universally the case during the contact period, however, and numerous accounts describe the active, albeit supporting, role of men in the harvest process (Boas 1921, n.d.; Adam Dick, personal communication 1998).

Along much of the Northwest Coast, specialized digging sticks were made for the turning of the soil and the excavation of estuarine roots, especially from the wood of the western yew (*Taxus brevifolia*), which is characterized by unusually high tensile strength; crabapple (*Pyrus fusca*) wood was also sometimes used. According to Kwakwa̱ka̱'wakw and Nuu-chah-nulth consultants, root plows varied in length, and Boas (n.d.: 166) reports that two distinct, highly specialized tools were used for this purpose, alternatively for *Trifolium* and *Potentilla*, with the *Potentilla* plow being "a little thinner and one span shorter than the clover digging-stick" (see also Turner et al. 1983). Specimens available in ethnological collections have averaged roughly 1.3 meters in length from end to end, arched with a maximum arc depth of roughly 25 cm, which provided a rounded base for efficient prying. When a woman dug roots with these sticks, as Turner et al. (1983: 18) recount "She leans on it, twisting it when going into the sand, always watching how far down it goes"; women would continue to dig shallow holes around the perimeter of a cluster of roots until, with one final deep push of the digging stick, the entire cluster could be pried out and the roots removed from the overturned sod. Digging sticks included an articulated handle segment, often with a knob at its terminal end, while they were characterized by a sharp or slightly splayed digging end (Boas 1921:146–47). According to Boas (1921: 149–50) and others, these digging sticks were constructed and waterproofed by specialists in a procedure that was carried out over the course of several days. Digging sticks of this type have been reported along the length of the Northwest Coast, contributing to the overall picture of a region-wide cultivation "complex" at contact (Turner and Kuhnlein 1982). By the late nineteenth century, some estuarine cultivators appear to have made the transition to metal tools, but many elders reportedly still kept and used their aged yew-wood root ploughs, blackened with age and wear. Adam Dick (personal communication 1998) reports seeing very old root plows when he was a child:

> They had a stick called *k'ellákw*—it was yew wood . . . three-cornered . . .
> like a three-cornered file . . . when they're breaking the earth. They've got

a knob on the end, where you're hanging onto, when you're breaking the ground. . . . They're all different sizes. . . . It was really sharp [on the tip], like a pencil. I remember what it's like. I'd play with that. I used to dig with that. It's hard because when you push it down you put all your weight on it, when you're breaking [and] softening that ground, that *t'akí'lakw.*

In Kwak'wala, the term "*k'ellákw*" is typically used in a verb form—it is a term that translates "to break up" or "to break up the soil"; in its noun form, it alluded to these specialized tools (Daisy Sewid-Smith, personal communication 1998). Likewise, the Nuu-chah-nulth had an etymologically distinct verb "*t'ikwa*" that alluded solely to the use of these specialized digging sticks for digging estuarine *Potentilla, Trifolium,* and *Lupinus* roots (Sapir and Swadesh 1939).

When gardens were distant from village sites, families would relocate to small collections of houses next to large garden sites. Such transhumance to permanent, owned root-digging plots was noted by numerous ethnographers, including Boas (1921, 1934), Drucker (n.d. Box 1: 2/2), and Newcombe (n.d. 24/6: 1551), and was depicted as a precontact phenomenon by each. These small houses were constructed at the sites of individual gardens to shelter harvesters, as harvesting took place over the course of several days at each garden site (Figure 11.1).[24]

Boas (1921: 190–91) suggests that, during harvest season, a single woman typically worked for several days selectively digging rhizomes from the family's individual plot inside the larger garden site. Such authors as Turner et al. (1983: 18) have reported similar, multiple-day harvests for the Nuu-chah-nulth. Though Boas (n.d.: 167) notes that "a good garden [subplot] may yield a hundred bundles of roots" per harvest, it is unclear how many roots were contained in each bundle. Sometimes, during the harvest, tied bundles of rhizomes were suspended in the digging house. Women reportedly had distinctive knots that they would use to tie their bundles, so that they could be discerned from other women's harvest (Turner et al. 1983).

According to Boas (n.d.) and others, harvests began in the morning, with the women digging with the sun at their back. Women kneeled on mats while digging, pulling roots from the "moving soil" and placing them in spruce root baskets to await further cleaning and processing. This process was repeated over the course of several days:

This is continued until the whole of the garden bed has been dug over. Every evening the clover roots are spread out on mats and covered over with others. In the morning, the covering is taken off, if the sky is clear. The clover dried outside is considered better than clover dried in the house. And finally the baskets are covered with grass, and are taken home, where the roots are preserved in a cool place. [Boas n.d.: 166]

Some groups appear to have timed harvests on the flats so that the morning work could begin with the receding tide, allowing plenty of time before the returning tide might submerge portions of the work area. Harvests at the gardens downstream from Kingcome village traditionally began in the morning: "The river would push us down to the flats [at the beginning of the outgoing tide]. People would dig [*Potentilla*, *Trifolium*, and *Lupinus*] for many hours. And when the tide starts coming up then we paddle up again with the tide" (Adam Dick, personal communication 1998).

Boas (1921: 618–19; n.d.) describes individuals leaving the "digging-house" at the end of the harvest, taking canoes full of roots as well as the boards from the digging house exterior with them to the winter village. As they departed for the final time, they directed an incantation at their digging houses:

> Look upon my wife and me, and protect us, so that nothing may happen to us, friend! and wish that we may come back to live in you happily, O house! when we come next year to dig cinquefoil. Good-bye![25]

If gardens were outside of the visual range of one's village, it was considered taboo to look back at the house and gardens after uttering this blessing and departing.

Antiquity

In light of the evidence presented so far, one is led to ask: Did these gardens exist prior to European contact? Such authors as Boas (1934) depict these practices as precontact, while more recent authors have tended to assume that these plant cultivation methods reflect a combination of endemic practices and European agricultural influences emanating from the early colonial period. Yet indigenous consultants, past and present, have asserted that the cultivation methods described here *were* practiced well before Europeans arrived. Certainly, explicit references to estuarine gardens can be found in a number of the region's earliest ethnographic records, but so too are references to the cultivation of the introduced potato (*Solanum tuberosum*). Unlike potato cultivation, however, estuarine root cultivation is mentioned prominently in many Northwest Coast oral traditions regarding the distant, and indeed prehuman past. Detailed environmental and cultural knowledge is encoded in etymologically distinct terms for plants, soils, tools, units of property tenure, and many other aspects of estuarine cultivation; some of the languages for which this is true were no longer in common use by the time ethnographers recorded these facts (Deur 2000; Turner and Deur 1999). Estuarine root use and cultivation appear to be at the center of elaborate ceremonial traditions, which suggests considerable antiquity for the practices described by ethnographic accounts.

In an effort to assess the antiquity of these practices, Deur (2000) conducted an archaeological investigation of particularly well-preserved gardens, which he identified in the course of reconnaissance surveys of documented garden sites. One garden became the focus of particularly thorough investigation, including an excavation of the garden site soils. This site, *ts'isakis* or "[place with] soil," sits in Clayoquot Sound on the west coast of Vancouver Island, near the border of the traditional territories of the Ahousaht and Tla-o-qui-aht Nuu-chah-nulth. Nuu-chah-nulth elders have identified *ts'isakis* as a former garden site (Bouchard and Kennedy 1990; Mary Hayes, personal communication 1999). The site consists of a small, level salt marsh; two parallel rock walls appear to have expanded an ancestral, naturally occurring salt marsh in the manner described in the preceding pages (Figure 11.5).

Excavations at the site, adjacent to the rock walls and in the upslope marsh, corroborated ethnographic testimony regarding garden plot maintenance, and tentatively suggest that these rockworks served to expand the salt-marsh surface prior to European contact. Soils adjacent to both walls proved to be amorphous and texturally diverse, as one would expect in a site that had been subject to regular churning, weeding, and root digging. Adjacent to both walls, sediment patterns were not characteristic of any known natural sedimentary process within salt-marsh environments. Moreover, laboratory analysis indicated that the gravel and sediments underlying the well-churned soil, on the apparent ancestral beach surface, represented a "buried soil" that had been covered in rapid, singular events. Oxidizable carbon ratio (OCR) dating of soil horizons indicated that the soils adjacent to both walls were deposited rapidly atop this ancestral beach surface, between 375 (+/– 11) and 471 (+/– 14) RCYBP, or between 1479 and 1575 A.D., and appear to have been churned extensively thereafter. On the basis of soil texture discontinuities and OCR dates in the upper portions of units adjacent to these features, the abandonment of intensive root-ground maintenance was dated to a time between 206 (+/– 6) and 49 (+/– 1) RCYBP, or between 1744 and 1901 A.D. This fits expected dates of abandonment, roughly correlating with the approximate dates of initial colonial occupation, epidemic-induced demographic collapse, and the replacement of traditional root foods with introduced crops. Future corroboration of these soils with more conventional dating techniques is currently planned, but the evidence available to date provides strong support of indigenous elders' testimony regarding both the methods and the precontact antiquity of traditional estuarine gardening on the Northwest Coast.

Conclusions

We come full circle, returning to where this chapter began, with Franz Boas, the "father of American anthropology," meticulously chronicling the traditional gardening practices of the Kwakwaka'wakw. In light of what Boas knew then, and in light of what more we know now, it seems puzzling that Boas and his academic peers were so eager to depict Northwest Coast peoples as noncultivators. Certainly, many of the criteria now widely accepted for the presence of "plant cultivation"—such as vegetative propagation, soil enhancement, weeding, and hydrological modifications—are well represented in the estuarine root-cultivation practices of the region's indigenous inhabitants. The practices that characterize Northwest Coast estuarine cultivation are remarkably similar to the "wetland cultivation" methods employed by cultivators in other coastal areas of the world, such as Oceania and the American tropics; they augmented the plant food output of nutrient-rich, seasonally flooded environments to augment diets rich in aquatic animal protein (Deur 2000). While engineered in some manner, these cultivation methods evidenced a subtle appreciation of shoreline environmental systems. Particularly when constructed in less than optimum sites, such as rocky, high-gradient estuaries, Northwest Coast peoples' gardens were—in Mathewson's (1985) terms— "physiomimetic," being modified in such a way that they mimicked, and even improved upon, the ideal natural conditions that fostered the growth of edible estuarine roots.

Together, these practices appear to have allowed the peoples of the precolonial Northwest Coast to place large, productive, and predictable concentrations of starchy foods within their territories, and adjacent to their villages. Not only did estuarine cultivation increase the cumulative output of starchy root foods as human demands on native plants increased over time, but it also would have decreased the time that was required to seek and obtain these foods, thereby reducing scheduling conflicts with the harvests of other significant resources. Arguably, these practices may have co-evolved with the emergence of increasingly large and sedentary settlements on the Northwest Coast over the course of centuries or millennia. In light of this evidence, we must now exchange one enigma for another: instead of a region that appears enigmatically devoid of cultivation, we now seem to behold a region with subsistence practices that were enigmatically misrepresented within the anthropological literature.

The genesis of the region's "noncultivating" designation can be traced to the peculiar circumstances of the colonial period. Early explorers' and anthropologists' observations of indigenous peoples were brief, focused on the visibly exotic aspects of indigenous ceremonialism, and uninformed by what we now know about the dramatic demographic collapses and dietary changes that accompanied (and may have even preceded) European resettlement. The region's tremendous "natural abundance" was too often overstated; we now recognize that the patterns of soils and vegetation described by early explorers were often not "natural," but now seem to be the intentional outcomes of generations of human engineering. Being largely the object of women's labor, estuarine gardening was given short shrift in the accounts of the region's early anthropologists, most of whom were men, and most of whom relied primarily on male consultants. Moreover, Northwest Coast estuarine gardening did not involve overt seed propagation or geometric plantings of familiar domesticates. Estuarine root gardening thus defied conventional nineteenth-century notions of cultivation, and provided the historical particularists of the Boasian school with a prominent ethnological anomaly with which to tear asunder prevailing models of cultural evolution (Deur and Turner, this volume). Thus, ironically, the region's most prominent early anthropologists—Boas, Sapir, Newcombe, Drucker, and others—simultaneously documented many aspects of the estuarine cultivation practices outlined here, while categorically denying the presence of plant cultivation on the Northwest Coast.

This is not to say that the reconstruction of traditional estuarine gardening would have been easy to document in the early years of anthropological inquiry on the Northwest Coast. Estuarine cultivation practices appear to have been rapidly swept away over the course of the nineteenth century, in the wake of epidemics, mass relocations, land alienation, cultural changes, and the replacement of native plants with introduced starchy crops, most notably the potato. Only in a few isolated enclaves was the knowledge of traditional gardening intentionally kept alive. Exposure to colonizing Europeans did not, as some authors paradoxically suggest, provide rapidly dwindling Northwest Coast populations with the revelation that plant cultivation was technically possible, thereby resulting in their rapid intensification of native plant resource output.[26] Quite the contrary, the dramatic reduction in Northwest Coast peoples' demands on estuarine root grounds and other plant resources on the heels of the European invasion arguably *reduced* the intensity with which these environments were modified. By the mid-nineteenth century, by some ethnographic accounts, estuarine cultivation was already being eclipsed by new modes of subsistence. The ethnographic information that has been available to twentieth-century researchers on these plant-management methods might only illuminate a small fragment of the traditional practices used to enhance these estuarine environments prior to European contact.

By necessity, then, the contemporary study of estuarine gardening practices must involve some degree of reconstruction. While ethnographic accounts of estuarine gardening, recorded in distant times and places, have described these practices with remarkable consistency, most individual accounts are understandably fragmentary; it is perhaps not surprising that only a handful of contemporary indigenous elders have detailed, if second-, third-, or fourth-hand, recollections of these practices. Yet clearly, the consistency of ethnographic accounts of estuarine cultivation methods along the coast hints that, at one time, the practice was widespread. Clear documentation of owned and intensively managed estuarine gardens can be found for much of the central Northwest Coast, particularly among the Nuxalk, Kwakwa̱ka̱'wakw, and Nuu-chah-nulth. Fragmentary ethnographic references of these management practices can be found from Southeast Alaska to western Washington, but additional research would be required to confirm the presence of this entire gardening "complex" in the northern and southern portions of the ethnographic Northwest Coast. Clearly, the fact that the Nuxalk, Kwakwa̱ka̱'wakw, and Nuu-chah-nulth have the best-documented traditions of estuarine cultivation may simply reflect the disproportionately large amount of ethnobotanical research that has been conducted with these peoples in recent years. As Adam Dick (personal communication 1998) asserted, in reference to the British Columbia coast:

Every village had their own gardens. You know, *every* different village had gardens like the ones we had at [Kingcome Inlet] . . . because that's what we ate. Not only these people here [the Kwakwa̱ka̱'wakw], but all over the whole coast.

Indeed, ethnographic and folkloric evidence clearly suggests that the peoples on the southern edge of the Northwest Coast, as far south as the Oregon and northern California coast, also appear to have relied heavily on these estuarine roots for their dietary starches and may have engaged in some degree of plot tending (e.g., Harrington n.d.; Jacobs n.d.; Kniffen 1939: 387). In some cases, we may yet be able to reconstruct aspects of these practices; in other cases, these practices may be lost to time, and to the two centuries of resettlement that have so dramatically transformed the social and natural landscapes of this coast. Regardless, it is clear: the peoples of the Northwest Coast modified estuarine landscapes in repeated and purposeful ways to enhance the output of certain culturally preferred species. They owned root plots, augmented the soil, transplanted and replanted rootlets, weeded out competing plants, and sometimes made changes in the hydrology of individual garden plots. They produced starchy roots in a manner that complemented a diet already rich in marine proteins and fats. They were cultivators. The revisionary work of a recent generation of scholars has provided us with a perspective on non-Western cultivation practices that is more inclusive, and less tied to the intellectual prejudices of colonial Europe. Clearly, in this light, it is time to thoughtfully reevaluate claims that plant cultivation was not practiced on the pre-European Northwest Coast. The region's estuaries, with their pronounced biological and cultural significance, seem a logical place to begin.

Notes

I wish to acknowledge the assistance of many knowledgeable individuals, who each made important contributions to this work. The research presented here would not have been possible without the kind assistance of Chief Adam Dick (*Kwaxistala*) and Kim Recalma-Clutesi (*Ogwilogwa*) (Kwakwa̱ka̱'wakw) and Dr. Nancy Turner. I also received valuable assistance from Dr. Daisy Sewid-Smith (*Mayanilth*), Dora Sewid, George Dawson, and John Bubba Moon (Kwakwa̱ka̱'wakw); Dr. Richard Atleo (*Umeek*), Chief Earl Maquinna George, and Mary Hayes (Nuu-chah-nulth); as well as Drs. Kent Mathewson, Fred York, Rebecca Saunders, Wayne Suttles, Sandra Peacock, Paige Raibmon, Dana Lepofsky, James MͨDonald, Barry Gough, Leslie Johnson, Denis St. Clair, and Steve Acheson. Funding from a number of sources has aided in this research, including research grants from Sigma Xi, the Louisiana State University Graduate School, and the Jacobs Fund; archaeological research in Nuu-chah-nulth country was carried out under the supervision of Dr. Richard Atleo (*Umeek*) and with the formal consent of the Ahousaht and Tla-o-qui-aht First Nations as well as the British Columbia Archaeology Branch.

1. "Rhizomes" are thick stem segments that expand laterally from the main plant, usually below ground. They allow the plant to expand horizontally, producing leaves above and roots below at some distance from the original plant. Rhizomatous plants

are common in the intertidal salt marshes of the world, as the rhizomes provide plants with the capacity to respond to rapidly shifting sediments and to effectively colonize unvegetated tidal flats.

2. Less frequently, these estuarine root plots reportedly contained such root vegetables as sea milkwort (*Glaux maritima*), and Pacific hemlock-parsley or "wild carrot" (*Conioselinum pacificum*). Like documentary references to "wild celery" and "Indian potatoes," the term "wild carrot" has a long history on the Northwest Coast, and is a source of much confusion. Compton (1993b) suggests that the plant designated as "wild carrot" was primarily *Conioselinum pacificum*. It is clear, however, that ethnographers and explorers—unfamiliar with the native plants of the region—referred to several other plants by this generic common name because of similarities to the root, leaf, or flower of domestic carrots. Other plants termed "wild carrot" included, significantly, *Potentilla anserina* ssp. *pacifica*. *Conioselinum pacificum* primarily grows in the upper estuary, but can also occur within meadows beyond the range of tidewater. The plant was often subject to indigenous cultivation, but as many references to the management of this plant do not refer specifically to estuarine garden sites, I have omitted most of these references from the current chapter.

3. Two articles have been written on the dietary significance of the two most common plants in these gardens, *Potenilla anserina* ssp. *pacifica* and *Trifolium wormskjoldii*, among Northwest Coast peoples—Turner and Kuhnlein (1982) and Kuhnlein et al. (1982). Root cultivation receives brief mention in these articles, but these practices are largely interpreted as post-contact phenomena.

4. These two plants, incidentally, have a history of cultivation elsewhere. In Eurasia, clovers (*Trifolium*) have been cultivated as a fallow cover crop, for animal feed, and as a supplementary component of the human diet. In contexts such as the Middle East, intensive use of *Trifolium* for human subsistence was eclipsed upon the arrival of other cultigens, and only the larger legumes have persisted (Miller 1992: 44, 50). Similarly, rhizomatous colonies of *Potentilla anserina* served as a cultivated crop in the Hebrides Islands and elsewhere in the northern British Isles prior to the introduction of the potato; the cultivation and consumption of *Potentilla* was temporarily readopted during the famines that followed the potato blight (Hedrick 1972: 451).

5. Indeed, Lepofsky (1985; personal communication, 1997) found that access to prime estuarine root grounds was a better predictor of the relative wealth and prestige of different villages among the Nuxalk than access to animal resource sites, such as productive salmon streams. Salmon, she concludes, were comparatively ubiquitous, while the control of these coveted but unevenly distributed estuarine plant resources provided households and villages with surplus wealth that, over time, shaped their dealings with neighboring tribes and enhanced their relative status (see also Lepofsky et al. 1985). There is much truth in Hayden's (1990) observation that competitive feasting brought about an impetus for the intensification of these gardens, but I would contend that this is only one of several factors to which one should attribute causality (see also Hayden 2001). Certainly, "infrastructural, structural, and superstructural" factors, many extraneous to competitive feasts, were all at play; Harris (1979); see also Deur and Turner, this volume.

6. This is significant, as Jewitt resided at Nootka Sound, very close to the perennial anchorage used by explorers since the time of Cook. The Yuquot or "Friendly Cove" area is quite rugged and exposed to the open ocean; salt-marsh plants are relatively scarce and estuarine cultivation areas were primarily concentrated many kilometers away, at the heads of the inlets that enter into Nootka Sound. This geographical quality of the Nootka Sound area no doubt contributed to the concealment of estuarine cultivation from the European gaze.

7. In the Chinookan realm, where these estuarine roots were scarce but wapato was a significant wetland food source, Boas (1901) found tales of wapato serving a similar function during crashes in the availability of salmon. See Darby, this volume.

8. Clearly, there are strong parallels between these practices and the management of terrestrial resources described in the other chapters of this volume. See especially Turner and Peacock, and Suttles, this volume. Indeed, it is quite possible that some of the ethnographic accounts of the ownership and cultivation of "wild carrot," chocolate lily, and other plants mentioned in other chapters (e.g., Suttles, this volume) refer to estuarine gardens, but the location of many of these plant communities was not made clear within the original ethnographic sources.

9. While the accounts of cultivation methods described here are drawn extensively from archival and published accounts of estuarine gardening on the Northwest Coast, this material has been augmented by the contemporary accounts of a small number of Kwakwaka'wakw elders. A small number of elders still recall that cultivation was reportedly widespread, and that "individual and family stewardship of these gardens was a factor in their [enhancement] and sustainable use" (Craig and Smith 1997: 36). The material presented here is informed by my extensive communication with Adam Dick, hereditary Chief *Kwaxistala* of the Dzewadenux Kwakwaka'wakw of Kingcome Inlet. As part of his childhood training for his role as hereditary chief, Adam Dick was given detailed instruction on traditional methods of estuarine root management by elders born during the mid-nineteenth century, and was also involved with maintaining a few remnant gardens with his maternal grandmother, *Wotkineyga*. In addition to conducting numerous interviews with Adam Dick, I have had the opportunity to visit former garden sites near Kingcome Village, along with Nancy Turner, as his guest, and to work with him in the partial reconstruction of a traditional garden on an intermittent estuarine channel near his home. These experiences have provided a wealth of contextual information on traditional plant management not available through other sources. Other knowledgeable Kwakwaka'wakw individuals, including Dr. Daisy Sewid-Smith (*Myanilth*) and Kim Recalma-Clutesi (*Ogwilogwa*), have also provided me with valuable information from their investigations of traditional land management that are mentioned here.

10. As Alert Bay Indian agent R. H. Pidcock reported in May of 1896, "On my visit to [Kingcome village] my attention was called by a number of Indians of the fact that all the land which they have cleared and cultivated and from which they obtain a large quantity of clover roots which they use as an article of food, has been taken from them by the white settlers who have recently acquired land at the head of Kingcome inlet. They had torn up some of the posts of a settler named McKay who was fencing in a portion of the land they claim, and I had some difficulty in getting them to allow him to go on with his fencing. There are about five acres altogether of this land which they have cleared, and as a large quantity of the clover root is annually dug here in the month of October, I am afraid that there will be some trouble with the settlers who have preempted the land unless some compensation is made them. . . . As nearly if not all, the small plots are claimed by the women of the several families, I think a small present to each would settle the difficulty, as the plots are a source of revenue to them from the sale of the roots to other tribes" (Kwagiulth Agency, n.d., Pidcock to Vowell, 19 May 1896; Volume 1648/BCA reel number B-1915, Alert Bay, Letterbook, 1891–1899).

11. While there is physical evidence of past pruning of forest-edge vegetation alongside ethnographically documented garden sites, I have been unable to find ethno-

graphic references to pruning in the vicinity of estuarine garden sites (Deur 2000). *Potentilla* and *Trifolium* growth in small marsh sites is impaired when forest vegetation is encroaching on the salt marsh edge, a fact that is readily apparent in the field and presumably reflects competition for sunlight. The relative paucity of references to pruning may be due in no small part to the fact that pruning would not be required on the large tidal-flat gardens that were still utilized in the colonial period. Only smaller, more peripheral gardens would have been crowded by adjacent forest vegetation, and the majority of these secondary sites appear to have been abandoned prior to the arrival of ethnographers.

12. Selective vegetative propagation of plants with desirable traits always raises the potential for the isolation of genetically distinctive "domesticates." The evidences for the emergence of domesticated *Potentilla* and *Trifolium* are few, but they are provocative. Numerous Northwest Coast peoples have mentioned two types of *Trifolium wormskjoldii*, sharply differentiated by the size and appearance of roots and leaves, although botanists investigating contemporary plant specimens along the coast have apparently not observed this distinction. References to such distinctions can be found most abundantly in the literature addressing the Kwakwa̱ka̱'wakw (Boas 1921; Hunt 1922; Newcombe n.d. 24/6: 1552, 59:10). Adam Dick (personal communication 1999) also asserted that he knew of two related types of edible tidewater *Trifolium*, one larger than the other, but he was not certain whether this difference was a product of environmental variability or an inherent property of the plant itself. The Ditidaht Nuu-chah-nulth also recognized two types of estuarine *Trifolium* (Turner et al. 1983). Turner et al. (1983) and others have noted readily apparent phenotypic bifurcation between rhizomatous colonies of *Potentilla* in the vicinity of formerly cultivated sites (but no wide variability in rhizome characteristics), which may possibly reflect the impress of long-term human intervention. As Turner (1995) and Suttles (1951b) have suggested, some explorers' descriptions of tidewater "Indian potatoes" on this coast, with long, finger-like, white starchy segments probably refers to a plant other than *Solanum tuberosum*. This description would better fit a root segment of *Potentilla* or *Trifolium* that had been enhanced through environmental improvements or long-term vegetative selection. One might interpret certain references to "Indian carrot" similarly (Drucker n.d.). The origins of these variations are entirely unknown, and may reflect environmental variation, genetic variation, or quite possibly both. It remains an intriguing, but as yet unexamined possibility that prolonged tending and selective vegetative propagation facilitated genetic discontinuities within populations of these plants. These domesticates or "proto-domesticates" would have been largely lost once human selection and vegetative propagation of distinctive, desirable plants ceased, as was common throughout the Americas during colonial reoccupation (Sauer 1952). At present, the potential past domestication of Northwest Coast estuarine plants remains strictly conjectural.

13. Vegetative transplanting and replanting are implicit, if unexamined, components of Boas's (1934, n.d.) accounts of estuarine gardening. Boas makes it clear that traditional root gardens were regularly constructed by the complete removal of all vegetation down to a level rock-and-soil surface. Though Boas did not say so explicitly, his narrative makes it clear that surfaces stripped bare in newly constructed gardens would have been subsequently revegetated with plants from the surrounding tidal marsh.

14. This site and its placename were also documented by Drucker (n.d.: Box 2, File 23(1): 12) who was, ironically, among the most vocal proponents of the view that Northwest Coast peoples did not cultivate plants.

15. See also Arima et al. (1991: 190).

16. Nuu-chah-nulth consultants of the late twentieth century related a tale that suggested that logs around the boundaries of plots had served as the foundation for the subsequent production of rock-work enclosures in some cases. At the place called "wa7uus," at the mouth of the Cypre River on Clayoquot Sound, individually owned subplots were all contiguously on the tidal flats: "the extent of each owned plot was marked by poles laid on the ground . . . there used to be strong disagreements over the boundaries of these cinquefoil beds, and some owners moved the poles to extend their own boundaries. This occurred until a strong man named hinkaa7at . . . of the Ahousahts, placed large rocks on the poles to prevent them from being moved" (Bouchard and Kennedy 1990: 377). It is not clear when this event took place, or what its significance in the origin or widespread use of rock barrier walls in the Nuu-chah-nulth world might be. This tale may have served to explain why wa7uus, a place with low-gradient tidal flats, ultimately came to have stone barriers in a manner that was more typical of high-gradient tidal flats.

17. Moderation of soil temperature caused by wintertime estuarine inundation of gardens and adjacent tidal flats may also play a role in portions of the range being discussed here. Benefiting from regular inputs of detritus from upstream and estuarine sources (which was regularly churned into garden plots) the nutrient composition of these mounded soils is particularly conducive to rapid plant growth. The author's nonsystematic tests of estuarine garden soils suggest that they exhibit a nutrient mix much higher in nitrogen, phosphorous, potassium, and such trace elements as calcium than do unmodified marsh soils.

18. Once again, ethnographic references to "wild carrot" gardening mentioned in other chapters (e.g., Suttles, this volume) may be instructive on estuarine root gardening methods, such as Marian Smith's (n.d. 5/3: 11; 1950) finding that the Nooksack Coast Salish intentionally kept the "soil . . . loose and easy to dig" around root plots.

19. The Nuu-chah-nulth also had distinct terms for numerous soil modification techniques: "to dig," "to dig by scratching away dirt from the surface with a digging stick," "to poke into the ground with a digging stick," "to cover with soil" (Sapir and Swadesh 1939).

20. The Skokomish of Puget Sound also reported that the roots of *Potentilla* would squirm when they were dug. Other accounts mention that this power caused the roots to squirm when cooked. This was due to the fact that the plant contains a spirit; this spirit reportedly could give certain women the power to identify the most productive wild patches of *Potentilla* roots (Elmendorf 1960: 127). Elmendorf (1960 [1992]: 127), a student of Alfred Kroeber, derided this suggestion, saying that the women's root-digging duties were not as demanding or as technically complex as men's hunting and fishing, and that any able-bodied woman could be a plant gatherer. Both of his informants were men.

21. In some cases, gardens also were referred to by a host of terms that translate as "[places of] logs manually laid crosswise." Not surprisingly, this term provides an apt physical description of certain garden sites described above, including those with mounded soil, or reinforced by log barriers.

22. It is possible, but at present I have been unable to determine whether the term for certain gardens used by the linguistically related Nuu-chah-nulth "*tlh'ayaqak*" is related to the Kwakwala term. This term was recorded by Sapir (1913–14) and others in reference to garden sites.

23. There are also a number of descriptive terms that Boas (1934) mentions for individual plots within estuarine root gardens, describing plot attributes, such as "patch of sandy beach," "place of long roots," or "having everything just right [for

growing roots]." Also, it is important to note that the one garden that Boas mapped in exacting detail appears to have initially served to expand a small, ancestral patch of naturally occurring roots, so that, within a single larger garden, there are individual plots named "place having [naturally occurring] roots" and "[places of] human-manufactured soil."

24. Daisy Sewid-Smith (personal communication 1998) reports that her mother was born in one of the digging houses mapped by Boas on the Nimpkish River tidal flats.

25. Such a plea for protection might corroborate other claims that rhizomes served as a "risk-reducing" resource, although it is not clear whether this "protection" was conceptualized as being dietary, economic, ceremonial, or some combination of the three.

26. Indeed, by the time that early anthropologists began asking about plant management, the indigenous population of the Northwest Coast had been reduced to perhaps 5 to 10 percent of their precontact population (Boyd 1985), and most had adopted the potato at the expense of native starchy crops, due to direct and indirect colonial influences (see McDonald, this volume, and Moss, this volume).

PART III

Conclusions

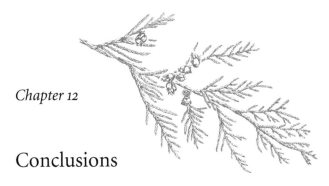

Chapter 12

Conclusions

DOUGLAS DEUR AND NANCY J. TURNER

At the beginning of the twenty-first century, much is still unknown about past indigenous Northwest Coast plant-management practices, and must be inferred from partial evidence. Yet despite the relative scarcity of information regarding plant use in this part of the world, it is obvious that plants were a substantial, integral part of Northwest Coast life. In many indigenous communities they continue to be. Plants have served as fundamental sources of foods, medicines, ceremonial substances, and objects, and have provided materials for constructing any number of goods, such as tools, containers, canoes, clothing, and housing, as well as for fuel for heating, cooking, and food processing. Indigenous peoples named and utilized over 300 plant species for a variety of purposes up and down the coast, with over 100 of these as foods, and many of them have been harvested in substantial quantities, from specific habitats and specific sites.

As plants have been valued resources throughout the region, plant-site ownership and proprietorship appears to have been widespread. While the exact methods and intensities of plant resource management did vary along the Northwest Coast, from place to place, from time to time, from village to village, from people to people, and from resource to resource, it is apparent that the practices described in this volume were found in some form throughout the entire culture area. Multiple, complementary methods of cultivation— weeding, pruning, tilling, vegetative transplanting, and sowing, as well as burning—were found among peoples geographically, linguistically, and culturally distant from one another along the coast. Cumulatively, accounts of intensive plant use and cultivation, drawn together from diverse sources, provide compelling evidence of the importance of plants and plant management techniques in the life of Northwest Coast peoples prior to European contact. Although it was not widely recognized by early European settlers, nor later by most anthropologists, Northwest Coast Aboriginal peoples certainly were not merely passive occupants and opportunistic users of their environments.[1] Their management practices, however, often left only subtle or fleeting imprints on the land. As such, the region's many human-modified plant communities were not visually apparent to newcomers, whose perceptions were shaped by notions of domesticated environments of the European style, with seed-planted crops, ploughed fields, grazing areas for livestock, and geometric enclosed gardens containing rows of individual domesticated plant species.

In this book we have provided a conceptual framework for understanding Northwest Coast peoples' many methods of plant cultivation and tending—for "keeping it living." Unlike some plants cultivated elsewhere in the world, most of the plants cultivated traditionally on the Northwest Coast were locally native; like most cultivators, however, the people manipulated localized environments in a number of ways with the intention of enhancing the productivity of their culturally significant plants. As we continue to document the vast quantities of plants once used by Northwest Coast peoples and the apparently sustainable nature of traditional plant harvests, the logic of "keeping it living" as an enduring and intentional strategy becomes increasingly apparent.

Admittedly, evidence of these plant-management techniques has been scattered. Too often, written evidence has been relegated to residual or parenthetical accounts and to passing references in ethnographic writings, or can only be found hidden in the journals of explorers, settlers, or casual observers. However, as can be seen from the diverse chapters in this book, as well as from previous works they incorporate, many examples are there for our consideration: from the camas maintenance traditions of the Coast Salish, to the burning practices of the Stó:lō and many other peoples to increase berry production; from the Haida and their neighbors planting tobacco in garden plots, to the intensive maintenance of intertidal root grounds among the Kwakwa̱ka'wakw, Nuu-chah-nulth and others; from the Chinook peoples' proprietorship of wapato beds to the Tsimshian management of berries and fruit-bearing trees. Knowledge of some of these practices persists in the oral traditions of the First Peoples of this coast; by listening to their accounts, past and present, we may yet learn more about their traditional management of plants. And, as portions of this volume have demonstrated, evidence from scientific sources—such as archaeological investigations and botanical analyses—often corroborates the accounts of indigenous peoples.

It now seems indisputable that the peoples of the coast were in some manner "cultivating" plants. This revelation is not based solely, however, on the fact that many practices were overlooked in the past. In addition, the prevailing definitions of cultivation have grown more inclusive over the years, and less tied to the assumptions and practices of European peoples. "Cultivation" is a problematic term, but now represents a range of plant-management practices, situated on a much broader continuum of human–environment interaction (Smith, this volume). As defined by Ford (1985a,b), Harlan (1975), and others, cultivation involves modifying plants or plant communities in some manner in order to promote the growth and productivity of culturally important plants, thus providing humans with easier access to, or larger quantities of, foods and other plant resources. Arguably, peoples would begin to cultivate plants in these ways, not to "revolutionize" their production of dietary staples, initially, but simply to increase the availability, accessibility, and predictability of culturally preferred plant species. Such activities as weeding, tending, pruning, tilling, transplanting, sowing, and burning represented the tools that facilitated this effort. And, to be sure, Northwest Coast peoples engaged in all of these activities.

Although contradicting a century of anthropological orthodoxy, the findings of this volume should not come as a complete surprise. Increasingly, we find that peoples who combine hunting, fishing, and gathering with low-intensity plant cultivation are common around the world. And, as many researchers now recognize, even if this state is seen as "intermediate" between more widely recognized categories—"hunter-gatherers" and "agriculturalists"—it often represents a "stable state" unto itself, a lifestyle characterized by staple fishing and hunting that are augmented by the enhanced production of certain plant foods. Indeed, in some ethnographic and archaeological contexts, this "intermediate state" appears to have been the norm, often persisting unchanged for millennia (Smith, this volume). Pure "agriculturalists" or pure "hunter-gatherers" may therefore never have been valid categories with which to categorize many human societies. As Smith (this volume) and others have suggested, these categories created false dichotomies. On the Northwest Coast, they have obscured much and revealed little, often with calamitous results for the First Nations themselves.

The scholars who claimed that the Northwest Coast's resource abundance precluded agriculture were therefore only partially correct. Within patchy but resource-rich environments such as those found on the Northwest Coast, there is little incentive for "agricultural revolutions" of the sort envisioned by V. G. Childe (1951) and others. The localized abundance of marine and riverine resources on the Northwest Coast likely *facilitated* the emergence of certain plant-management strategies by concentrating people together in large villages within finite territories and limited local plant resources. In turn, however, this localized resource abundance may have precluded coastal people's "transformation" into a people primarily dependent on agricultural practices for sustenance (Ames, this volume; Deur 1999). In light of the localized abundance of salmon and other staple animal resources on the Northwest Coast, it would have made little sense to abandon these resources in favor of an agricultural lifestyle based on the growing of annual crop plants.

If the peoples of this coast, living on the productive coastal margins, had no particular motive to abandon marine animals as a source of staple food, they nonetheless had strong incentives—dietary, socioeconomic, medicinal, and ceremonial—to maintain and enhance the productivity of their plant resources. Cumulatively, the peoples of the Northwest Coast were motivated for reasons that were at once, in Marvin Harris's (1979) terms, "infrastructural," "structural," and "superstructural." This was particularly true of resources they harvested in large quantities.

Plant foods contributed substantially to the diet of Northwest Coast peoples, providing not only dietary diversity, but also essential vitamins and minerals, some carbohydrates, and the dietary fiber required for proper digestion (Kuhnlein and Turner 1991; Turner 1995). In addition, plant medicines helped maintain peoples' health and well-being. Plant materials—fibers, wood products, and others—were vital in providing the necessities of life, including the means for procurement, processing, and storage of animal food resources. By intensifying conveniently located patches of food plants, the peoples of the coast also reduced the risk of fluctuations of other dietary resources such as salmon. Salmon populations do exhibit wide vacillations and it is clear that this affected indigenous peoples and their survival strategies (Schalk 1977). Abundant references in traditional oral narratives to plants serving as virtually the only foods during famines, as well as traditional "prayers," or "words of praise"[2] to plants and plant sites asking for protection, are suggestive evidence of this function (Boas 1930; Turner and Davis 1993).

Further, the peoples of this coast had strong socioeconomic motives for plant management. Plant materials were valued items of trade and exchange and thus served, practically and symbolically, as wealth. Plant products, therefore, contributed in numerous ways to the general wealth and prestige of the chiefs and their households. Furthermore, these products were often convertible into other forms of wealth through exchange, as a form of currency. Large quantities of plant resources were acquired or changed hands through trade, or were distributed widely during ceremonial occasions (Boas 1921; Turner and Loewen 1998). Boxes of Pacific crabapples and highbush cranberries, for example, were common and highly prized gifts at feasts and potlatches, as were preserved soapberries, salal berry cakes, dried seaweed, camas bulbs, wapato tubers, silverweed and springbank clover rhizomes, and bog cranberries. Even the green shoots of thimbleberry and salmonberry were exchanged and shared among Northwest Coast peoples. Such plant products as cedar canoes, cedarwood boxes, cedar-bark blankets, mats, and clothing, yew wood implements, cattail and tule mats, nettle fiber twine and nets, and baskets of many types also featured prominently in the economic exchanges and property of peoples throughout the region (Turner 1996a, 1998; Turner and Loewen 1998).

Moreover, the peoples of this coast had ceremonial or "religious" motives for plant management. This aspect of plant use is little documented and little understood by contemporary scholars. One ceremonially prized plant, "wild celery" (*Lomatium nudicaule*), was, and is still today, widely used and sought for medicinal and ceremonial purposes, and its seeds used as gifts. Some contemporary Northwest Coast peoples are careful to leave wild celery seeds behind, or to scatter seeds (when gathering medicines) to ensure its continuation, and this seems a likely candidate as a species that was managed and whose range has been extended through past human intervention. The attribution of a "spirit" within all of nature's creations, and of the powers of plants to affect human lives and human well-being, is another reflection of, and rea-

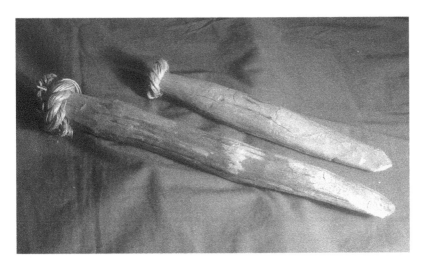

Yew wood wedges, Royal British Columbia Museum collection (Photo by
R. D. Turner)

son for, peoples' stewardship of the plants they depend upon (Turner and Atleo
1998). One reason given for not girdling red-cedar trees when harvesting their
bark is that the tree, conceived of as a sentient being with a "spirit," would
die and nearby trees would curse the perpetrator of this act. Similarly, devil's
club, a highly respected and powerful plant considered sacred by many, is carefully and selectively harvested from the main central stalk by Haida, Tlingit,
and others in order not to destroy the original "mother" plant (Walker 1999).
Hence, "keeping it living" appears to have represented a philosophical
approach to culturally significant plants and plant communities that was common to Northwest Coast peoples; it has a moral and ethical basis that translates into a practical one.

In light of the evidence presented in this volume, there is a pressing need
to reevaluate anthropological orthodoxy regarding the complete absence of
plant cultivation on this coast. We must reevaluate the basis for Drucker's claims
(1951: 81) that "The plants just grew by themselves" when there is such abundant evidence to the contrary, including that, significantly, among the Nuu-
chah-nulth, with whom Drucker conducted his research. And we must turn
a very critical eye to claims within the anthropological literature that the only
hint of human proprietorship was to be found in "small plots of problematical tobacco" (Murdock 1934: 223).

The origins of these errors and biases are revealing objects of study in their
own right. The cultural prejudices and gender biases, and the academic agendas of the majority culture in the nineteenth and twentieth centuries were
mutually reinforcing, and together served to undermine actual indigenous
knowledge of, and claims on, the natural environment. The vast majority of
documentary information from the Northwest Coast cultures was recorded

by men of European background and culture—explorers, traders, settlers, colonial officials, and anthropologists. It is perhaps not surprising, then, that activities such as root-vegetable tending and harvesting, or berry picking and the maintenance of berry grounds—all primarily "women's work"—have been given comparatively little attention. This is not necessarily a condemnation of these chroniclers; they, like all of us, were constrained by the blinders imposed by their own cultural, educational and experiential perspectives. They were also limited by the practical and logistical problems of access to such knowledge. Nonetheless it is time, now, to re-evaluate the evidence, and to explore the implications of these reevaluations.

In this light, the veracity of the colonial literature on Northwest Coast plant use appears particularly suspect, for it embodied not only the cultural biases of the time, but also manifested the territorial, military, economic, and political ambitions of the colonial elite. Colonial attempts to sever the ties between Native peoples and their resource base are among the first and most potent steps in the legal and physical removal of Native peoples from their lands (Wishart 1994). As such, the appropriation of resources is a prerequisite of colonial enterprises and resettlement in a land. The designation of the peoples of the Northwest Coast as "noncultivators," and indeed the denial of their use and proprietorship of plant resources generally, have had many profound implications for these peoples. This designation has undercut land claims on terrestrial resource sites. In turn this has facilitated the dispossession of lands, the displacement of Native peoples, and ultimately the loss of many traditional practices and much traditional knowledge (Deur 1997, 2000, 2002a,b). The claims of nineteenth-century colonial surveyors that Northwest Coast peoples "have no aboriginal plant which they cultivate" revealed as much—and perhaps more—of the authors' agendas and Eurocentric biases as they revealed the "ground truths" of indigenous plant management (R. Brown 1873: 50). When men of great political influence made similar pronouncements, they became potent tools of dispossession. British Columbia Land Commissioner Gilbert James Sproat's (1868 [1987]: 8) claim that indigenous land use was limited to "plucking wild fruits, and cutting a few trees," or British Columbia Governor James Douglas's (1859), characterization of Aboriginal people as mere "wandering denizen[s] of the forest" revealed much more than mere ethnocentric bias. Rather, these assertions served to devalue and dismiss indigenous peoples' competing claims to the lands and resources of the entire region, lands and resources that were coveted by the fledgling frontier societies on both sides of the international border. Together, colonial claims regarding traditional resource use provided a representational facade that served the extractive ambitions of colonial and frontier economies and the expansionist territorial agendas of colonial governments and the nation-state.

Northwest Coast First Peoples' concepts of land and resource ownership differed substantially from those of the European newcomers to the Northwest Coast. Seen through European eyes, Aboriginal peoples' ownership of the land was regarded as alien, much as their land-management techniques were, and

E. S. Curtis, *Gatherer on the Beach,* about 1915. The woman is Virginia Tom of Hesquiaht. (British Columbia Archives: D-08316)

both were generally considered neither valid nor legitimate. In most cases, the newcomers would recognize only large, permanent settlements and highly visible agricultural modification of crop plants as criteria for land ownership and retention. This demonstrates the important connections between land ownership, land title, and the question of cultivation and land management. As a consequence, the finely tuned mechanisms for managing and conserving resources that were embedded in indigenous institutions for ownership and proprietorship were disrupted when traditional territories were taken over by European systems of land use and tenure.

The emphasis on agriculture and enclosure of lands as a prerequisite for land title in colonial correspondence and treaty texts reflects a broader philosophical attitude toward the land and land use, one which was shared by many Europeans at the time. The relationship between private ownership and Western modes of "cultivation" was deeply embedded in European cultural traditions, and was upheld with almost religious conviction, as reflected in enlightenment philosopher John Locke's (1988: 291) words, originally written in 1690:

> God gave the World to Men in Common; but since he gave it them for their benefit, and the greatest Conveniences of Life they were capable to draw from it, it cannot be supposed that he meant it should always remain common and uncultivated. He gave it to the use of the Industrious and Rational (and Labour was to be his Title to it). [Locke 1988: 291]

As this volume demonstrates, coastal First Peoples expended considerable labor in the maintenance and modification of their plant communities. That this labor was not recognized or acknowledged by casual observers served as justification for appropriation.

The primacy given to Western forms of agriculture within colonial policy, and the lack of understanding of traditional ownership and resource management is reflected in the fourteen Fort Victoria Treaties imposed upon the First Nations on Vancouver Island between 1850 and 1854 by Governor James Douglas. There is little variation in the format and terms of the treaties.[3] The logic underlying the treaties was also uniform, reflecting the colonial experiences across the entirety of Canada and beyond. As the Secretary of the Hudson's Bay Company wrote to James Douglas regarding the division of indigenous peoples lands on Vancouver Island,

> in your negotiations with [the chiefs of the tribes] . . . you are to consider the natives as the rightful possessors of such *lands only as they are occupied by cultivation, or had houses built on* [emphasis ours], at the time when the Island came under the undivided sovereignty of Great Britain in 1846. All other land is to be regarded as waste, and applicable to the purposes of colonization . . . " [Archibald Barclay to James Douglas, May 16, 1850, cited by Arnett 1999: 32]

In each reserve on Canada's west coast, the outcome of the treaties was to prove similar in each case. The boundaries of the traditional territory to be surrendered were given in detail, and the following conditions were specified:

> The condition of our[4] understanding of this sale is this, that our village sites and enclosed fields are to be kept for our own use, for the use of our children, and for those who may follow after us; and the land shall be properly surveyed hereafter. It is understood, however, that the land itself, with these small exceptions, becomes the entire property of the white people forever; it is also understood that we are at liberty to hunt over the unoccupied lands, and to carry on our fisheries as formerly. [Government, British Columbia 1875: 5]

Wilson Duff (1969) points out that these early treaties reflect more of the European conceptions of land and property than the realities of the Aboriginal peoples with whom they were negotiated. According to Duff (1969: 3), "A treaty, of the kind discussed here, is a white man's certificate of a transaction . . . to read a treaty is to understand the white man's conception (or at least his rationalization) of the situation as it was and of the transaction that took place."

Nevertheless, these early treaties contain an implicit recognition that Aboriginal peoples exercised some form of ownership over the land that had to be extinguished by colonial authorities before settlement could occur (Fisher 1992: 67; Tennant 1990).[5] This recognition was shallow and short-lived, however, as later administrators would effectively deny the existence of Aboriginal title in British Columbia, and also the applicability of the Royal Proclamation of 1763 that asserted the need to negotiate with Aboriginal peoples for the surrender of lands. The result was a system whereby reserves were allotted by executive action, rather than by negotiation (Bartlett 1990). The unresolved issues of Aboriginal title to land in British Columbia are a legacy which is only now being addressed through modern-day treaty negotiations. Within the United States, the resolution of these long-standing disputes remains even more elusive, but is being slowly and sporadically explored within litigation concerning the interpretation of specific treaties and within "government-to-government" negotiations between American Indian tribes and individual federal agencies with regard to sites of enduring cultural significance.

As land issues are revisited along the Northwest Coast today, the practices of "keeping it living" described in this volume continue to be highly relevant in efforts to understand and promote restoration ecology, cultural restoration, and biological conservation. Today, the knowledge of Northwest Coast peoples and the practices they undertook to enhance a range of species and habitats must be considered, for example, in developing land policy and planning. The rationale for this is based both on ethical and legal grounds, as Tribes and First Nations increasingly assert and renegotiate their treaty rights with the governments of the United States and Canada. The recent work of the Scientific Panel for Sustainable Forest Practices in Clayoquot Sound (1995) provides one model of how traditional resource-management knowledge can both shape and be accommodated within contemporary resource-planning efforts. This panel has been co-chaired by the First Nations scholar Dr. Richard Atleo (author of this volume's foreword) and wildlife biologist Dr. Fred Bunnell, with panel members from both the academic and Nuu-chah-nulth communities. Working together, the panel generated many recommendations for land management policy that will embody the kinds of information incorporated in this volume into forest management planning in British Columbia's Clayoquot Sound. The experiences of panels of this sort point to and indicate the importance of identifying human-modified landscapes and culturally valued areas, rather than simply focusing attention on discrete locations containing commonplace archaeological sites. This in turn has allowed land managers to account for the cultural roots and environmental outcomes of diverse traditional environmental practices, including many that are addressed in the current volume. These cultural landscapes include traditional management sites—river estuaries encompassing former clover and silverweed gardens, for example, or tended berry patches that might have been pruned or burned in the past. Such landscapes may still be of cultural and dietary significance to contemporary indigenous peoples, and may yet continue to play important cultural, economic, and dietary roles into the future. Moreover, forest and fisheries management has been aided considerably by an understanding the role of human management in the production of environments that were until recently viewed as completely "natural."

Much more needs to be done in this regard, but the first step is simple recognition that such resource sites and traditional management areas existed in the past and that, in some cases, both the modified landscapes and their cultural significance still persist today. Future research may allow us to identify these areas in greater detail and with greater sensitivity to their genesis and traditional significance. Further, we may gain a better appreciation of their antiquity and former scale, using such methods as charcoal and pollen analyses, dendrochronology, or the chemical and physical analysis of soils. This work will be aided by more thorough historical research, employing early maps and written records, and through linguistic investigations of, for example, terms pertaining to traditional management techniques and concepts. Ongoing collaborative work with knowledgeable contemporary Aboriginal elders, such as those who advised us in this volume, is invaluable. Experimental modeling and monitored replication of traditional management techniques will also yield important new insights into the mechanics and outcomes of the many traditional methods used to maintain and enhance the output of plants on the Northwest Coast.

In presenting this work—in highlighting some of the research on Northwest Coast plant management and real examples of the practices traditionally employed in "keeping it living"—we hope to stimulate discussion, as well as a reevaluation of traditional assumptions regarding Northwest Coast traditional lifeways. It is our hope that we may foster a better understanding of the

Chief Adam Dick at Kingcome River estuary, near the site of his family's root gardens, with Kim Recalma-Clutesi and Johnny Moon in the boat (Photo by N. J. Turner)

importance of plant resources, particularly of plant foods, on the Northwest Coast. And we hope to provide some hint of the depth of knowledge and the amount of energy traditionally directed toward the maintenance and enhancement of these resources. For too long these practices have been belittled and overlooked. It is time for a considered and critical reevaluation.

Notes

1. See endnote 11 in Turner, Smith, and Jones, this volume.

2. Daisy Sewid-Smith explained that the Kwak'wala term translated by Boas as "prayer" would be more aptly translated as "words of praise" (Sewid-Smith et al. 1998).

3. Notably, the template for the so-called "Douglas treaties" was drawn from the wording of the 1840 Treaty of Waitangi negotiated by the British government with the Maori of New Zealand, where similar philosophies and misunderstandings were imposed (Arnett 1999: 31).

4. "Our" here refers to the chiefs who were invited to mark their X's on the treaty on behalf of the people in their communities.

5. It is notable that these Treaties were developed by James Douglas because he *did* partially recognize indigenous peoples' ownership concepts, although his reasons for this recognition were focused on the colonial perspective of the need for peaceful land takeover. In March, 1861, he wrote to the Secretary of State for the Colonies, asking that the remaining land of Vancouver Island be purchased from the native peoples: "As the native Indian population of Vancouver Island have distinct ideas of property in land, and mutually recognize their several exclusive possessory rights in certain districts, they would not fail to regard the occupation of such portions of the Colony by white settlers, unless with the full consent of the proprietary tribes, as national wrongs; and the sense of injury might produce a feeling of irritation against the settlers, and perhaps disaffection to the Government that would endanger the peace of the country. . . . Knowing their feelings on that subject, I made it a practice up to the year 1859 to purchase the native rights in the land. . . . but since that time it has not been within my power to continue it. . . ." (Victoria, 25th March, 1861, Governor Douglas to the Secretary of State for the Colonies, Government, British Columbia 1875: 19).

Bibliography

Acheson, S. R. 1991. In the Wake of the ya'åats' xaatgáay ['Iron People']: A Study of Changing Settlement Strategies among the Kunghit Haida. Ph.D. Dissertation. Oxford, UK: Oxford University.

Agee, J. K. 1993. *Fire Ecology of the Pacific Northwest Forests.* Covelo, CA: Island Press.

Agee, J. K., and L. Smith. 1984. Subalpine Tree Establishment after Fire in the Olympic Mountains, Washington. *Ecology* 65: 810–19.

Alaback, P., and J. Pojar. 1997. Vegetation from Ridgetop to Seashore. In P. K. Schoonmaker, B. von Hagen, and E. C. Wolf, eds., *The Rainforests of Home: Profile of a North American Bioregion.* Covelo, CA and Washington, DC: Island Press.

Allaire, L. 1984. A Native Mental Map of Coast Tsimshian Villages. In M. Seguin, ed., *The Tsimshian: Images of the Past, Views for the Present,* pp. 82–98. Vancouver: University of British Columbia Press.

Alvarez de Williams, A. 1983. Cocopa. In A. Ortiz, ed., *Handbook of North American Indians, Volume 10: The Southwest,* pp. 99–112. Washington, DC: Smithsonian Institution.

Ames, K. M. 1985. Hierarchies, Stress, and Logistical Strategies among Hunter-Gatherers in Northwestern North America. In T. D. Price and J. Brown, eds., *Prehistoric Hunter-Gatherers,* pp. 155–80. Orlando, FL: Academic Press.

———. 1988. Early Holocene Forager Mobility Strategies on the Southern Columbia Plateau. In J. A. Willig, C. M. Aikens, and J. Fagan, eds., *Early Human Occupation in Western North America.* Carson City: Nevada State Museum Papers, vol. 21.

———. 1991. Sedentism, a Temporal Shift or a Transitional Change in Hunter-Gatherer Mobility Strategies. In S. Gregg, ed., *Between Bands and States: Sedentism, Subsistence, and Interaction in Small Scale Societies,* pp. 108–33. Carbondale: Southern Illinois University Press.

———. 1991a. The Archaeology of the *Longue Durée*: Temporal and Spatial Scale in the Evolution of Social Complexity on the Southern Northwest Coast. *Antiquity* 65: 935–45.

———. 1994. The Northwest Coast: Complex Hunter-Gatherers, Ecology, and Social Evolution. *Annual Reviews of Anthropology* 23: 209–29.

———. 1995. Chiefly Power and Household Production on the Northwest Coast. In T. D. Price and G. Feinman, eds., *Foundations of Social Inequality,* pp. 155–87. New York: Plenum Press.

———. 1997. Review of "Dale R. Croes. The Hoko River Archaeological Site Complex: the Wet/dry Site (45CA213), 3000–1700 BP." *Antiquity* 71: 236–38.

———. 1998. Economic Prehistory of the North British Columbia Coast. *Arctic Anthropology* 35(1): 68–87.

———. 2000. Kennewick Man: Cultural Affiliation Report. Chapter 2, *Review of the Archaeological Data*. United States Department of the Interior, National Park Service. *www.cr.nps.gov/aad/kennewick/AMES.HTM.*

Ames, K. M., W. Cornett, and S. Hamilton. 1995. Archaeological Investigations (1991–1994) at 45CL (Cathlapotle), Clark County, Washington: A Preliminary Report. (Document in the possession of Kenneth Ames, Portland State University, Portland, OR).

Ames, K. M., and A. G. Marshall. 1980–81. Villages, Demography and Subsistence Intensification on the Southern Columbia Plateau. *North American Archaeologist* 2: 25–52.

Ames, K. M, and H. D. G. Maschner. 1999. *Peoples of the Northwest Coast: Their Archaeology and Prehistory*. London: Thames and Hudson.

Anderson, E. N. 1969. *Plants, Man, and Life*. Berkeley: University of California Press.

———. 1988. *The Food of China*. New Haven, CT: Yale University Press.

———. 1996. *Ecologies of the Heart: Emotion, Belief and the Environment*. New York: Oxford University Press.

Anderson, M. K. 1993a. *The Experimental Approach to Assessment of the Potential Ecological Effects of Horticultural Practices by Indigenous Peoples on California Wildlands*. Ph.D. Dissertation. Berkeley: Wildland Resource Science, University of California.

———. 1993b. Native Californians as Ancient and Contemporary Cultivators. In T. C. Blackburn and M. K. Anderson, eds., *Before the Wilderness: Environmental Management by Native Californians*. (Ballena Press Anthropological Papers No. 40), pp. 151–74. Menlo Park, CA: Ballena Press.

———. 1993c. California Indian Horticulture: Management and Use of Redbud by the Southern Sierra Miwok. *Journal of Ethnobiology* 11(1): 145–57.

———. 1996. Tending the Wilderness. *Restoration and Management Notes* 14(2): 154–66.

Anonymous. 1849. Colonization of Vancouver Island. *The (London) Times*. May 4, 1849, pp. 18–19.

Anonymous. 1898. Revival of the Wapato. *The (Portland) Oregonian*. Article in Scrapbook 35, Oregon Historical Society Archives. Portland, OR.

Arima, E. Y., D. St. Claire, L. Clamhouse, J. Edgar, C. Jones, and J. Thomas. 1991. *Between Ports Alberni and Renfrew: Notes on West Coast Peoples*. Ottawa, ON: Canadian Ethnology Service, Mercury Series Paper 121; and Hull, PQ: Canadian Museum of Civilization.

Arnett, C. 1999. *The Terror of the Coast. Land Alienation and Colonial War on Vancouver Island and the Gulf Islands, 1849–1863*. Vancouver, BC: Talon Books.

Arnold, J. E. 1992. Complex Hunter-Gatherer-Fishers of Prehistoric California: Chiefs, Specialists and Maritime Adaptations of the Channel Islands. *American Antiquity* 57: 60–84.

———. 1995. Transportation Innovation and Social Complexity among Maritime Hunter-Gatherer Societies. *American Anthropology* 97(4): 733–47.

———. 1996. The Archaeology of Complex Hunter-Gatherers. *Journal of Archaeological Method and Theory* 3(2): 77–125.

———. 2001. Social Evolution and the Political Economy in the Northern Channel Islands. In J. Arnold, ed., *The Origins of a Pacific Coast Chiefdom*, pp. 287–96. Salt Lake City: University of Utah Press.

Asch, D., and N. Asch. 1978. The Economic Potential of *Iva annua* and Its

Prehistoric Importance in the Lower Illinois Valley. In R. I. Ford, ed., *The Nature and Status of Ethnobotany*. Anthropological Papers 67, pp. 301–41. Ann Arbor: University of Michigan Museum of Anthropology.

———. 1985. Prehistoric Plant Cultivation in West-Central Illinois. In R. I. Ford, ed., *Prehistoric Food Production in North America*. Anthropological Papers 75, pp. 149–204. Ann Arbor: University of Michigan Museum of Anthropology.

Babcock, M. 1967. *Camas—descriptions of getting and preparing—from informants of Tsartlip Reserve (W. Saanich), Vancouver Island*. Unpublished manuscript, cited with permission of author; copy in possession of N. Turner.

Bach, E., ed. 1992. *Ninuyems Xenaksiala: Stories from Kitlope and Kemano by Gordon Robertson*. Unpublished manuscript. Amherst, MA: Department of Linguistics, University of Massachusetts.

Balée, W. 1994. *Footprints in the Forest. Ka'apor Ethnobotany—the Historical Ecology of Plant Utilization by an Amazonian People*. New York: Columbia University Press.

Bamforth, D. B., and P. Bleed. 1997. Technology, Flaked Stone Technology, and Risk. In C. M. Barton and G. A. Clark, eds., *Rediscovering Darwin: Evolutionary Theory in Archaeological Explanation*, pp. 109–40. Archaeological Papers of the American Anthropological Society, Volume 7.

Barbeau, M. 1929. *Totem Poles of the Gitksan, Upper Skeena River, British Columbia*. National Museum of Canada, Bulletin No. 61; Anthropological Series No. 12. Ottawa, ON: Department of Mines.

Barnes, G. L. 1993. *China, Korea and Japan: The Rise of Civilization in East Asia*. London: Thames and Hudson.

Barnett, H. G. 1939. Culture Element Distributions: IX Gulf of Georgia Salish. *Anthropological Records*. Berkeley: University of California Press.

———. 1955. *The Coast Salish of British Columbia*. Eugene: University of Oregon Monographs. Studies in Anthropology, 4.

Bartlett, R. H. 1990. *Indian Reserves and Aboriginal Lands in Canada, a Homeland: A Study in Law and History*. Saskatoon: University of Saskatchewan, Native Law Centre.

Barrett, S. W. 1980. Indians and Fire. *Western Wildlands* 6(3): 17–21

Bates, D., T. Hess, and V. Hilbert. 1994. *Lushootseed Dictionary*. Seattle: University of Washington Press.

Beaglehole, J. C., ed. 1967. *The Journals of Captain James Cook on his Voyages of Discovery; the Voyage of the "Resolution" and "Discovery," 1776–1780*. Parts 1 and 2. Cambridge, UK: Hakluyt Society.

Bean, L. J., and H. W. Lawton. 1993. Some Explanations for the Rise of Cultural Complexity in Native California with Comments on Proto-Agriculture and Agriculture. In T. C. Blackburn and M. K. Anderson, eds., *Before the Wilderness: Environmental Management by Native Californians*, pp. 27–54. (Ballena Press Anthropological Papers No. 40). Menlo Park, CA: Ballena Press.

Beckwith, B. R. 2002. Colonial Eden or Indigenous Cultivated Landscape: Reconstructing Nineteenth Century Camas Meadows on Southern Vancouver Island. In P. J. Burton, ed., *Garry Oak Ecosystem Restoration: Progress and Prognosis*, pp. 64–72. Proceedings of the Third Annual Meeting of the B.C. Chapter of the Society for Ecological Restoration, April 27–29, 2002, University of Victoria, B.C. Chapter of the Society for Ecological Restoration, Victoria, British Columbia.

———. 2004. *"The Queen Root of This Clime": Ethnoecological Investigations of Blue Camas* (Camassia leichtlinii *(Baker) Wats.,* C. quamash *(Pursh) Greene;*

Liliaceae) and its Landscapes on SouthernVancouver Island, British Columbia. Unpublished PhD dissertation. Victoria, B.C.: Department of Biology, University of Victoria.

Bender, B. 1978. Gatherer-Hunter to Farmer: A Social Perspective. *World Archaeology* 10: 204–22.

———. 1985. Prehistoric Developments in the American Mid-Continent and in Brittany, Northwest France. In T. D. Price and J. Brown, eds., *Prehistoric Hunter-Gatherers*, pp. 21–58. Orlando, FL: Academic Press.

Benedict, R. 1934. *Patterns of Culture.* Boston: Houghton Mifflin Co.

Berkes, F. 1999. *Sacred Ecology: Traditional Ecological Knowledge and Resource Management.* Philadelphia, PA: Taylor & Francis.

———., ed. 1989. *Common Property Resources: Ecology and Community-Based Sustainable Development.* London and New York: Belhaven Press.

Bettinger, R. L. 1977. Aboriginal Human Ecology in Owens Valley: Prehistoric Change in the Great Basin. *American Antiquity* 42: 3–17.

———. 1991. *Hunter-Gatherers: Archaeological and Evolutionary Theory.* New York: Plenum Press.

Betts, R. 1985. *In Search of York.* Boulder, CO: Colorado Associated University Press.

Beynon, W., and M. Barbeau. n.d. The Marius Barbeau and William Beynon Field Notes. Ottawa, ON: Archives, Canadian Centre for Folk Culture Studies, National Museum of Man.

Biddle, N., and P. Allen 1962. *The Journals of the Expedition Under the Command of Capts. Lewis and Clark.* New York: Heritage Press, Easton Press.

Binford, L. R. 1968. Post-Pleistocene Adaptations. In S. R. Binford and L. R. Binford, eds., *New Perspectives in Archaeology*, pp. 313–41. Chicago: Aldine Publishing Company.

———. 2001. *Constructing Frames of Reference: An Analytical Method for Archaeological Theory Building Using Ethnographic and Environmental Data Sets.* Berkeley: University of California Press.

Bishop, C. 1927. Journal of the Ship *Ruby*, by Captain Charles Bishop. T. C. Elliott, ed. *Oregon Historical Quarterly* 28(3): 258–80.

Blackburn, T. C., and M. K. Anderson, eds. 1993. *Before the Wilderness: Environmental Management by Native Californians.* Menlo Park, CA: Ballena Press.

Blackman, M. 1990. Haida: Traditional Culture. In W. Suttles, ed., *Handbook of North American Indians, Volume 7: Northwest Coast*, pp. 240–60. Washington, DC: Smithsonian Institution.

Black, M. J. 1994. Plant Dispersal by Native North Americans of the Canadian Subarctic. In R. Ford, ed., *The Nature and Status of Ethnobotany*, 2nd edition, pp. 255–62. Ann Arbor: Anthropological Papers, Museum of Anthropology, University of Michigan, No. 67.

Blaut, J. M. 1993. *The Colonizer's Model of the World: Geographical Diffusionism and Eurocentric History.* New York: Guilford.

Boas, F. 1889. The Houses of the Kwakiutl Indians, British Columbia. *Proceedings of the United States National Museum for 1888* 11: 197–312.

———. 1891. The LkungEn. *60th Report of the British Association for the Advancement of Science for 1890*, pp. 563–92. London, UK.

———. 1894a. *Chinook Texts.* Bureau of American Ethnology Bulletin 20. Washington, DC: Smithsonian Institution.

———. 1894b. Indian Tribes of the Lower Fraser River. *64th Report of the British Association for the Advancement of Science for 1890*, pp. 4534–63. London, UK.

———. 1889. The Indians of British Columbia: Tlingit, Haida, Tsimshian, Kotomaga. Fifth Report on the Northwestern Tribes of Canada. *Report of the British Association for the Advancement of Science*, pp. 797–93. London, UK.

———. 1895. *Indianische Sagen von der Nord-Pacifischen Küste Amerikas*. Berlin, GDR: A. Asher.

———. 1901. *Kathlamet Texts*. Bureau of American Ethnology, Bulletin 26. Washington DC: Smithsonian Institution.

———. 1905. *Kwakiutl Texts*. (*Publications of the Jessup North Pacific Expedition*, vol. 3). (Originally published in separate installments, in 1902 and 1905). Leiden, The Netherlands: E. J. Brill.

———. 1908. *Kwakiutl Texts*. Publications of the Jessup North Pacific Expedition 10(1), 1. Leiden, The Netherlands: E. J. Brill.

———. 1909. *The Kwakiutl of Vancouver Island*. (Publications of the Jessup North Pacific Expedition 5(2) / *Bulletin of the American Museum of Natural History* 8) 2: 301–522.

———. 1910. *Kwakiutl Tales*. Columbia University Contributions to Anthropology, vol. 2. New York: Columbia University Press.

———. 1916. *Tsimshian Mythology*. 31st Annual Report, Bureau of American Ethnology, pp. 27–1037. Washington, DC: Smithsonian Institution.

———. 1921. *Ethnology of the Kwakiutl*. Bureau of American Ethnology 35th Annual Report, Parts 1 and 2. Washington, DC: Smithsonian Institution.

———. 1928. *Anthropology and Modern Life*. New York: Dover.

———. 1929. Metaphorical Expression in the Language of the Kwakiutl Indians. In *Versameling van Opstellen door Oud-Leeringen en Bevriende Vakgenooten Opgedragen ann Mgr. Prof. Dr. Jos. Schrijen, 3 Mei 1929*. Chartres, France, pp. 147–53. (Reprinted in F. Boas, ed., *Race, Language, and Culture*, pp. 232–39. Chicago: University of Chicago Press, 1940).

———. 1930. *The Religion of the Kwakiutl Indians*. New York: Columbia University Press. (reprinted in 1969, New York: AMS Press Inc.).

———. 1933. Relationships Between North-West America and North-East Asia. In D. Jenness, ed. *The American Aborigines: Their Origins and Antiquity*. Toronto: University of Toronto Press. (Reprinted in 1940 in F. Boas, ed., *Race, Language and Culture*, pp. 344–55. Chicago: University of Chicago Press).

———. 1934. *Geographical Names of the Kwakiutl Indians*. Columbia University Contributions to Anthropology, 20. New York: Columbia University Press.

———. 1940. *Race, Language and Culture*. Chicago and New York: Macmillan.

———. 1947. Kwakiutl Grammar with a Glossary of the Suffixes. H. Boas Yampolsky, ed. *Transactions of the American Philosophical Society* 37(3): 204–377.

———. 1966. *Kwakiutl Ethnography*. H. Codere, ed. Chicago: University of Chicago Press.

———. n.d. Notes on Kwakiutl Plant Use. (unpublished ms.). Item #1915, Philadelphia: American Philosophical Society Archives.

Bogucki, P. 1995. Prelude to Agriculture in North-central Europe. In *Before Farming: Hunter-Gatherer Society and Subsistence*, MASCA Research Papers in Science and Archaeology, Supplement to Volume 12, pp. 105–66. Philadelphia: University of Pennsylvania Museum of Archaeology and Anthropology.

Boit, J. 1960. *Voyage of the Columbia Around the World with John Boit, 1790–1793*. D. O. Johansen, ed. Portland, OR: Beaver Books.

Boserup, E. 1965. *The Conditions of Agricultural Growth: The Economics of Agrarian Change under Population Pressure*. Chicago: Aldine Publishing Company.

Botkin, D. B. 1990. *Discordant Harmonies: A New Ecology for the Twenty-first Century*. Oxford, UK: Oxford University Press.

Bouchard, R., and D. Kennedy. 1990. *Clayoquot Sound Indian Land Use.* Report prepared for: MacMillan Bloedel Limited, Fletcher Challenge Canada and British Columbia Ministry of Forests. Victoria, BC: British Columbia Indian Language Project. [B.&K. Key to consultants: PW Peter Webster; LS Luke Swan; SS Stanley Sam; GL George Louie; JS James Swam; JW Jesie Webster; JT Joe Tom; MT Mike Tom; AP Alice Paul; MH Mary Hayes; MJ Margaret Joseph; BA Ben Andrews.]

Boyd, R. T. 1985. *The Introduction of Infectious Diseases Among the Indians of the Pacific Northwest, 1774–1874.* Unpublished Ph.D. dissertation. Seattle: Department of Anthropology, University of Washington.

———. 1986. Strategies of Indian Burning in the Willamette Valley. *Canadian Journal of Anthropology* 5 (1): 65–86.

———, ed. 1999. *Indians, Fire, and the Land in the Pacific Northwest.* Corvallis: Oregon State University Press.

Boyd, R. T., and Y. P. Hajda. 1987. Seasonal Population Movement along the Lower Columbia River: The Social and Ecological Context. *American Ethnologist* 14: 309–26.

Braidwood, R. 1952. From Cave to Village. *Scientific American* 187: 62–66.

———. 1960. The Agricultural Revolution. *Scientific American* 203: 130–48.

Brayshaw, T. C. 1996. *Trees and Shrubs of British Columbia.* Vancouver: University of British Columbia Press.

British Columbia, Government. 1871–1981. *Revised Statutes of British Columbia.* 1924 C.93 s.1. Victoria, BC: Government Printing Office.

———. 1875. *Papers Connected with the Indian Land Question, 1850–1875.* Victoria, BC: Government Printer (Reprinted 1987, as *Indian Land Question, 1850–1875, 1877*).

———. 1885. Sessional Papers. Victoria, BC: Government Printer.

———. 1887. Sessional Papers. Victoria, BC: Government Printer.

British Columbia Archives. MS-1077. Newcombe Family Papers. Vol. 35, File 8, Frame 833. Charles Newcombe, Haida-Ksaan Notebook, 1902. Victoria, BC: British Columbia Archives and Records.

British Columbia Ministry of Forests. 1988. *Biogeoclimatic Zones of British Columbia 1988.* (Annotated map.) Victoria, BC: Research Branch, Ministry of Forests.

———. 1997. *Culturally Modified Trees of British Columbia.* Victoria, BC: Research Branch, Ministry of Forests.

Brody, H. 2000. *The Other Side of Eden: Hunters, Farmers and the Shaping of the World.* Vancouver, BC: Douglas & McIntyre.

Bronson, B. 1977. The Earliest Farming: Demography as Cause and Consequence. In C. Reed, ed., *Origins of Agriculture*, pp. 23–48. The Hague, The Netherlands: Mouton Publishers.

Brookfield, H. C. 1972. Intensification and Deintensification in Pacific Agriculture: A Theoretical Approach. *Pacific Viewpoint* 13: 30–48.

Broughton, J. M. 1994a. Declines in Mammalian Foraging Efficiency During the Late Holocene, San Francisco Bay, California. *Journal of Anthropological Archaeology* 13: 371–401.

———. 1994b. Late Holocene Resource Intensification in the Sacramento Valley, California: The Vertebrate Evidence. *Journal of Archaeological Science* 21: 501–14.

———. 1997. Widening Diet Breadth, Declining Foraging Efficiency, and Prehistoric Harvest Pressure: Ichthyofaunal Evidence from the Emeryville Shellmound. *Antiquity* 71: 845–58.

Brown, D. 1997. *Archaeological Investigations at the Somenos Creek Site (DeRw 18).* Permit report 1994-122. On file, British Columbia Archaeology Branch, Victoria.

Brown, R. 1868. On the Vegetable Products, Used by the Northwest American Indians as Food and Medicine, in the Arts, and in Superstitious Rites. *Transactions of the Edinburgh Botanical Society* 9: 378–96.

———. 1873–1876. *The Races of Mankind: Being a Popular Description of the Characteristics, Manners and Customs of the Principal Varieties of the Human Family.* 4 vols. London, UK: Cassell, Petter, and Galpin.

Buck, C. D. 1949. *A Dictionary of Selected Synonyms in the Principal Indo-European Languages.* Chicago: University of Chicago Press.

Burch, E. S. Jr., and L. J. Ellanna, eds. 1994. *Key Issues in Hunter-Gatherer Research.* Oxford, UK: Berg Publishers.

Burley, D. V. 1980. Marpole. *Anthropological Reconstructions of a Prehistoric Northwest Culture Type.* Department of Archaeology Publication, No. 8. Burnaby, BC: Simon Fraser University.

———. 1983. Cultural Complexity and Evolution in the Development of Coastal Adaptation among the Micmac and Coast Salish. In R. J. Nash, ed., *The Evolution of Maritime Cultures on the Northeast and Northwest Coasts of America,* pp. 157–72. Department of Archaeology Publication, No. 11. Burnaby, BC: Simon Fraser University.

Burley, D. V., and C. Knusel. 1989. Burial Patterns and Archaeological Interpretation: Problems in the Recognition of Ranked Society in the Coast Salish Region. Paper presented at the Circum-Pacific Prehistory Conference. Seattle, WA.

Burtchard, G. C. 1988. *Patterns in the Intensive Utilization of Camas: Regional Implications from the Calispell and Willamette Valleys.* Paper presented to the 41st annual Northwest Anthropological Conference, Tacoma, WA.

Butler, V. L. 2000. Resource Depression on the Northwest Coast of North America. *Antiquity* 74: 649–61.

Butzer, K. 1990. The Indian Legacy in the American Landscape. In M. P. Conzen, ed., *The Making of the American Landscape,* pp. 27–50. Boston: Unwin Hyman.

Calvert, S. G. 1970. *A Cultural Analysis of Faunal Remains from Three Archaeological Sites in Hesquiat Harbour, B.C.* Ph.D. Dissertation, Vancouver: Department of Anthropology, University of British Columbia.

Canada. n.d. RG 10 v.1022. Unpublished Documents in the Public Archives of Canada. Ottawa, ON.

Canada and British Columbia. 1916. *Report of the Royal Commission on Indian Affairs for the Province of B.C.* Victoria, BC: Government Printer.

Cannizzo, J. 1983. George Hunt and the Invention of Kwakiutl Culture. *The Canadian Review of Sociology and Anthropology* 20(1): 44–58.

Cannon, A. 1991. *The Economic Prehistory of Namu.* Burnaby, BC: Archaeology Press, Simon Fraser University.

———. 1998. Contingency and Agency in the Growth of Northwest Coast Maritime Economies. *Arctic Anthropology* 35(1): 57–67.

Carlson, K. T., ed. 2001. *A Stó:lo Coast Salish Historical Atlas.* Vancouver, BC: Douglas & McIntyre; Seattle: University of Washington Press; and Chilliwack, BC: Stó:lo Heritage Trust.

Carlson, K. T., and J. Lutz. 1997. Sto:lo People and the Development of the B.C. Wage Labour Economy. In K. T. Carlson, ed., *You Are Asked to Witness: The Sto:lo in Canada's Pacific Coast History,* pp. 109–24. Chilliwack, BC: Sto:lo Heritage Trust.

Carlson, R. L. 1996a. Coastal British Columbia in the Light of North Pacific Maritime Adaptations. *Arctic Anthropology* 35(1): 23–35.

———. 1996b. Introduction to Early Human Occupation in British Columbia. In R. L. Carlson and L. Dalla Bona, eds., *Early Human Occupation in British Columbia*, pp. 1–4. Vancouver: University of British Columbia Press.

Carniero, R. L. 1970. A Theory of the Origin of the State. *Science* 169: 733–38.

Castetter, E., and W. Bell. 1951. *Yuman Indian Agriculture*. Albuquerque: University of New Mexico Press.

Chatters, J. C. 1995. Population Growth, Climatic Cooling, and the Development of Collector Strategies on the Southern Plateau, Western North America. *Journal of World Prehistory* 9(3): 341–400.

Childe, V. G. 1951. *Man Makes Himself*. New York: The New American Library.

Clark, J. S., and Royall, P. D. 1995. Transformation of a Northern Hardwood by Aboriginal (Iroquois) Fire: Charcoal Evidence from Crawford Lake, Ontario, Canada. *The Holocene* 5: 1–9.

———. 1996. Local and Regional Sediment Charcoal Evidence for Fire Regimes in Presettlement North-eastern North America. *Journal of Ecology* 84: 365–82.

Clark, W. R., and D. W. Kroeker. 1993. Population Dynamics of Muskrats in Experimental Marshes at Delta, Manitoba. *Canadian Journal of Zoology* 71: 1620–28.

Codere, H. 1959. The Understanding of the Kwakiutl. In W. Goldschmidt, ed., *The Anthropology of Franz Boas*. American Anthropological Association Memoir No. 89, 61(5): Part 2. pp. 61–75.

Cohen, M. N. 1977. *The Food Crisis in Prehistory*. New Haven, CT: Yale University Press.

Cole, D., and B. Lochner, eds. 1993. *"To The Charlottes," George Dawson's 1878 Survey of the Queen Charlotte Islands*. Vancouver: University of British Columbia Press.

Collins, J. M. 1974. *Valley of the Spirits: The Upper Skagit Indians of Western Washington*. American Ethnological Society Monographs, 56. Seattle: University of Washington Press.

Compton, B. D. 1993a. *Upper North Wakashan and Southern Tsimshian Ethnobotany: The Knowledge and Usage of Plants and Fungi Among the Oweekeno, Hanaksiala, Haisla and Kitasoo Peoples of the Central and North Coast of British Columbia*. Ph.D. Dissertation, Vancouver: Department of Botany, University of British Columbia.

———. 1993b. The North Wakashan "Wild Carrots": Clarification of Some Ethnobotanical Ambiguity in Pacific Northwest Apiaceae. *Economic Botany* 47(3): 297–303.

Connolly, T. J., C. M. Hodges, et al. 1997. Cultural Chronology and Environmental History in the Willamette Valley, Oregon. Ellensburg, WA: 50th Annual Northwest Anthropological Conference.

Cook, J., and J. King. 1784. *A Voyage to the Pacific Ocean . . . performed under the discretion of Captains Cook, Clerke, and Gore, in his majesty's ship the Resolution and the Discovery in the years 1776, 1777, 1778, 1779, and 1780*. 3 vols. London: G. Nichol and T. Cadell.

Cotton, A. F. 1894. Survey of Toba Inlet, Powell Lake, and Kingcome Inlet. *British Columbia Sessional Papers* (Crown Land Surveys Series), Victoria, pp. 800–02.

Coupland, G. 1998. Maritime Adaptation and Evolution of the Developed Northwest Coast Pattern on the Central Northwest Coast. *Arctic Anthropology* 35(1): 36–56.

Cove, J. J. 1992. The Gitksan Traditional Concept of Land Ownership. In D. Miller,

W. Heber, J. Dempsey, and C. Beal, eds. *The First Ones: Readings in Indian/ Native Studies*, pp. 234–39. Saskatchewan Indian Federated College, Piapot Indian Reserve No. 75.

Cowan, C. W., and P. J. Watson, eds. 1992. *The Origins of Agriculture: An International Perspective.* Washington, DC: Smithsonian Institution.

Cox, R. 1957. *The Columbia River or scenes and adventures during a residence of six years on the western side of the Rocky Mountains among various tribes of Indians hitherto unknown; together with "A Journey across the American Continent."* (orig. published, 1832). E. I. Stewart and J. R. Stewart, eds. Norman: University of Oklahoma Press.

Craig, J. 1998. *Nature Was the Provider: Traditional Ecological Knowledge and Inventory of Culturally Significant Plants in the Atleo River Watershed, Ahousaht Territory, Clayoquot Sound, B.C.* M.Sc. Thesis, Victoria, BC: Department of Biology and School of Environmental Studies, University of Victoria.

Craig, J., and R. V. Smith. 1997. *"A Rich Forest": Traditional Knowledge, Inventory and Restoration of Culturally Important Plants and Habitats in the Atleo River Watershed.* Report to Ahousaht Band Council, Ahousaht, BC, and Long Beach Model Forest, Ucluelet, BC. Victoria, BC: University of Victoria

Crawford, G. 1992a. The Transitions to Agriculture in Japan. In A. Gebauer and T. D. Price, eds. *Transitions to Agriculture in Prehistory*, pp. 117–32. Madison, WI: Prehistory Press.

————. 1992b. Prehistoric Plant Domestication in East Asia. In C. W. Cowan and P. J. Watson, eds., *The Origins of Agriculture*, pp. 7–38. Washington, DC: Smithsonian Institution Press.

Cressman, L. S., D. L. Cole, W. A. Davis, T. M. Newman, and D. J. Scheans. 1960. Cultural Sequences at The Dalles, Oregon: A Contribution to Pacific Northwest Prehistory. *Transactions of the American Philosophical Society, New Series* 50(10).

Croes, D. R. 1991. Exploring Prehistoric Subsistence Change on the Northwest Coast. In D. R. Croes, ed., *Long-Term Subsistence Change in Prehistoric North America*, pp. 337–66. Research in Economic Anthropology, Supplement 6. Greenwich, CT: JAI Press.

————. 1995. *The Hoko River Archaeological Complex: The Wet/Dry Site (45CA213), 3,000–1,700 B.P.* Pullman, WA: Washington State University Press.

Croes, D. R., and S. Hackenberger. 1988. Hoko River Archaeological Complex: Modeling Prehistoric Northwest Coast Economic Evolution, pp. 19–86. Research in Economic Anthropology, Supplement 3. Greenwich, CT: JAI Press.

Dall, W. H. 1877. *Tribes of the Extreme Northwest.* Washington, DC: U.S. Government Printing Office.

Daly, R., and R. Vast. 2001. *Gitxsan and Wet'suwet'en Ownership and Management.* Vancouver: University of British Columbia Press.

Darby, M. C. 1996. *Wapato for the People: An Ecological Approach to Understanding the Native American Use of Sagittaria latifolia on the Lower Columbia River.* M.A. Thesis, Portland, OR: Department of Anthropology, Portland State University.

Darling, J. D. 1955. *The Effects of Culture Contact on the Tsimshian System of Land Tenure During the 19th Century.* M.A. Thesis, Vancouver: Department of Economics, Political Science and Sociology, University of British Columbia.

Davis, A., with B. Wilson, and B. Compton. 1995. *Salmonberry Blossoms in the New Year. Some Culturally Significant Plants of the Haisla Known to Occur Within the Greater Kitlope Ecosystems.* Kitamaat, BC: Nanakila Press.

Dawson, G. M. 1887a. Notes and Observations on the Kwakiool People of the

Northern Part of Vancouver Island and Adjacent Coasts, Made During the Summer of 1885. *Proceedings and Transactions, Royal Society of Canada* 5(2): 63–98.

———. 1887b. Map 1: Geological Map of the Northern Part of Vancouver Island and Adjacent Coasts. *Canadian Geological and Natural History Survey*. 1886 Annual Report. Montreal.

Day, G. M. 1953. The Indian as an Ecological Factor in the Northeastern Forest. *Ecology* 34(2): 329–46.

de Laguna, F. 1960. The Story of a Tlingit Community: A Problem in the Relationship between Archeological, Ethnological, and Historical Methods. Washington, DC: Smithsonian Institution Bureau of American Ethnology Bulletin 172.

———. 1972. *Under Mount Saint Elias: the History and Culture of the Yakutat Tlingit*. Washington, DC: Smithsonian Contributions to Anthropology, vol. 7.

Delgamuukw v. British Columbia. 11 December, 1997. File 23799 (S.C.C.).

Denevan. W. 1992. The Pristine Myth: The Landscape of the Americas in 1492. *Annals of the Association of American Geographers* 82(3): 369–85.

Densmore, F. 1939. *Nootka and Quileute Music*. Bureau of American Ethnology, Bulletin 124. Washington, DC: Smithsonian Institution.

Deur, D. 1997. Subsistence, Territorial Sovereignty, and Ethnographic Representations of the Northwest Coast. Paper presented at the Association of American Geographers 1997 Annual Meeting, Fort Worth, TX, April 5, 1997.

———. 1998. Estuarine Rhizome Cultivation on the Northwest Coast: A Critical Assessment. Paper presented at the Association of American Geographers 1998 Annual Meeting, Boston, MA, March 28, 1998.

———. 1999. Salmon, Sedentism and Cultivation: Toward an Environmental Prehistory of the Northwest Coast. In D. Goble and P. Hirt, eds., *Northwest Lands and Peoples: An Environmental History Anthology*, pp. 129–55. Seattle: University of Washington Press.

———. 2000. *A Domesticated Landscape: Native American Plant Cultivation on the Northwest Coast of North America*. Ph.D. Dissertation, Baton Rouge: Department of Geography and Anthropology, Louisiana State University.

———. 2002a. Plant Cultivation on the Northwest Coast: A Reconsideration. *Journal of Cultural Geography* 19(1): 9–35.

———. 2002b. Rethinking Precolonial Plant Cultivation on the Northwest Coast of North America. *The Professional Geographer* 54(2): 140–57.

———. 2002c. Huckleberry Mountain Traditional Use Study. Report to the USDI National Park Service, Columbia-Cascades Support Office, Seattle, WA, and the USDA Forest Service, Rogue River National Forest, Medford, OR.

Deur, D., and N. J. Turner. 1999. Plant Cultivation on the Central Northwest Coast?: First Nations Management of Estuarine Plant Resources. Newport, OR: 52nd Annual Northwest Anthropological Conference.

Diamond, J. 1994. How to Tame a Wild Plant. *Discover* September 1994, 100–06.

Doebley, J., A. Stec, and L. Hubbard. 1997. The Evolution of Apical Dominance in Maize. *Nature* 386: 485–88.

Donald, L. 1983. Was Nuu-chah-nulth-aht (Nootka) Society Based on Slave Labor? In E Tooker, ed., *The Development of Political Organization in Native North America*, pp. 108–19. Proceedings of the American Ethnological Society.

———. 1997. *Aboriginal Slavery on the Northwest Coast of North America*. Berkeley: University of California Press.

Doolittle, W. E. 1992. Agriculture in North America on the Eve of Contact: a Reassessment. *Annals of the Association of American Geographers* 82(3): 386–401.

———. 2000. *Cultivated Landscapes of Native North America*. Oxford, UK: Oxford University Press.

Dorweiler, J., A. Stec, J. Kermicle, and J. Doebley. 1993. Teosinte Glume Architecture 1: A Genetic Locus Controlling a Key Step in Maize Evolution. *Science* 262: 233–35.

Douglas, J. 1859. Dispatch from Governor Douglas to the Right Hon. Sir E. B. Lytton, Bart. (No. 114), Victoria, Vancouver's Island, March 14. P.16. *British Columbia. Papers Connected with the Indian Land Question 1850–1875, 1877*. Victoria: Government Printing Office.

Douglas, W. 1790. Voyage of the *Iphigenia*, Capt. Douglas, from Samboingan, to the North West Coast of America. In J. Meares, *Voyages Made in the Years 1788 and 1789, from China to the North West Coast of America. . . .* , pp. 287–372. London: Logographic Press.

Downie, M. W. 1859. Report of Trip to the Interior of British Columbia. In *Papers Relating to British Columbia*, pp. 71–74. Victoria: British Columbia Archives and Records Service.

Drucker, P. 1951. *The Northern and Central Nootkan Tribes*. Bureau of American Ethnology, Bulletin No. 144. Washington, DC: Smithsonian Institution.

———. n.d. Unpublished Field Notes: Nootka. Washington, DC: Smithsonian Institution, National Anthropological Archives.

Duff, W. 1952. *The Upper Stalo Indians of the Fraser River of B.C.* Anthropology in British Columbia, Memoir No. 1, Victoria: British Columbia Provincial Museum.

———, ed. 1959. *Histories, Territories, and Laws of the Kitwancool*. Anthropology in British Columbia, Memoir No. 4. Victoria: Department of Education, British Columbia Provincial Museum.

———. 1964. *The Indian History of British Columbia, Volume 1: The Impact of the White Man*. Anthropology in British Columbia Memoir No. 5. Victoria, BC: Provincial Museum of Natural History and Anthropology.

———. 1969. The Fort Victoria Treaties. *B.C. Studies* 3: 3–57.

Dunn, J. 1844. *History of the Oregon Territory and British North American Fur-trade: with an Account of the Habits and Customs of the Principal Native Tribes on the Northern Continent*. London: Edwards and Hughes.

Dunn, J. A., and A. Booth. 1990. Tsimshian of Metlakatla, Alaska. In W. Suttles, ed., *Handbook of North American Indians, Volume 7: Northwest Coast*, pp. 294–97. Washington, DC: Smithsonian Institution.

Easton, N. A. 1985. The Underwater Archaeology of Straits Salish Reef-netting. M.A. Thesis, Victoria, BC: Department of Anthropology, University of Victoria.

Edwards, G. T. 1979. Indian Spaghetti. *The Beaver*. Autumn 1979: 4–11.

———. 1980. Bella Coola: Indian and European Medicines. *The Beaver* Winter 1980: 5–11.

Eldridge, M., and S. Acheson. 1992. The Antiquity of Fish Weirs on the Southern Coast: A Response to Moss, Erlandson and Stuckenrath. *Canadian Journal of Archaeology* 16: 112–16.

Ellen, R. 1982. *Environment, Subsistence and System: The Ecology of Small Scale Social Formations*. Cambridge, UK: University of Cambridge Press.

———. 1998. Foraging, Starch Extraction and the Sedentary Lifestyle in the Lowland Rainforest of Central Seram. In T. Ingold, D. Riches, and J. Woodburn, eds., *Hunters and Gatherers*, vol. 1, pp. 117–34. Oxford, UK: Berg Publishers.

Elmendorf, W. W. 1960. *The Structure of Twana Culture*. Pullman: Washington State University Research Studies 28(3), Monograph Supplement 2.

————. 1992. *The Structure of Twana Culture.* Pullman: Washington State University Press (reprint of Elmendorf 1960).

Elster, J. 1986. *An Introduction to Karl Marx.* Cambridge, UK: Cambridge University Press.

Emmons, G. T. 1991. *The Tlingit Indians.* Edited with additions by F. de Laguna. Seattle: University of Washington Press; and New York: American Museum of Natural History.

Erlandson, J. M., M. A. Tveskov, and R. S. Byram. 1997. The Development of Maritime Adaptations on the Southern Northwest Coast of North American. *Arctic Anthropology* 35(1): 6–22.

Fahnestock, G. R., and Agee, J. K. 1983. Biomass Consumption and Smoke Production by Prehistoric and Modern Forest Fires in Western Washington. *Journal of Forestry* 81: 653–57.

Fetzer, P. S. 1950–51. Nooksack Ethnographic Field Notes. (In the possession of Wayne Suttles.)

Finney, B. P., I. Gregory-Eaves, S. V. Douglas, and J. P. Smol. 2002. Fisheries Productivity in the Northeastern Pacific Ocean Over the Past 2,200 Years. *Nature* 416: 729–33.

Fish, S. 1995. Mixed Agricultural Technologies in Southern Arizona and their Implications. In H. W. Toll, ed., *Soil, Water, Biology, and Belief in Prehistoric and Traditional Southwestern Agriculture*, pp. 101–16. New Mexico Archaeological Council Special Publication, No. 2, Albuquerque, NM.

Fisher, R. 1992. *Contact and Conflict: Indian-European Relations in British Columbia, 1774–1890.* 2nd ed. Vancouver: University of British Columbia Press.

Fiske, J. 1991. Colonization and the Decline of Women's Status: the Tsimshian Case. *Feminist Studies* 17: 509–53.

Fladmark, K. R. 1975. A Paleoecological Model for Northwest Coast Prehistory. *Archaeological Survey of Canada Paper, Mercury Series.* No. 43. Ottawa, ON: National Museum of Civilization.

Flannery, K. 1968. Archaeological Systems Theory and Early Mesoamerica. In B. Meggers, ed., *Anthropological Archaeology in the Americas*, pp. 67–87. Washington, DC: Anthropological Society of Washington.

————. 1973. The Origins of Agriculture. *Annual Review of Anthropology* 2: 271–310.

————. 1986. *Guila Naquitz.* Orlando, FL: Academic Press.

Folan, W. J. 1984. On the Diet of Early Northwest Coast Peoples. *Current Anthropology* 25: 123–24.

Ford, C. S. 1941. *Smoke From Their Fires: The Life of a Kwakiutl Chief.* New Haven, CT: Yale University Press.

Ford, R. I. 1979. Gathering and Gardening: Trends and Consequences of Hopewell Subsistence Strategies. In D. Brose and N. Greber, eds., *Hopewell Archaeology*, pp. 234–38. Kent, OH: Kent State University Press.

————. 1985a. The Processes of Plant Food Production in Prehistoric North America. In R. I. Ford, ed., *Prehistoric Food Production in North America*, pp. 1–18. University of Michigan Museum of Anthropology Papers No. 75. Ann Arbor, MI: Department of Anthropology, University of Michigan.

————. 1985b. *Prehistoric Food Production in North America.* University of Michigan Museum of Anthropology, Anthropological Papers No. 75. Ann Arbor: University of Michigan Press.

Forde, C. D. 1934. *Habitat, Economy, and Society.* London: Methuen.

Foucault, M. 1970. *The Order of Things: An Archaeology of the Social Sciences.* London: Tavistock.

Fowler, C. S., and N. J. Turner. 1999. Ecological/Cosmological Knowledge and Land Management among Hunter-Gatherers. In R. B. Lee and R. H. Daly, eds., *Cambridge Encyclopedia of Hunters and Gatherers*, pp. 419–25. Cambridge, UK: Cambridge University Press.

Freeman, M., and L. Carbyn, eds. 1988. *Traditional Knowledge and Renewable Resource Management in Northern Regions*. Edmonton: Boreal Institute for Northern Studies, University of Alberta.

French, D. n.d. Aboriginal Control of Huckleberry Yield in the Northwest. Paper presented at the 1957 American Anthropological Association Meetings, Chicago, IL. Reprinted in R. Boyd, ed., *Indians, Fire, and the Land in the Pacific Northwest*, pp. 31–35. Corvallis: Oregon State University Press.

Fritz, G. 1990. Multiple Pathways to Farming in Precontact Eastern North America. *Journal of World Prehistory* 4: 387–435.

Galois, R. 1994. *Kwakwa̱ka̱'wakw Settlements, 1775–1920: A Geographical Analysis and Gazetteer*. Vancouver: University of British Columbia Press.

Garfield, V. E. 1939. Tsimshian Clan and Society. *University of Washington Publications in Anthropology* 7(3): 167–340.

———. 1966. The Tsimshian and Their Neighbours. In V. E. Garfield and P. S. Wingert, eds., *The Tsimshian Indians and Their Arts*, pp. 3–70. Seattle: University of Washington Press.

Garrick, D. 1998. *Shaped Cedars and Cedar Shaping (Hanson Island, B.C.)*. Vancouver, B.C. Western Canada Wilderness Committee.

Gatschet, A. S. 1877. Texts, Sentences and Vocabulary of the Atfa'lati Dialect of the Tualatin Language of Willamette Valley, Northwestern Oregon. (Manuscript field notes, catalogue no. 472-a, National Anthropological Archives, Smithsonian Institution, Washington, DC).

Gavin, D. G. 2001. Estimation of Inbuilt Age in Radiocarbon-Derived Ages of Soil Charcoal for Fire History Studies. *Radiocarbon* 43: 27–44.

Gavin, D. G., L. B. Brubaker, and K. P. Lertzman. 2002. Holocene Fire History of a Coastal Temperate Rain Forest Based on Soil Charcoal Radiocarbon Dates. *Ecology* 84(1): 186–201.

Gavin, D. G., K. P. Lertzman, L. B. Brubaker, and E. Nelson. 1996. Long-Term Fire Histories in a Coastal Temperate Rainforest. *Bulletin of the Ecological Society of America* 77(3): 157.

———. 1997. Holocene Fire Size and Frequency in a Coastal Temperate Rainforest. *Bulletin of the Ecological Society of America* 78(4): 93.

Gibbs, G. 1855. *Report to Captain Mc'Clellan, on the Indian Tribes of the Territory of Washington. Report of Explorations for a Route ... from St. Paul to Puget Sound, by I. I. Stevens*, pp. 402–34. (Reprinted 1972 by Ye Galleon Press, Fairfield, WA.)

———. 1863. *Alphabetical Vocabularies of the Clallam and Lummi*. New York: Cramoisy Press.

———. 1877. Tribes of Western Washington and Northwestern Oregon. In J. W. Powell, ed., *Contributions to North American Ethnology*. Washington, DC: U.S. Geographical and Geological Survey of the Rocky Mountain Region 1(2): 157–361.

Gibson, J. 1978. European Dependence upon American Natives: The Case of Russian America. *Ethnohistory* 25(4): 359–85.

Gilbert, H. 1990. Productivite végétal dans un marais intertidal d'eau douce. *Canadian Journal of Botany* 68: 825–56.

Gilmore, M. R. 1931. Dispersal by Indians a Factor in the Extension of Discontinuous Distribution of Certain Species of Native Plants. *Papers of the Michigan Academy of Science, Arts and Letters* 13: 89–94.

Gisday Wa and Delgam Uukw. 1989. *The Spirit in the Land. The Opening Statement of the Gitksan and Wet'suwet'en Hereditary Chiefs in the Supreme Court of British Columbia, May 11, 1987.* Gabriola, BC: Reflections.

Goldman, Irving. 1975. *The Mouth of Heaven: An Introduction to Kwakiutl Religious Thought.* New York: John Wiley and Sons.

Goldschmidt, W. R., and T. H. Haas. 1946. *Possessory Rights of the Natives of Southeastern Alaska, a Report to the Commissioner of Indian Affairs.* Washington, DC: U.S. Department of the Interior.

Golovin, P. N. 1979. *The End of Russian America: Captain P. N. Golovin's Last Report, 1862.* Translated by Basil Dmytryshyn and E.A.P. Crownhart-Vaughan. Portland, OR: Western Imprints, Oregon Historical Society.

Golovin, P. N. 1983. *Civil and Savage Encounters, the Worldly Travel Letters of an Imperial Russian Navy Officer 1860–1861.* Translated and annotated by Basil Dmytryshyn and E.A.P. Crownhart-Vaughan. Portland, OR: Western Imprints, Oregon Historical Society.

Gottesfeld, A. S., R. W. Mathewes, and L. M. J. Gottesfeld. 1991. Holocene Debris Flows and Environmental History, Hazelton Area. *Canadian Journal of Earth Science* 28: 1583–1593.

Gottesfeld, L. M. J. 1993. *Plants, Land and People: A Study of Wet'suwet'en Ethno-botany.* M.A. Thesis, Edmonton: Department of Anthropology, University of Alberta.

———. 1994a. Aboriginal Burning for Vegetation Management in Northwest British Columbia. *Human Ecology* 22(2): 171–88.

———. 1994b. Wet'suwet'en Ethnobotany: Traditional Plant Uses. *Journal of Ethnobiology* 14(2): 185–210.

———. 1994c. Conservation, Territory, and Traditional Beliefs: An Analysis of Gitksan and Wet'suwet'en Subsistence, Northwest British Columbia, Canada. *Human Ecology* 22 (No. 4): 443–65.

Gould, F. W. 1942. A Systematic Treatment of the Genus *Camassia* Lindl. *American Midland Naturalist* 28: 712–42.

Gould, R. 1985. "Now Let's Invent Agriculture . . .": A Critical Review of Concepts of Complexity Among Hunter-Gatherers. In T. D. Price and J. Brown, eds., *Prehistoric Hunter-Gatherers,*. pp. 427–35. Orlando, FL: Academic Press.

Grant, W. C. ca. 1848. *Report on Vancouver Island.* Unpublished report to Governor James Douglas, Fort Victoria.

Griffen, P. B. 1989. Hunting, Farming, and Sedentism in a Rain Forest Foraging Society. In S. Kent, ed., *Farmers as Hunters: The Implications of Sedentism*, pp. 60–70. Cambridge, UK: Cambridge University Press.

Grissino-Mayer, H. D. 1995. *Tree-Ring Reconstructions of Climate and Fire History at El Malpais National Monument, New Mexico.* Ph.D. dissertation. Tucson: University of Arizona.

Grumet, R. S. 1975. Changes in Coast Tsimshian Redistributive Activities in the Fort Simpson Region of British Columbia, 1788–1862. *Ethnohistory* 22(4): 295–318.

Gunther, E. 1927. *Klallam Ethnography.* Seattle: University of Washington Publications in Anthropology 1(5): 171–314.

———. 1973. *Ethnobotany of Western Washington; The Knowledge and Use of Indigenous Plants by Native Americans.* Rev. ed., first published 1945. Seattle: University of Washington Press.

Gwaganad. 1990. Speaking in the Haida Way. In V. Andruss, C. Plant, J. Plant, and E. Wright, eds., *Home! A Bioregional Reader*, pp. 49–52. Gabriola Island, BC: New Society Publishers.

Hadley, K. S. 1999. Forest History and Meadow Invasion at the Rigdon Meadows Archaeological Site, Western Cascades, Orego*n. Physical Geography* 20: 116–33.

Haeberlin, H. K., and E. Gunther. 1930. *The Indians of Puget Sound.* Seattle: University of Washington Publications in Anthropology 4(1): 1–81.

Hajda, Y. P. 1984. *Regional Social Organization in the Greater Lower Columbia, 1792–1830.* Ph.D. Dissertation, Seattle: Department of Anthropology, University of Washington.

Hallett, D. J. 2001. *Holocene Fire History and Climate Change in Southern British Columbia, based on high-resolution analyses of sedimentary charcoal.* Ph.D. dissertation. Burnaby, BC: Simon Fraser University.

Hallett, D. J., D. S. Lepofsky, R. W. Mathewes, and K. P. Lertzman. 2003. 11,000 Years of Fire History and Climate in the Mountain Hemlock Rainforests of Southwestern British Columbia Based on Sedimentary Charcoal. *Canadian Journal of Forest Research* 33: 292–312.

Halliday, W. 1910. *1909 Annual Report: Kwagiulth Agency.* Department of Indian Affairs. (Unpublished ms. on microfilm, British Columbia Provincial Museum.) Ottawa, ON: Canada Department of Indian Affairs.

Halperin, R. H. 1994. *Cultural Economies, Past and Present.* Austin: University of Texas Press.

Hamilton, E. H., and H. K. Yearsley. 1987. Revegetation After Burning in the Sub-Boreal Spruce Zone. Abstract in *B.C. Vegetation Working Group Newsletter* 1(2): 8.

Hanson, D. K. 1991. *Late Prehistoric Subsistence in the Strait of Georgia Region of the Northwest Coast.* Ph.D. Dissertation, Burnaby, BC: Department of Archaeology, Simon Fraser University.

Hardin, G. 1968. The Tragedy of the Commons. *Science* 162: 1243–48.

Harlan, J. R. 1975. *Crops and Man.* Madison, WI: American Society of Agronomy.

———. 1995. *The Living Fields: Our Agricultural Heritage.* Oxford, UK: Oxford University Press.

Harlan, J. R., J. De Wet, and E. Price. 1973. Comparative Evolution of Cereals. *Evolution* 27: 311–25.

Harrington, J. P. n.d. Tlingit Ethnographic Notes, vol. 1, no. 11. Unpublished ms., Washington, DC: National Anthropological Archives, Smithsonian Institution.

Harris, C. 1997. *The Resettlement of British Columbia: Essays on Colonialism and Geographical Change.* Vancouver: University of British Columbia Press.

Harris, D. R. 1977. Alternative Pathways Toward Agriculture. In C. A. Reed, ed., *Origins of Agriculture*, pp. 179–244. The Hague, The Netherlands: Mouton.

———. 1989. An Evolutionary Continuum of People–Plant Interaction. In D. R. Harris and G. Hillman, eds., *Foraging and Farming: the Evolution of Plant Exploitation*, pp. 11–26. London: Unwin Hyman.

———. 1990. Settling Down and Breaking Ground: Rethinking the Neolithic Revolution. *Twaalfde Kroon-Voordracht.* Amsterdam, The Netherlands: Stichting Nederlands Museum voor Anthropologie en Praehistorie.

———. 1996a. Introduction: Themes and Concepts in the Study of Early Agriculture. In D. R. Harris, ed., *The Origins and Spread of Agriculture and Pastoralism in Eurasia*, pp. 1–11. Washington, DC: Smithsonian Institution.

———. 1996b. Domesticatory Relations of People, Plants and Animals. In R. Ellen and K. Fukui, eds., *Redefining Nature*, pp. 437–63. Oxford, UK: Berg Publishers.

Harris, M. 1968. *The Rise of Anthropological Theory.* New York: Thomas Y. Crowell.

———. 1979. *Cultural Materialism: The Struggle for a Science of Culture.* New York: Vintage.

Harris, R. C. 1997. *The Resettlement of British Columbia: Essays on Colonialism and Geographical Change.* Vancouver: University of British Columbia Press.

Harrison, C. 1925. *Ancient Warriors of the North Pacific: The Haidas, their Laws, Customs and Legends, with some Historical Account of the Queen Charlotte Islands.* London: H. F. & G. Witherby.

Haskins, L. 1934. *Wild Flowers of the Pacific Coast.* Portland, OR: Binfords and Mort.

Hayden, B. 1981. Research and Development in the Stone Age: Technological Transitions among Hunter-Gatherers. *Current Anthropology* 22: 519–48.

———. 1990. Nimrods, Piscators, Pluckers, and Planters: The Emergence of Food Production. *Journal of Anthropological Anthropology* 9: 31–69.

———. 1995. A New Overview of Domestication. In T. D. Price and A. B. Gebauer, eds., *Last Hunters—First Farmers,* pp. 273–300. Santa Fe, NM: School of American Research.

———. 1996. Pathways to Power: Principles for Creating Socioeconomic Inequalities. In T. D. Price and G. M. Feinman, eds., *Foundations of Social Inequality,* pp. 15–85. New York: Plenum Press.

———. 2001. Fabulous Feasts: A Prolegomenon to the Importance of Feasting. In M. Dietler and B. Hayden, eds., *Feasts: Archaeological and Ethnographic Perspectives on Food, Politics, and Power,* pp. 23–64. Washington, DC: Smithsonian Institution Press.

Hebda, R. J., and R. W. Mathewes. 1984. *Postglacial History of Cedar and Evolution of Northwest Coast Native Cultures.* Paper presented to the 35th annual Northwest Anthropological Conference, Burnaby, BC: Simon Fraser University.

Hebda, R., and C. Whitlock. 1997. Environmental History. In P. K. Schoonmaker, B. von Hagen, and E. C. Wolf, eds., *The Rainforests of Home, Profile of a North American Bioregion,* pp. 227–54. Washington, DC: Island Press.

Hedrick, U. P. 1972. *Sturtevant's Edible Plants of the World.* New York: Dover.

Herskovitz, M. J. 1953. *Franz Boas: The Science of Man in the Making.* New York: Charles Scribner's Sons.

Hesse, B. 1984. These Are Our Goats: The Origins of Herding in West Central Iran. In J. Clutton-Brock and C. Grigson, eds., *Animals and Archaeology,* vol. 3, pp. 243–64. Oxford, UK: British Archaeological Reports, International Series 202.

Hewes, G. W. 1947. *Aboriginal Use of Fishery Resources in Northwestern North America.* Ph.D. Dissertation, Berkeley: Department of Geography, University of California.

———. 1973. Indian Fisheries Productivity in Pre-contact Times in the Pacific Salmon Area. *Northwest Anthropological Research Notes* 7(2): 133–55.

Higgs, E. S. 1972. The Origins of Animal and Plant Husbandry. In E. S. Higgs, ed., *Papers in Economic Prehistory,* pp. 3–15. Cambridge, UK: Cambridge University Press.

———. 1975. Paleoeconomy. In E. S. Higgs, ed., *Paleoeconomy,* pp. 1–7. Cambridge, UK: Cambridge University Press.

Hill-Tout, C. 1900. Notes on the Sk°qo´ýmic of British Columbia, a Branch of the Great Salish Stock of North America. *70th Report of the British Association for the Advancement of Science for 1900,* pp. 472–549 (Appendix II). London, UK.

———. 1903. Ethnological Studies of the Mainland Halkome´lem, a Division of the Salish of British Columbia. *72nd Report of the British Association for the Advancement of Science for 1902,* pp. 355–449. London, UK.

———. 1904. Ethnological Report on the Stsee´lis and Sk°aúlits Tribes of the Halkome´lem Division of the Salish of British Columbia. *Journal of the Anthropological Institute of Great Britain and Ireland* 34: 311–76.

———. 1907. Report of the Ethnology of the South-eastern Tribes of Vancouver Island, British Columbia. *Journal of the Royal Anthropological Institute of Great Britain and Ireland* 37: 306–74.

———. 1978. *The Salish People. The Local Contribution of Charles Hill-Tout.* Vol. III: *The Mainland Halkomelem,* ed. Ralph Maud. Vancouver, BC: Talonbooks.

Hitchcock, L. C., and A. Cronquist. 1973. *Flora of the Pacific Northwest.* Seattle: University of Washington Press.

Hobler, P. M. 1983. Settlement Location Determinants: An Exploration of Some Northwest Coast Data. In R. J. Nash, ed., *The Evolution of Maritime Cultures on the Northeast and Northwest Coasts of America,* pp. 149–56. Publication Number 11. Burnaby, BC: Simon Fraser University, Department of Archaeology.

Hole, F. 1996. The Context of Caprine Domestication in the Zagros Region. In D. Harris, ed., *The Origins and Spread of Agriculture and Pastoralism in Eurasia,* pp. 263–81. Washington, DC: Smithsonian Institution Press.

Howat, J. P. 1929. Records of Some Early Transactions at Fort Simpson, B.C. *The Beaver* March, 155–56.

Howay, F. W. 1920. The Voyage of the *Hope. Washington Historical Quarterly* 11(1): 3–28.

———. 1929. Potatoes: Records of Some Early Transactions at Fort Simpson, British Columbia. *The Beaver,* March, pp. 155–56.

———. 1942. The Introduction of Intoxicating Liquors amongst the Indians of the Northwest Coast. *British Columbia Historical Quarterly* 6(3): 157–69.

Howay, F. W., and T. C. Elliott. 1942. Extract from the Journal of Thomas Manby, Master of the *Chatham. Oregon Historical Quarterly* 43(4): 318–27.

Howell, T. 1903. *A Flora of Northwest America.* Vol. 1. Portland, OR: C. C. Lane.

Hu, P., D. B. Hannaway, and H. W. Youngberg. 1992. *Forage Resources of China.* Corvallis, OR: Beijing Agricultural University and Oregon State University.

Hudson's Bay Company Archives. 1832–70. *Papers Relating to Fort Simpson.* Winnipeg, MN: Public Archives of Manitoba.

Huelsbeck, D. R. 1988. The Surplus Economy of the Central Northwest Coast. In B. L. Isaac, ed., *Prehistoric Economies of the Northwest Coast,* pp. 149–77. Research in Economic Anthropology, Supplement 3. Greenwich, CT: JAI Press.

Hunn, E. S., and J. Selam. 1990. Nch'i-Wána, *"The Big River": Mid-Columbia Indians and Their Land.* Seattle: University of Washington Press.

Hunn, E. S., and Williams, N. M. 1982. Introduction. In N. M. Williams and E. S. Hunn, eds., *Resource Managers: North America and Australian Hunter-Gatherers,* pp. 1–16. Boulder, CO: Westview Press.

Hunt, G. 1922. Notes on Kwakiutl Plant Use. (Unpublished ms., C.F. Newcombe notes.) Victoria, BC: British Columbia Archives and Records Service.

Hyatt, M. 1990. *Franz Boas, Social Activist: The Dynamics of Ethnicity.* Contributions to the Study of Anthropology, Number 6. Westport, CT: Greenwood.

Imamura, K. 1996. *Prehistoric Japan: New Perspectives on Insular East Asia.* Honolulu: University of Hawai'i Press.

Inglis, G. B., D. R. Hudson, B. K. Rigsby, and B. Rigsby. 1990. Tsimshian of British Columbia Since 1900. In W. Suttles, ed., *Handbook of North American Indians, Volume 7: Northwest Coast.* Washington, DC: Smithsonian Institution.

Inglis, J. T., ed. 1993. *Traditional Ecological Knowledge: Concepts and Cases.* Ottawa, ON: International Program on Traditional Ecological Knowledge.

Jackson, D., ed. 1962. *Letters of the Lewis and Clark Expedition, with Related Documents, 1783–1854.* Urbana, IL: University of Illinois Press.

Jacobs, E. D. n.d. *Tillamook Field Notebooks and Research Notes.* Unpublished ms., Seattle: Melville Jacobs Collection, University of Washington Archives.

———, ed. 1990. *Nehalem Tillamook Tales.* (Northwest Reprints.) Corvallis: Oregon State University Press.

Jacobs, M. 1945. *Kalapuyan Texts: Santiam Kalapuya Myth Texts.* Seattle: University of Washington Publications in Anthropology, vol. 2.

Jacobsen, G. L., and R.H.W. Bradshaw. 1981. The Selection of Sites for Paleo-vegetation Studies. *Quaternary Research* 16: 80–96.

Jarman, M., G. Bailey, and H. Jarman, eds. 1982. *Early European Agriculture: Its Foundation and Development.* Cambridge, UK: Cambridge University Press.

Jefferson, C. 1973. *Some Aspects of Plant Succession in Oregon Estuarine Salt Marshes.* Ph.D. Dissertation, Corvallis: Oregon State University.

Jenness, D. 1934–35. *The Saanich Indians of Vancouver Island.* Manuscript (No. VII-G-8M) in Canadian Ethnology Service Archives, Canadian Museum of Civilization, Ottawa, ON, and Hull, PQ.

Jewitt, J. R. 1807. *A Journal Kept at Nootka Sound, by John R. Jewitt. One of the Surviving Crew of the Ship* Boston, *of Boston; John Salter, Commander Who Was Massacred on 22d of March, 1803. Interspersed with Some Account of the Natives, Their Manners and Customs.* Printed by the author, Boston, MA. Reprinted 1987, H. Stewart, ed. Seattle: University of Washington Press; and Vancouver, BC: Douglas & McIntyre.

Jochim, M. A. 1976. *Hunter-Gatherer Subsistence and Settlement, A Predictive Model.* New York: Academic Press.

Johannesen, C. L., W. Davenport, A. Millet, and S. McWilliams. 1971. The Vegetation of the Willamette Valley. *Annals of the Association of American Geographers* 61 (2): 286–306.

Johnson, M., ed. 1992. *Lore: Capturing Traditional Environmental Knowledge. Dene Cultural Institute, Hay River, Northwest Territories.* Ottawa, ON: International Development Research Centre.

Jones, J. 2002. *"We Looked After All the Salmon Streams." Traditional Heiltsuk Cultural Stewardship of Salmon and Salmon Streams: A Preliminary Study.* Unpublished M.A. Thesis, Victoria, British Columbia: School of Environmental Studies, University of Victoria.

Kane, P. 1967. *Wanderings of an Artist.* Rutland, VT: Charles E. Tuttle.

Keeley, L. 1995. Protoagricultural Practices among Hunter-Gatherers: A Cross Cultural Survey. In T. D. Price and A. Gebauer, eds., *Last Hunters—First Farmers*, pp. 243–72. Santa Fe, NM: School of American Research.

Keely, P. B. 1980. *Nutrient Compositon of Selected Important Plant Foods of the Pre-Contact Diet of the Northwest Coast Native Peoples.* Unpublished M.Sc. Thesis, Seattle: Nutritional Sciences and Textiles, University of Washington.

Kelly, R. L. 1991. Sedentism, Sociopolitical Inequality, and Resource Fluctuations. In S. Gregg, ed., *Between Bands and States: Sedentism, Subsistence, and Interaction in Small Scale Societies*, pp. 135–60. Carbondale, IL: Southern Illinois University Press.

———. 1995. *The Foraging Spectrum: Diversity on Hunter-Gatherer Lifeways.* Washington, DC: Smithsonian Institution Press.

Kelly, W. H. 1977. *Cocopa Ethnography.* Tucson: Anthropological Papers of the University of Arizona, 29.

Kennedy, D. 1993. *Looking for Tribes in All the Wrong Places: An Examination of the Central Coast Salish Social Network.* M.A. Thesis, Victoria, BC: Department of Anthropology, University of Victoria.

Kennedy, D., and R. Bouchard. 1983. *Sliammon Life, Sliammon Lands.* Vancouver, BC: Talonbooks.

Kew, M. 1992. Salmon Availability, Technology, and Cultural Adaptation in the Fraser River Watershed. In B. Hayden, ed., *A Complex Culture of the British Columbia Plateau: Traditional Stl'atl'imx Resource Use*, pp. 177–221. Vancouver: University of British Columbia Press.

Kirch, P. V. 1984. *The Evolution of the Polynesian Chiefdoms.* Cambridge, UK: Cambridge University Press.

Knight, R. 1996. *Indians at Work: An Informal History of Native Labour in British Columbia, 1858–1930.* Rev. ed. Vancouver, BC: New Star Books.

Kniffen, F. 1939. Pomo Geography. *University of California Publications in American Archaeology and Ethnology* 36: 353–400.

Krause, A. 1956. *The Tlingit Indians.* E. Gunther, trans. Seattle: University of Washington Press.

Kroeber, A. L. 1939. *Cultural and Natural Areas of Native North America.* Berkeley: University of California Press.

———. 1962. *A Roster of Civilizations and Cultures.* Viking Publications in Anthropology No. 33. New York: Wenner-Gren.

Kubiak-Martens, L. 1996. Evidence for Possible Use of Plant Foods in Paleolithic and Mesolithic Diet from the Site of Calowanie in the Central Part of the Polish Plain. *Vegetation History and Archaeobotany* 5: 33–38.

Kuhnlein, H. V., and N. J. Turner. 1987. Cow-parsnip (*Heracleum lanatum* Michx.): An Indigenous Vegetable of Native People of Northwestern North America. *Journal of Ethnobiology* 6(2): 309–24.

———. 1991. *Traditional Plant Foods of Canadian Indigenous Peoples. Nutrition, Botany and Use.* In S. Katz, ed., *Food and Nutrition in History and Anthropology*, vol. 8. Philadelphia, PA: Gordon and Breach Science Publishers.

Kuhnlein, H. V., N. J. Turner, and P. D. Kluckner. 1982. Nutritional Significance of Two Important "Root" Foods (Springbank Clover and Pacific Silverweed) used by Native People on the Coast of British Columbia. *Ecology of Food and Nutrition* 12(2): 89–95.

Kwawkewlth Agency. n.d. Letterbooks of the Kwawkewlth Agency. (Unpublished documents, on file with University of British Columbia Libraries Special Collections, Vancouver.)

Lakatos, I. 1970. Falsification and the Methodology of Scientific Research Programmes. In I. Lakatos and H. Musgrave, eds., *Criticism and the Growth of Knowledge*, pp. 91–196. Cambridge, UK: Cambridge University Press.

LaLande, J. and R. Pullen. 1999. Burning for a "Fine and Beautiful Open Country": Native Uses of Fire in Southwestern Oregon. In R. Boyd, ed., *Indians, Fire and the Land in the Pacific Northwest*, pp. 255–76. Corvallis: Oregon State University Press.

Lampman, B. H. 1946. *The Coming of the Pond Fishes.* Portland, OR: Binfords and Mort.

Langdon, S. J. 1979. Comparative Tlingit and Kaigani Adaptation to the West Coast of the Prince of Wales Archipelago. *Ethnology* 18(2): 101–19.

Lantz, T. 2001. The Population Ecology and Ethnobotany of Devil's Club (*Oplopanax horridus* (Sm.) Torr. & A. Gray ex. Miq. Araliaceae). M.Sc. Thesis. Victoria, BC: Department of Biology and School of Environmental Studies, University of Victoria.

Lantz, T., and N. J. Turner. 2003. Traditional Phenological Knowledge (TPK) of Aboriginal Peoples in British Columbia. *Journal of Ethnobiology* 23(2): 263–86.

Larsen, C. P. S., and G. M. MacDonald. 1994. Lake Morphometry, Sediment Mixing and the Selection of Sites for Fine Resolution Palaeoecological Studies. *Quaternary Science Reviews* 12: 781–92.

Lawrence, S. 1972. The Haida Potato. *Raincoast Chronicles* 1(1): 44.

Lazenby, R. A., and P. McCormack. 1985. Salmon and Malnutrition on the Northwest Coast. *Current Anthropology* 26: 379–84.

Lee, R. B. 1968a. What Hunters Do for a Living, or, How to Make Out on Scarce Resources. In R. B. Lee and I. Devore, eds., *Man the Hunter*, pp. 30–48. Chicago: Aldine Publishing Company.

———. 1968b. Introduction: Problems in the Study of Hunter-gatherers. In R. B. Lee and I. Devore, eds., *Man the Hunter*, pp. 3–12. Chicago: Aldine Publishing Company.

Lee, R. B. and I. Devore, eds. 1968. *Man the Hunter*. Chicago: Aldine Publishing Company.

Lepofsky, D. n.d. The Northwest. In P. Minnis, ed., *Plants and People in Ancient North America*. Washington, DC: Smithsonian Institution Press.

Lepofsky, D., E. Heyerdahl, K. Lertzman, D. Schaepe, and B. Mierendorf. 2003. Historical Meadow Dynamics in Southwest British Columbia: A Multidisciplinary Approach. *Conservation Ecology* (now *Ecology and Society*) 7(3): 5; *www.consecol.org/vol7/iss3/art5*.

Lepofsky, D., K. Lertzman, D. Hallett, and R. Mathewes. 2005. Climate Change and Culture Change on the Southern Coast of British Columbia 2400–1200 cal. B.P.: An Hypothesis. *American Antiquity* 70: 267–93.

Lepofsky, D., and S. Peacock. 2004. A Question of Intensity: Exploring the Role of Plant Foods in Northern Plateau Prehistory. In B. Prentiss and I. Kuijt, eds., *Complex Hunter-Gatherers: Evolution and Organization of Prehistoric Communities on the Plateau of Northwestern North America*, pp. 115–39. Salt Lake City: University of Utah Press.

Lepofsky, D., N. J. Turner, and H. V. Kuhnlein. 1985. Determining the Availability of Traditional Wild Plant Foods: An Example of Nuxalk Foods, Bella Coola, British Columbia. *Ecology of Food and Nutrition* 16: 223–41.

Lertzman, K. P. 1992. Patterns of Gap-Phase Replacement in a Sub-alpine Old Growth Forest. *Ecology* 73: 657–69.

Lertzman, K., D. Gavin, D. Hallett, L. Brubaker, D. Lepofsky, and R. Mathewes. (2002) Long-term Fire Regime Estimated from Soil Charcoal in Coastal Temperate Rainforests. *Conservation Ecology* (now *Ecology and Society*) 6(2): 5; *www.consecol.org/vol6/iss2/art5*.

Lertzman, K. P., D. Gavin, D. Hallett, D. Lepofsky, L. Brubaker, R. Mathewes, and E. Nelsen. n.d. Long-term Fire Histories in Coastal Temperate Rainforest. Manuscript in K. P. Lertzman's possession.

Lertzman, K. P., and C. J. Krebs. 1991. Gap-Phase Structure of a Subalpine Old-Growth Forest. *Canadian Journal of Forest Research* 12: 1730–41.

Lertzman, K. P., G. Sutherland, A. Inselberg, and S. Saunders. 1996. Canopy Gaps and the Landscape Mosaic in a Temperate Rainforest. *Ecology* 77: 1254–70.

Lewis, A. B. 1906. Tribes of the Columbia Valley and the Coast of Washington and Oregon. *Memoirs of the American Anthropological Association* 1(2): 147–209.

Lewis, H. T. 1973. Patterns of Indian Burning in California. *Ecology and Ethnohistory*. Socorro, CA: Ballena Press Anthropological Papers No. 1.

———. 1977. Maskuta: The Ecology of Indian Fires in Northern Alberta. *The Western Canadian Journal of Anthropology* 7(1): 15–52.

———. 1978. Traditional Indian Uses of Fire in Northern Alberta. In D. E. Dube, ed. *Fire Ecology in Resource Management,* pp. 61–62. Edmonton, AL: Northern Forest Research Centre, Canadian Forestry Service, Environment Canada.

———. 1982. *A Time for Burning.* Occasional Publication No. 17, Edmonton: Boreal Institute for Northern Studies, University of Alberta.

Lewis, H. T., and T. A. Ferguson. 1988. Yards, Corridors, and Mosaics: How to Burn a Boreal Forest. *Human Ecology* 16(1): 57–77.

Lewis, M. 1814. *The Lewis and Clark Expedition,* N. Biddle and P. Allen, eds. Philadelphia, PA: J. B. Lippincott Co., 1961.

Liljeblad, S., and K. Fowler. 1986. Owens Valley Paiute. In W. D'Azevedo, ed. *Handbook of North American Indians, Volume 11: Great Basin,* pp. 412–34. Washington, DC: Smithsonian Institution.

Lillard, C. 1987. Introduction. In G. M. Sproat, *The Nootka. Scenes and Studies of Savage Life.* C. Lillard, ed. Victoria, BC: Sono Nis Press.

Linton, R. 1936. *The Study of Man: An Introduction.* New York: Appleton-Century-Crofts.

Locke, J. 1988. *Two Treatises of Government.* Peter Laslett, ed. Cambridge, UK: Cambridge University Press.

Loewen, D. C. 1998. *Ecological, Ethnobotanical, and Nutritional Aspects of Yellow Glacier Lily,* Erythronium grandiflorum *Pursh (Liliaceae) in Western Canada.* M.Sc. Thesis, Victoria, BC: Department of Biology and School of Environmental Studies, University of Victoria.

Lopatin, Ivan A. 1945. *Social Life and Religion of the Indians of Kitimat, British Columbia.* Social Science Series 26. Los Angeles: University of Southern California.

Lowie, R. 1920. *Primitive Society.* New York: Harper and Brothers.

———. 1937. *The History of Ethnological Theory.* New York: Ferrar and Rinehart.

Lutz, J. 1995. Preparing Eden: Aboriginal Land Use and European Settlement. Paper presented to the 1995 Meeting of the Canadian Historical Association, Université de Québec à Montréal, PQ.

Lyman, R. L. 1991. *Prehistory of the Oregon Coast: The Effects of Excavation Strategies and Assemblage Size on Archaeological Inquiry.* Orlando, FL: Academic Press.

MacDonald, F. A. 1929. A Historical Review of Forest Protection in British Columbia. *Forestry Chronicle* 5: 31–35.

Mackie, J. D. 1964. *A History of Scotland.* Harmondsworth, UK: Penguin Books, Ltd.

Maclachlan, M., ed. 1998. *Fort Langley, 1827–1830.* Vancouver: University of British Columbia Press.

MacNeish, R. S. 1958. *Preliminary Archaeological Investigations in the Sierra de Tamaulipas, Mexico.* Transactions, Vol. 48, Part 6. Philadelphia, PA: American Philosophical Society.

———. 1967. A Summary of the Subsistence. In D. Byers, ed. *Environment and Subsistence, Volume 1: The Prehistory of the Tehuacán Valley,* pp. 290–309. Austin: University of Texas Press.

———. 1992. *The Origins of Agriculture and Settled Life.* Norman: University of Oklahoma Press.

Mangelsdorf, P., R. MacNeish, and G. Willey. 1964. Origins of Agriculture in Middle America. In R. West, ed. *Handbook of Middle American Indians, Volume 1: Natural Environment and Early Cultures,* pp. 427–45. Austin: University of Texas Press.

Maschner, H. D. G. 1991. The Emergence of Cultural Complexity on the Northern Northwest Coast. *Antiquity* 65: 924–34.

————. 1992. The Origins of Hunter-Gatherer Sedentism and Political Complexity: A Case Study from the Northern Northwest Coast. Unpublished Ph.D. dissertation, Santa Barbara: Department of Anthropology, University of California.

Mathewson, K. 1985. Taxonomy of Raised and Drained Fields: A Morphogenetic Approach. In I. S. Farrington, ed., *Prehistoric Intensive Agriculture in the Tropics*, pp. 835–45. Oxford, UK: British Archaeological Reports, International Series 232.

Matson, R. G. 1976. The Glenrose Cannery Site. *Canadian National Museum, Archaeological Survey of Canada Paper, Mercury Series No. 52,* Ottawa, ON and Hull, QC: National Museum of Civilization.

————. 1983. Intensification and the Development of Cultural Complexity: The Northwest versus the Northeast Coast. In R. J. Nash, ed., *The Evolution of Maritime Cultures on the Northeast and Northwest Coasts of America,* pp. 124–48. Department of Archaeology Publication No. 11. Burnaby, BC: Simon Fraser University.

————. 1992. The Evolution of Northwest Coast Subsistence. In D. E. Croes, R. A. Hawkins, and B. L. Isaac, eds. *Long-Term Subsistence Change in Prehistoric North America,* pp. 367–430. Research in Economic Anthropology, Supplement 6. Greenwich, CT: JAI Press.

Matson, R. G., and G. Coupland. 1995. *The Prehistory of the Northwest Coast.* Orlando, FL: Academic Press.

Maud, R., ed. 1978. *The Salish People: The Local Contribution of Charles Hill-Tout.* (4 Vols.) Vancouver, BC: Talonbooks.

Mayne, R. C. 1862. *Four Years in British Columbia and Vancouver Island.* Toronto, ON: S. R. Publishers Limited (reprinted in Johnson Reprint Collection, 1969).

McDonald, J. A. 1983. An Historic Event in the Political Economy of the Tsimshian: Information on the Ownership of the Zimacord District. In P. Tennant, ed. *British Columbia: A Place for Aboriginal Peoples, A Special Issue of B.C. Studies* 57: 24–37

————. 1985. *Trying to Make a Life: the Historic Political Economy of Kitsumkalum.* Ph.D. Dissertation, Vancouver: Department of Anthropology and Sociology, University of British Columbia.

————. 1987. The Marginalization of a Cultural Ecology: The Seasonal Cycle of Kitsumkalum. In B. Cox, ed. *Native Peoples, Native Lands.* (Carleton Library Series), pp. 109–218. Ottawa, ON: Macmillan.

————. 1986. *A Report on the Lagybaaw of Kitsumkalum.* Kitsumkalum, BC: Kitsumkalum Social History Research Project, Report No. 10.

————. 1990. Bleeding Day and Night: The Construction of the Grand Trunk Pacific Railway Across Tsimshian Reserve Lands. *Canadian Journal of Native Studies* 10 (1): 33–69.

————. 1994. Social Change and the Creation of Underdevelopment: A Northwest Coast Case. *American Ethnologist* 21 (1): 152–75.

————. 2002. Land and Resource Governance. Final Project Report. B.C. Capacity Initiative Project. Kitsumkalum, BC: Kitsumkalum Treaty Office.

McDonald, J. A., with the assistance of first Nations Education Centre, Coast Mountain School District. 2003. *People of the Robin: the Tsimshian of Kitsumkalum: A Resource Book for The Kitsumkalum Education Committee and The Coast Mountain School District 82, Terrace.* Edmonton, AB: CCI Press.

McDougall, A. S., B. R. Beckwith, and C. Y. Maslovat. 2004. Defining Conservation Strategies with Historical Perspectives: A Case Study from a Degraded Oak Grassland Ecosystem. *Conservation Biology* 18(2): 1–11.

McIlwraith, T. F. 1948. *The Bella Coola Indians*, vols. I, II. Toronto, ON: University of Toronto Press.

McKenna-McBride Royal Commission. 1913–1916. Evidence, Exhibits, Applications, and Reports of the McKenna-McBride Royal Commission. (Unpublished documents, on file with the British Columbia Archives and Records Service, Victoria, BC).

McNeary, S. A. 1976. *When Fire Came Down: Social and Economic Life of the Niska*. Ph.D. Dissertation. Bryn Mawr, PA: Department of Anthropology, Bryn Mawr College.

Meidinger, D., and J. Pojar, eds. 1991. *Ecosystems of British Columbia*. Special Report Series, 6. Victoria: B.C. Ministry of Forests.

Meillassoux, C. 1972. From Reproduction to Production. *Economy and Society* 1: 93–105.

Meilleur, B. A. 1979. Speculations on the Diffusion of *Nicotiana quadrivalvis* Pursh to the Queen Charlotte Islands and Adjacent Alaskan Mainland. *Syesis* 12: 101–04.

Menzies, A. 1923. *Menzies' Journal of Vancouver's Voyage, April to October 1792*. C. F. Newcombe, ed. Victoria, BC: W. H. Cullin.

Miller, J. 1997. *Tsimshian Culture: A Light Through the Ages*. Lincoln: University of Nebraska Press.

Miller, N. F. 1992. The Origins of Plant Cultivation in the Near East. In C. W. Cowan and P. J. Watson, eds., *The Origins of Agriculture: An International Perspective*, pp. 39–58. Washington, DC: Smithsonian Institution.

Millon, R. F. 1955. Trade, Tree Cultivation, and the Development of Private Property in Land. *American Anthropologist* 57: 698–712.

Mills, B. J. 1986. Prescribed Burning and Hunter-Gatherer Subsistence Systems. *Haliska'i: UNM Contributions to Anthropology* 5: 1–26.

Millspaugh, S. H., and C. Whitlock. 1995. A 750-Year Fire History Based on Lake Sediment Records in Central Yellowstone National Park, USA. *The Holocene* 5: 283–92.

Minore, D. 1972. *The Wild Huckleberries of Oregon and Washington—A Dwindling Resource*. Portland, OR: USDA Forest Service Research Paper 143, Pacific Northwest Forest and Range Experiment Station.

Mitchell. D. 1971. Archaeology of the Gulf of Georgia Area, A Natural Region and its Culture Types, *Syesis* 4, Supplement 1. Victoria, BC.

———. 1988. Changing Patterns of Resources Use in the Prehistory of Queen Charlotte Strait, British Columbia. In D. Croes, ed., *Prehistoric Economies of the Northwest Coast*, pp. 245–92. Research in Economic Anthropology, Supplement 3. Greenwich, CT: JAI Press.

———. 1990. Prehistory of the Coasts of Southern British Columbia and Northern Washington, In W. Suttles, ed. *Handbook of North American Indians, Volume 7: The Northwest Coast*, pp. 340–58. Washington, DC: Smithsonian Institution.

Mitchell, M. R. 1968. *A Dictionary of Songish, a Dialect of Straits Salish*. M.A. Thesis, Victoria, BC: Department of Linguistics, University of Victoria.

Moran, E. 1993. *Through Amazonian Eyes: The Human Ecology of Amazonian Populations*. Iowa City: University of Iowa Press.

———. 1996. Nurturing the Forest: Strategies of Native Amazonians. In R. Ellen and K. Fukui, eds. *Redefining Nature*. Oxford, UK: Berg Publishers.

Moseley, M. E. 1975. *The Maritime Foundations of Andean Civilization*. Menlo Park, CA: Cummings.

Moss, M. L. 1985. Phosphate Analysis of Archaeological Sites, Admiralty Island, Southeast Alaska. *Syesis* 17: 95–100.

———. 1989. *Archaeology and Cultural Ecology of the Prehistoric Angoon Tlingit*. Ph.D. Dissertation, Santa Barbara: Department of Anthropology, University of California.

———. 1993. Shellfish, Gender and Status on the Northwest Coast: Reconciling Archeological, Ethnographic, and Ethnohistoric Records of the Tlingit. *American Anthropologist* 95: 631–52.

———. 1998. Northern Northwest Coast Regional Overview. *Arctic Anthropology* 35(1): 88–111.

Moss, M. L. and J. M. Erlandson. 1992. Forts, Refuge Rocks, and Defensive Sites: the Antiquity of Warfare along the North Pacific Coast of North America. *Arctic Anthropology* 29(2): 73–90.

Moss, M. L., J. M. Erlandson, and R. Stuckenrath. 1989. The Antiquity of Tlingit Settlement on Admiralty Island, Southeast Alaska. *American Antiquity* 54(3): 534–43.

———. 1990. Wood Stake Fish Weirs and Salmon Fishing on the Northwest Coast: Evidence from Southeast Alaska. *Canadian Journal of Archaeology* 14: 143–58.

Moulton, G. E., ed. 1983. *The Journals of the Lewis and Clark Expedition*. Lincoln: University of Nebraska Press.

Moziño, J. M. 1970. *Noticias de Nutka: An Account of Nootka Sound in 1792*. I. H. Wilson Engstrand, ed. and trans. Seattle: University of Washington Press.

Mudie, P. J., S. Greer, J. Brakel, J. H. Dickson, C. Schinkel, R. Peterson-Welsh, M. Stevens, N. J. Turner, M. Shadow, and R. Washington. Forensic Palynology and Ethnobotany of *Salicornia* Species (Chenopodiaceae) in Northwest Canada and Alaska. *Canadian Journal of Botany,* in press.

Murdock, G. P. 1934. *Our Primitive Contemporaries*. New York: MacMillan Co.

———. 1967. *Ethnographic Atlas*. Pittsburgh: University of Pittsburgh Press.

Nelson, J. 1983. *The Weavers*. Vancouver, BC: Pacific Educational Press.

Netting, R. M. 1986. *Cultural Ecology*, 2nd ed. Prospect Heights, IL: Waveland.

Neumann, A., R. Holloway, and C. Busby. 1989. Determination of Prehistoric Use of Arrowhead (*Sagittaria*, Alismataceae) in the Great Basin of North America by Scanning Electron Microscopy. *Economic Botany* 43(3): 287–96.

Newcombe, C.F. n.d. Notes on Northwest Coast Ethnography. (Unpublished field notes on microfilm, Charles Newcombe Notes Collection, ca. 1897–1916). Victoria: Royal British Columbia Museum.

———, ed. 1923. Menzies' Journal of Vancouver's Voyage, April to October 1792. *British Columbia Archives Memoir* 5(8), Victoria, BC.

Newsom, L., and M. Scarry. In press. Homegardens and Mangrove Swamps: Pineland Archaeobotanical Research. In K. Walker and B. Marquardt, eds. *The Archaeology of Pineland*. University of Florida, Gainesville: Institute of Archaeology and Paleoenvironmental Studies, Monograph 4.

Newton, R. G., and M. L. Moss. 1984. *The Subsistence Lifeway of the Tlingit People: Excerpts of Oral Interviews*. Juneau, AK: USDA Forest Service, Alaska Region Report No. 179.

Niblack, A. P. 1888. *Coast Indians of Southern Alaska and Northern British Columbia*. Washington, DC: Report to the National Museum.

Nicholas, G. P. 1999. A Light but Lasting Footprint: Human Influences on the Northeastern Landscape. In M. L. Levine, M. S. Nassaney, and K. E. Sassaman, eds., *The Archaeological Northeast*, pp. 25–38. Greenwich, CT: Bergin and Garvey.

Nikiforuk, A. 1991. *The Fourth Horseman: A Short History of Epidemics, Plagues, Famine and Other Scourges*. Toronto, ON: Viking.

Norton, H. H. 1979a. Evidence for Bracken Fern as a Food for Aboriginal Peoples of Western Washington. *Economic Botany* 33(4): 384–96.

———. 1979b. The Association Between Anthropogenic Prairies and Important Food Plants in Western Washington. *Northwest Anthropological Research Notes* 13(2): 175–200.

———. 1981. Plant Use in Kaigani Haida Culture: Correction of an Ethnohistorical Oversight. *Economic Botany* 35(4): 434–49.

Norton, H. H., E. S. Hunn, C .S. Martinson, and P. B. Keely. 1984. Vegetable Food Products of the Foraging Economies of the Pacific Northwest. *Ecology of Food and Nutrition* 14: 219–28.

Oberg, K. 1973. *The Social Economy of the Tlingit Indians.* Seattle: University of Washington Press.

Olson, R. L. 1940. The Social Organization of the Haisla of British Columbia. *University of California Anthropological Records* 2(5): 169–200.

Oregon Department of Fish and Wildlife. n.d. *Statistics of Waterfowl Use Days 1980–1994.* Sauvies Island: Oregon Fish and Wildlife Office, unpublished manuscript.

Paul, P. K., P. C. Paul, E. Carmack, and R. Macdonald. 1994. *The Care-Takers: The Re-emergence of the Saanich Indian Map.* Indigenous Science Series, Report No. 1. Sidney, BC: Institute of Ocean Science, Department of Fisheries and Oceans.

Peacock, S. L. 1998. *Putting Down Roots: The Emergence of Wild Plant Food Production on the Canadian Plateau.* Ph.D. Dissertation, Victoria, BC: Department of Geography and School of Environmental Studies, University of Victoria.

Peacock, S. L., and N. J. Turner. 2000. "Just Like a Garden": Traditional Plant Resource Management and Biodiversity Conservation on the British Columbia Plateau. In P. Minnis and W. Elisens, eds., *Biodiversity and Native America*, pp. 133–79, Norman: University of Oklahoma Press.

Pickett, S. T. A., V. T. Parker, and P. L. Fiedler. 1992. The New Paradigm in Ecology: Implications for Conservation Biology above the Species Level. In P. L. Fiedler and S. K. Jain, eds., *Conservation Biology: The Theory and Practice of Nature Conservation*, pp. 65–88. New York: Chapman and Hall.

Piddocke, S. 1965. The Potlatch System of the Southern Kwakiutl: A New Perspective. *Southwestern Journal of Anthropology* 21: 244–64.

Piperno, D., A. Renere, A. Holst, and P. Hansell. 2000. Starch Grains Reveal Early Root Crop Horticulture in the Panamanian Tropical Forest. *Nature* 407: 894–97.

Piperno, D., I. Holst, L. Wessel-Beaver, and T. Andres. 2002. Evidence for the Control of Phytolith Formation in *Cucurbita* Fruits by the Hard Rind (*Hr*) Genetic Locus: Archaeological and Ecological Implications. *PNAS* 99: 10923–28.

Pojar, J., and A. MacKinnon. 1994. *Plants of the Pacific Northwest Coast: Washington, Oregon, British Columbia and Alaska.* Victoria, BC: Ministry of Forests, and Vancouver, BC: Lone Pine Publishing.

Pokotylo, D. L., and P. D. Froese. 1983. Archaeological Evidence for Prehistoric Root Gathering on the Southern Interior Plateau of British Columbia: A Case Study from Upper Hat Creek Valley. *Canadian Journal of Archaeology* 7(2): 127–57.

Popper, K. R. 1959. *The Logic of Scientific Discovery.* London: Hutchinson.

Posey, D. 1985. Indigenous Management of Tropical Forest Ecosystems: The Case of the Kayapó Indians of the Brazilian Amazon. *Agroforestry Systems* 3: 139–58.

Powell, F. W. 1917. Memoir of Hall Jackson Kelley. *Oregon Historical Quarterly* 17(4): 271–95.

Price, T. D., and J. Brown, eds. 1985. *Prehistoric Hunter-Gatherers: The Emergence of Cultural Complexity.* Orlando, FL: Academic Press.

Price, T. D., and G. M. Feinman, eds. 1996. *Foundations of Social Inequality.* New York: Plenum Press.

Pritchard, J. C. 1977. *Economic Development and the Disintegration of Traditional Culture among the Haisla.* Ph.D. Dissertation, Vancouver: Department of Anthropology and Sociology, University of British Columbia.

Pyne, S. J. 1993. Keeper of the Flame: A Survey of Anthropogenic Fire. In P. J. Crutzen and J. G. Goldammer, eds., *Fire in the Environment,* pp. 245–66. New York: John Wiley & Sons.

Ray, V. F. 1938. Lower Chinook Ethnographic Notes. *University of Washington Publications in Anthropology* 7(2): 29–165. Seattle: University of Washington Press.

———. 1989. Boas and the Neglect of Commoners. In T. McFeat, ed., *Indians of the North Pacific Coast,* pp. 159–65. Seattle: University of Washington Press.

Reed, C. A., ed. 1977. *Origins of Agriculture.* The Hague, The Netherlands: Mouton.

Richardson, A. 1982. The Control of Productive Resources on the Northwest Coast of North America. In N. M. Williams and E. S. Hunn, eds., *Resource Managers: North American and Australian Hunter-Gatherers,* pp. 93–120. American Association for the Advancement of Science Symposium No. 67. Boulder, CO: Westview.

Rindos, D. 1984. *The Origins of Agriculture: An Evolutionary Perspective.* New York: Academic Press.

Robinson, G. 1973. Talk on Haisla Culture (7 Dec.). *Indian Education Newsletter,* vols. 4–6 (Feb./March 1974). Vancouver: Indian Education Resources Center, University of British Columbia.

Robinson, S. P. 1983. *Men and Resources on the Northern Northwest Coast of North America, 1785–1840: A Geographical Approach to the Maritime Fur Trade.* Ph.D. Dissertation, London, ON: Department of Geography, University of London.

Rochefort, R. M., R. L. Little, A. Woodward, and D. L. Peterson. 1994. Changes in Sub-alpine Tree Distribution in Western North America: A Review of Climatic and Other Causal Factors. *The Holocene* 4: 89–100.

Roemer, H. 1992. Proposed Ecological Reserve: Timbercrest Estates Subdivision near Somenos Lake, Duncan. Unpublished ms. on file. Victoria: B.C. Parks, Province of British Columbia.

Ross, A. 1966. *Adventures of the First Settlers on the Columbia River.* Ann Arbor, MI: University Microfilms, Inc.

Rowley-Conway, P., and M. Zvelebil. 1989. Saving It for Later: Storage by Prehistoric Hunter-Gatherers in Europe. In P. P. Halstead and J. O'Shea, eds., *Bad Year Economics: Cultural Responses to Risk and Uncertainty,* pp. 40–56. Cambridge, UK: Cambridge University Press.

Ryder, J. M., and B. Thomson. 1986. Neoglaciation in the Southern Coast Mountains of British Columbia: Chronology Prior to the Late Neoglacial Maximum. *Canadian Journal of Earth Science* 23: 273–87.

Sahlins, M. D. 1968. Notes on the Original Affluent Society. In R. B. Lee and I. Devore, eds., *Man the Hunter,* pp. 85–89. Chicago: Aldine.

———. 1972. *Stone Age Economics.* Chicago: Aldine-Atherton.

Sahlins, M. D., and E. R. Service, eds. 1960. *Evolution and Culture.* Ann Arbor: University of Michigan Press.

Said, E. 1985. *Orientalism.* London: Penguin Books.

———. 1993. *Culture and Imperialism.* New York: Alfred A. Knopf.

Saleeby, B., and R. M. Pettigrew. 1983. Seasonality of Occupation of the Ethno-historically-documented Villages of the Lower Columbia River. In R. E. Greengo, ed., *Prehistoric Places on the Southern Northwest Coast*, pp. 169–93. Seattle: University of Washington, Washington State Museum.

Sapir, E. 1913–14. Unpublished notes: Nootka, Notebook 17, p. 23a, Dec. 1913–Jan. 1914, Roll 23 Microfilm from American Philosophical Society, copy in possession of Denis St. Clair, Victoria, BC.

Sapir, E., and M. Swadesh. 1939. *Nootka Texts: Tales and Ethnological Narratives, with Grammatical Notes and Lexical Materials.* Linguistic Society of America. Philadelphia: University of Pennsylvania Press.

Sauer, C. O. 1936. American Agricultural Origins: A Consideration of Nature and Culture. In R. Lowie, ed., *Essays in Anthropology Presented to A. L. Kroeber in Celebration of His Sixtieth Birthday, June 11, 1936*, pp. 278–97. Berkeley: University of California Press.

———. 1952. *Agricultural Origins and Dispersals.* New York: American Geographical Society.

Schalk, R. F. 1977. The Structure of an Anadromous Fish Resource. In L. R. Binford, ed. *For Theory Building in Archaeology*, pp. 207–49. Orlando, FL: Academic Press.

———. 1981. Land Use and Organizational Complexity among Foragers of Northwestern North America. In S. Koyama and D. H. Thomas, eds., *Affluent Foragers: Pacific Coasts East and West*, pp. 53–76. *Senri Ethnological Studies No. 9*, Osaka, Japan: National Museum of Ethnology.

———. 1987. Estimating Salmon and Steelhead Usage in the Columbia Basin before 1850: The Anthropological Perspective. *The Northwest Environmental Journal* 2(2): 1–29.

———. 1988. *The Evolution and Diversification of Native Land Use Systems on the Olympic Peninsula: A Research Design.* Seattle: University of Washington Institute of Environmental Studies.

Schlick, M. D. 1994. *Columbia River Basketry: Gift of the Ancestors, Gift of the Earth.* Seattle: University of Washington Press.

Schoonmaker, P. K., B. von Hagen, and E. C. Wolf, eds. 1997. *The Rain Forests of Home: Profile of a North American Bioregion.* Covelo, CA and Washington, DC: Island Press.

Scientific Panel for Sustainable Forest Practices in Clayoquot Sound. 1995. *First Nations' Perspectives on Forest Practices in Clayoquot Sound.* Report 3, Victoria, BC.

Service, E. R. 1963. *Profiles in Ethnology.* New York: Harper and Row.

Sewid-Smith, D. (Mayanilh). 1979. *Prosecution or Persecution.* Cape Mudge, BC: Nu-yum-balees Society.

Sewid-Smith, D. (Mayanilh), Chief A. Dick (Kwaxsistala), and N. J. Turner. 1998. The Sacred Cedar Tree of the Kwakwa̲ka̲'wakw People. In M. Bohl, ed., *Stars Above, Earth Below: Native Americans and Nature*, pp. 189–209. Pittsburgh, PA: Carnegie Museum of Natural History.

Shipek, F. 1989. An Example of Intensive Plant Husbandry: The Kumeyaay of Southern California. In D. Harris and G. Hillman, eds., *Foraging and Farming: The Evolution of Plant Exploitation*, pp. 159–67. London: Unwin Hyman.

Skarsten, M. O. 1964. *George Drouillard, Hunter and Interpreter for Lewis and Clark and Fur Trader, 1807–1810.* Glendale, CA: A. H. Clark Co.

Siemens, A. H. 1983. Wetland Agriculture in Pre-Hispanic Mesoamerica. *Geographical Review* 73: 166–81.

———. 1990. *Between the Summit and the Sea: Central Veracruz in the Nineteenth Century.* Vancouver: University of British Columbia Press.

Simpson, G. 1968. *Fur Trade and Empire, George Simpson's Journal, Entitled*

Remarks Connected with the Fur Trade in the Course of a Voyage from York Factory to Fort George and Back to York Factory 1824–1825. F. Merk, ed. Cambridge, MA: Belknap Press, Harvard University Press.

Sluyter, A. 1999. The Making of the Myth in Postcolonial Development: Material–Conceptual Landscape Transformation in Sixteenth-Century Veracruz. *Annals of the Association of American Geographers* 89(3): 377–401.

Smith, B. D. 1984. *Chenopodium* as a Prehistoric Domesticate in Eastern North America: Evidence from Russell Cave, Alabama. *Science* 226: 165–67.

———. 1985a. The Role of *Chenopodium* as a Domesticate in Pre-Maize Garden Systems of the Eastern United States. *Southeastern Archaeology* 4: 51–72.

———. 1985b. *Chenopodium berlandieri* ssp. *jonesianum*: Evidence for a Hopewellian Domesticate from Ash Cave, Ohio. *Southeastern Archaeology* 4: 107–33.

———. 1995. *The Emergence of Agriculture.* New York: W. H. Freeman.

———. 1997a. The Initial Domestication of *Cucurbita pepo* in the Americas 10,000 Years Ago. *Science* 276: 932–34.

———. 1997b. Reconsidering the Ocampo Caves and the Era of Incipient Cultivation in Mesoamerica. *Latin American Antiquity* 8: 342–83.

———. 2001. Low-Level Food Production. *Journal of Archaeological Research* 9: 1–43.

———. 2002. *Rivers of Change.* Washington, DC: Smithsonian Institution Press.

Smith, E. A. 1991. *Inujjuamiut Foraging Strategies: Evolutionary Ecology of an Arctic Hunting Economy.* New York: Aldine De Gruyter.

Smith, H. I. 1928. Materia Medica of the Bella Coola and Neighbouring Tribes of British Columbia. Ottawa, ON: *National Museum of Canada Bulletin.* No. 56.

———. 1997. *Ethnobotany of the Gitksan Indians of British Columbia.* B. Compton, B. Rigsby, and M. L. Tarpent, eds. Ottawa, ON, and Hull, PQ: Canadian Museum of Civilization.

Smith, M. W. 1950. The Nooksack, the Chilliwack, and the Middle Fraser. *Pacific Northwest Quarterly* 41(4): 330–41.

———. n.d. Nooksack Field Notes. Unpublished ms. Victoria: Royal British Columbia Provincial Museum.

Smith, R. Y. 1997. *"Hishuk ish Ts'awalk" All Things are One: Traditional Ecological Knowledge and Forest Practices in Ahousaht First Nation's Traditional Territory, Clayoquot Sound, B.C.* M.A. Thesis. Peterborough, ON: Departments of Canadian Studies and Native Studies, Trent University.

Speth, W. W. 1977. The Anthropogeographic Theory of Franz Boas. *Anthropos* 73: 1–31.

Spier, L., and E. Sapir. 1930. Wishram Ethnography. *University of Washington Publications in Anthropology* 3(3): 151–300.

Spinden, H. J. 1917. The Origin and Distribution of Agriculture in America. *International Congress of Americanists, Proceedings* 19: 269–76.

Sponsel, L. E. 1989. Farming and Foraging: A Necessary Complementarity in Amazonia? In S. Kent, ed., *Farmers as Hunters: The Implications of Sedentism,* pp. 37–45. Cambridge, UK: Cambridge University Press.

Sproat, G. M. 1868. *The Nootka: Scenes and Studies of Savage Life.* London: Smith, Elder. Reprint, Victoria, BC: Sono-Nis Press, 1987.

Spurgeon, T. 2001. *Wapato* (Sagittaria latifolia) *in Katzie Traditional Territory, Pitt Meadows, British Columbia.* Unpublished M.A. Thesis. Burnaby, BC: Department of Archaeology, Simon Fraser University.

Stern, B. J. 1934. *The Lummi Indians of Northwest Washington.* New York: Columbia University Press.

Steward, J. H. 1930. Irrigation Without Agriculture. *Papers of the Michigan Academy of Science, Arts, and Letters* 12: 149–56.

———. 1933. *Ethnography of the Owens Valley Paiute*, pp. 233–350. Berkeley: University of California Publications in American Archaeology and Ethnology, No. 33.

———. 1938. *Basin-Plateau Aboriginal Sociopolitical Groups.* Smithsonian Institution Bureau of American Ethnology Bulletin 120. Washington, DC: Government Printing Office.

———. 1949. Cultural Causality and Law: A Trial Formulation of the Development of Early Civilizations. *American Anthropologist* 51: 1–27.

Stewart, H. 1984. *Cedar: Tree of Life to the Northwest Coast Indians.* Vancouver, BC: Douglas & McIntyre.

Stokes, M. A., and T. L. Smiley. 1996. *An Introduction to Tree-Ring Dating.* Tucson: University of Arizona Press.

Stoltman, J., and D. Baerreis. 1983. The Evolution of Human Ecosystems in the Eastern United States. In H. Wright, ed., *Late-Quaternary Environments of the United States,* vol. 2, pp. 252–65. Minneapolis: University of Minnesota Press.

Stuckey, R. L. 1994. *Sagittaria:* The Distribution of its Species in Relation to Physiography and Glaciation in Eastern North America. *American Journal of Botany* 81(6), supplement 4.

Suttles, W. 1947–52. Field Notes from work with speakers of Northern Straits, Nooksack, and Northern Lushootseed. (In possession of Wayne Suttles.)

———. 1951a. *The Economic Life of the Coast Salish of Haro and Rosario Straits.* Ph.D. Dissertation, Seattle: Department of Anthropology, University of Washington.

———. 1951b. The Early Diffusion of the Potato Among the Coast Salish. *Southwestern Journal of Anthropology* 7(3): 272–88.

———. 1954. Post-Contact Culture Change among the Lummi Indians. *British Columbia Historical Quarterly* 18: 29–101.

———. 1955. *Katzie Ethnographic Notes.* Anthropology in British Columbia Memoir No. 2, edited by Wilson Duff. Victoria: Department of Education and British Columbia Provincial Museum.

———. 1956–62. Musqueam Linguistic and Ethnographic Field Notes. (Unpublished document in possession of Wayne Suttles.)

———. 1960. Affinal Ties, Subsistence, and Prestige among the Coast Salish. *American Anthropologist* 62: 296–305.

———. 1962. Variation in Habitat and Culture on the Northwest Coast. *Proceedings of the 34th International Congress of Americanists,* pp. 522–36. Vienna, Austria.

———. 1968. Coping with Abundance: Subsistence on the Northwest Coast. In R. B. Lee and I. Devore, eds., *Man the Hunter,* pp. 56–68. Chicago: Aldine.

———. 1974. *Coast Salish and Western Washington Indians I: The Economic Life of the Coast Salish of Haro and Rosario Straits.* New York: Garland Publishing.

———. 1987. *Coast Salish Essays.* Vancouver, BC: Talonbooks.

———. 1990. Introduction. In W. Suttles, ed., *Handbook of North American Indians, Volume 7: Northwest Coast,* pp. 1–15. Washington, DC: Smithsonian Institution.

———. 2004. *Musqueam Reference Grammar.* Vancouver: University of British Columbia Press.

———, ed. 1990. *Handbook of North American Indians, Volume 7: Northwest Coast.* Washington, DC: Smithsonian Institution.

Swan, J. G. 1857. *The Northwest Coast; Or, Three Years' Residence in Washington*

Territory. New York: Harper. Reprint, Seattle: University of Washington Press, 1972.

———. 1879. Official Report of James G. Swan . . . an Account of the Cruise of the U.S. Revenue Cutter *Wolcott*, in Alaska During the Summer of 1875. Appendix to Report upon the Customs District, Public Service, and Resources of Alaska Territory, by William Gouverneur Morris, pp. 143–50. 45th Congress 3rd Session, Senate Executive Document No. 59. Originally published 1877. Port Townsend, WA: Port Townsend Argus.

Swanton, J. G. 1905. *Contributions to the Ethnology of the Haida.* Leiden, The Netherlands: E. J. Brill.

Tanimoto, T. 1989. Promotion of Flowering and Seed Germination in Chinese Arrowhead *Sagittaria trifolia* var. *edulis. Japanese Journal of Breeding* 39: 345–52.

Tennant, P. 1990. *Aboriginal Peoples and Politics: The Indian Land Question in British Columbia, 1849–1989.* Vancouver: University of British Columbia.

Testart, A. 1982. The Significance of Food Storage among Hunter-Gatherers: Residence Patterns, Population Densities, and Social Inequities. *Current Anthropology* 23: 523–37.

Thom, R. M. 1987. The Biological Importance of Pacific Northwest Estuaries. *Northwest Environmental Journal* 3: 21–42.

Thoms, A. V. 1989. *The Northern Roots of Hunter-Gatherer Intensification: Camas and the Pacific Northwest.* Ph.D. Dissertation, Pullman, WA: Department of Anthropology, Washington State University.

Thoms, A. V., and G. C. Burtchard, eds. 1986. Calispell Valley Archaeological Project: Interim Report for 1984 and 1985. *Contributions in Cultural Resource Management* No. 10. Pullman, WA: Center for Northwest Anthropology, Washington State University.

Thornton, T. F. 1999. Tleikwaani, the "Berried" Landscape: The Structure of Tlingit Edible Fruit Resources at Glacier Bay, Alaska. *Journal of Ethnobiology* 19(1): 27–48.

Thwaites, R. G., ed. 1904a. Ross's Adventures of the First Settlers on the Oregon or Columbia River, 1810–1813. In *Early Western Travels 1748–1846*, vol. VII. Cleveland, OH: Arthur H. Clark.

———. 1904b. Franchere's Voyage to the Northwest Coast, 1811–1814. In R. G. Thwaites, ed., *Early Western Travels 1748–1846*, vol. VI. Cleveland, OH: Arthur H. Clark.

———. 1969. [1904–1905]. *Original Journals of the Lewis and Clark Expedition 1804–1806.* 8 vols. New York: Arno Press. (First published, New York: Dodd, Mead and Co.)

Timbrook, J., J. R. Johnson, and D. D. Earle. 1982. Vegetation Burning by the Chumash. *Journal of California and Great Basin Anthropology* 4(20): 163–86.

Turner, N. J. 1973. The Ethnobotany of the Bella Coola Indians of British Columbia. *Syesis* 6: 193–220.

———. 1975. *Food Plants of British Columbia Indians. Part 1: Coastal Peoples.* Victoria, BC: British Columbia Provincial Museum. Handbook 34.

———. 1991 "Burning Mountain Sides for Better Crops": Aboriginal Landscape Burning in British Columbia. *Archaeology in Montana* 32(2). Bozeman, MT: Montana Archaeological Society.

———. 1995. *Food Plants of Coastal First Peoples.* Victoria, BC: Royal British Columbia Museum; Vancouver: University of British Columbia Press. (Revised from 1975 edition, *Food Plants of British Columbia Indians.* Part 1. *Coastal Peoples.* Victoria: B.C. Provincial Museum).

————. 1996a. "Dans une Hotte." L'importance de la vannerie dans l'économie des peuples chasseurs-pêcheurs-cueilleurs du Nord-Ouest de l'Amérique du Nord. ("'Into a Basket Carried on the Back': Importance of Basketry in Foraging/Hunting/Fishing Economies in Northwestern North America.") *Anthropologie et Sociétiés.* Special Issue on Contemporary Ecological Anthropology. Theories, Methods and Research Fields. Montréal, Québec, 20(3): 55–84.

————. 1996b. *Food Plants of Interior First Peoples.* Vancouver: University of British Columbia Press.

————. 1997a. Traditional Ecological Knowledge. In P. K. Schoonmaker, B. Von Hagen, and E. C. Wolf, eds., *The Rain Forests of Home. Profile of a North American Bioregion*, pp. 275–98. Covelo, CA and Washington, DC: Island Press.

————. 1997b. "Le fruit de l'ours": Les rapports entre les plantes et les animaux dans les langues et les cultures amérindiennes de la Côte-Ouest" ("'The Bear's Own Berry': Ethnobotanical Knowledge as a Reflection of Plant/Animal Interrelationships in Northwestern North America.") In: *Recherches amérindiennes au Québec*, vol. 27(3–4), 1997. Special Edition on *Des Plantes et des Animaux: Visions et Pratiques Autochtones*, P. Beaucage, ed., pp. 31–48. Montréal, PQ: Université de Montréal.

————. 1997c. *Food Plants of Interior First Peoples.* Victoria, BC: Royal British Columbia Museum and Vancouver: University of British Columbia Press.

————. 1998. *Plant Technology of British Columbia First Peoples.* Victoria, BC: Royal British Columbia Museum; Vancouver: University of British Columbia Press.

————. 1999. "Time to Burn": Traditional Use of Fire to Enhance Resource Production by Aboriginal Peoples in British Columbia. In R. Boyd, ed., *Indians, Fire, and the Land in the Pacific Northwest*, pp. 185–218. Corvallis: Oregon State University Press.

————. 2003a. The Ethnobotany of "Edible Seaweed" (*Porphyra abbottiae* Krishnamurthy and related species; Rhodophyta: Bangiales) and its use by First Nations on the Pacific Coast of Canada. *Canadian Journal of Botany* 81(2): 283-93.

————. 2003b. "Passing on the News": Women's Work, Traditional Knowledge and Plant Resource Management in Indigenous Societies of Northwest North America. In Dr. P. Howard, ed., *Women and Plants: Case Studies on Gender Relations in Local Plant Genetic Resource Management*, pp. 133–149. London: Zed Books.

————. 2004. *Plants of Haida Gwaii.* Xaadaa Gwaay guud gina k'aws (Skidegate), Xaadaa Gwaayee guu giin k'aws (Massett). Winlaw, BC: Sono Nis Press.

Turner, N. J., and E. R. Atleo (Chief Umeek). 1998. Pacific North American First Peoples and the Environment. In H. Coward, ed. *Traditional and Modern Approaches to the Environment on the Pacific Rim, Tensions and Values*, pp. 105–24. Centre for Studies in Religion and Society. Albany, NY: State University of New York, Albany Press.

Turner, N. J., and M.A.M. Bell. 1971. The Ethnobotany of the Coast Salish Indians of Vancouver Island. *Economic Botany* 25(1): 61–104.

Turner, N. J., and H. Clifton. 2002. "The Forest and the Seaweed": Gitga'at Seaweed, Traditional Ecological Knowledge and Community Survival. Paper presented with Helen Clifton (Gitga'at Nation) at Workshop on Local Knowledge, Natural Resources and Community Survival: Charting a Way Forward. Sponsored by Forest Renewal B.C., Charles Menzies, Organizer, Prince Rupert, B.C., February 2002.

Turner, N. J., I. J. Davidson-Hunt, and M. O'Flaherty. 2003. Ecological Edges and

Cultural Edges: Diversity and Resilience of Traditional Knowledge Systems. *Human Ecology* 31(3): 439–63.

Turner, N. J., and A. Davis. 1993. "When Everything Was Scarce": The Role of Plants as Famine Foods in Northwestern North America. *Journal of Ethnobiology* 13(2): 1–28.

Turner, N. J., and D. E. Deur. 1999. "Cultivating the Clover": Managing Plant Resources on the Northwest Coast. Paper presented at the Society for Ethnobiology 1999 annual meeting, Oaxaca, Mexico.

Turner, N. J., and B. S. Efrat. 1982. *Ethnobotany of the Hesquiat Indians of Vancouver Island.* Victoria, BC: British Columbia Provincial Museum, Cultural Recovery Paper No. 2.

Turner, N. J., and R. J. Hebda. 1990. Contemporary Use of Bark for Medicine by Two Salishan Native Elders of Southeast Vancouver Island, Canada. *Journal of Ethnopharmacology* 29: 59–72.

Turner, N. J., L. M. J. Gottesfeld, H. V. Kuhnlein, and A. Ceska. 1992. Edible Wood Fern Rootstocks of Western North America: Solving an Ethnobotanical Puzzle. *Journal of Ethnobiology* 12(1): 1–34.

Turner, N. J., M. B. Ignace, and R. Ignace. 2000. Traditional Ecological Knowledge and Wisdom of Aboriginal Peoples in British Columbia. *Ecological Applications* 10(5): 1275–87. Special Issue on Traditional Ecological Knowledge, Ecosystem Science and Environmental Management, J. Ford and D. R. Martinez, eds.

Turner, N. J., and H. V. Kuhnlein. 1982. Two Important "Root" Foods of the Northwest Coast Indians: Springbank Clover (*Trifolium wormskioldii*) and Pacific Silverweed (*Potentilla anserina* ssp. *pacifica*). *Economic Botany* 36(4): 411–32.

———. 1983. Camas (*Camassia* spp.) and Riceroot (*Fritillaria* spp): Two Liliaceous "Root" Foods of the Northwest Coast Indians. *Ecology of Food and Nutrition* 13: 199–219.

Turner, N. J., and D. C. Loewen. 1998. The Original "Free Trade": Exchange of Botanical Products and Associated Plant Knowledge in Northwestern North America. *Anthropologica* XL: 49–70.

Turner, N. J., and R. L. Taylor. 1972. A Review of the Northwest Coast Tobacco Mystery. *Syesis* 5: 249–57.

Turner, N. J., J. Thomas, B. F. Carlson, and R. T. Ogilvie. 1983. *Ethnobotany of the Nitinaht Indians of Vancouver Island.* Victoria, BC: British Columbia Provincial Museum, Occasional Paper No. 24.

Turner, N. J., L. C. Thompson, M. T. Thompson, and A. Z. York. 1990. *Thompson Ethnobotany: Knowledge and Usage of Plants by the Thompson Indians of British Columbia.* Victoria, BC: Royal British Columbia Museum, Memoir No. 3; and Vancouver: University of British Columbia Press.

Vancouver, G. 1967. *A Voyage of Discovery to the North Pacific Ocean and Round the World.* Vol. III. New York: N. Israel and DaCapo Press. (Facsimile republication of 1798, 1801 London Editions.)

———. 1984. *A Voyage of Discovery to the North Pacific Ocean and Round the World 1791–1795 with an Introduction and Appendices.* 4 vols. W. K. Lamb, ed. London: The Hakluyt Society.

Vander Kloet, S. P. 1994. The Burning Tolerance of *Vaccinium myrtilloides* Michaux. *Canadian Journal of Plant Science* 74: 577–79.

Vayda, A. P. 1961. A Re-examination of Northwest Coast Economic Systems. *Transactions of the New York Academy of Sciences* Ser. 2, vol. 23(7): 618–24.

———. 1968. Economic Systems in Ecological Perspective: The Case of the Northwest Coast. In M. H. Fried, ed., *Readings in Anthropology,* vol. II: pp. 172–78. New York: Crowell.

Wagner, P. L. 1972. Persistence of Native Settlement in Coastal British Columbia. In J. Minghi, ed., *Peoples of the Living Lands,* pp. 13–27. British Columbia Geographical Series, No. 15. Vancouver, BC: Tantalus.

———. 1977. The Concept of Environmental Determinism in Cultural Evolution. In C. A. Reed, ed., *The Origins of Agriculture,* pp.49–74. The Hague, The Netherlands: Mouton.

———. 1996. *Showing Off: The Geltung Hypothesis.* Austin: University of Texas Press.

Walker, M. 1999. Shamanism and Traditional Ecological Knowledge. *Proceedings of the International Congress on Shamanism and Other Indigenous Spiritual Beliefs and Practices.* Moscow, Russia.

Warren, William J. 1860. Journal of Wm J. Warren, Sec'y N.W. Boundary Commission of an Expedition in Company with C.R.R. Kennerly, Surgeon and Naturalist, to the Haro Archipelago Jan., Feb, 1860. Appendix E of Archibald Campbell, ed., *Geographical Memoir of the Islands between the Continent and Vancouver Island in the Vicinity of the Forty Ninth Parallel of North Latitude.* (Manuscript with appendices by Dr. C. B. R. Kennerly, George Gibbs, Henry Custer, and W. J. Warren.) U.S. National Archives RG 76, Entry 198, Journals of Exploring Surveys.

Webster, P. 1983. *As Far as I Remember: Reminiscences of an Ahousat Elder.* Campbell River, BC: Campbell River Museum and Archives.

Webster, G. C., and J. Powell. 1994. Geography, Ethnogeography, and the Perspective of the Kwakwa̠ka̠'wakw. In R. Galois, ed., *Kwakwa̠ka̠'wakw Settlements, 1775–1920: A Geographical Analysis and Gazetteer,* pp. 4–11. Vancouver: University of British Columbia Press.

Weinstein, M. S., and M. Morrell. 1994. *Need Is Not a Number: Report on the Kwakiutl Marine Food Fisheries Reconnaissance Survey.* Campbell River, BC: Kwakiutl Territorial Fisheries Commission.

Wells, R., K. P. Lertzman, and S. Saunders. 1998. Old-Growth Definitions for the Forest of British Columbia. *Natural Areas Journal* 18: 280–94.

White, R. 1980. *Land Use, Environment, and Social Change: The Shaping of Island County, Washington.* Seattle: University of Washington Press.

———. 1992. *Land Use, Environment, and Social Change.* Seattle: University of Washington Press.

Whitlock, C., and M. A. Knox. 2002. Prehistoric Burning in the Pacific Northwest. In T. R. Vale, ed., *Fire, Native Peoples, and the Natural Landscape,* pp. 195–231. Washington, DC: Island Press.

Wilk, R. R., and W. L. Rathje 1982. Household Archaeology. *American Behavioral Scientist* (25)6: 631–40.

Wilken, G. C. 1987. *The Good Farmers: Traditional Agricultural Resource Management in Mexico and Central America.* Berkeley: University of California Press.

Willems-Braun, B. 1997. Buried Epistemologies: The Politics of Nature in (Post)Colonial British Columbia. *Annals of the Association of American Geographers* 87: 3–31.

Williams, N. M., and G. Baines, eds. 1993. *Traditional Ecological Knowledge: Wisdom for Sustainable Development.* Canberra: Australian National University, Centre for Resource and Environmental Studies.

Wilson, C. W. 1866. Report on the Indian Tribes Inhabiting the Country in the Vicinity of the 49th Parallel of North Latitude. *Transactions of the Ethnological Society of London* 4: 275–332.

Winterhalder, B., and C. Goland. 1997. An Evolutionary Ecology Perspective on Diet Choice, Risk, and Plant Domestication. In K. J. Gremillion, ed., *People, Plants, and Landscapes: Studies in Paleoethnobotany*, pp. 123–60. Tuscaloosa: University of Alabama Press.

Winterhalder, B., F. Lu, and B. Tucker. 1999. Risk-Sensitive Adaptive Tactics: Models and Evidence from Subsistence Studies in Biology and Anthropology. *Journal of Archaeological Research* 7: 301–48.

Wishart, D. J. 1994. *An Unspeakable Sadness: The Dispossession of the Nebraska Indians*. Lincoln: University of Nebraska Press.

Woodburn, J. 1968. An Introduction to Hadza Ecology. In R. B. Lee and I. Devore, eds., *Man the Hunter*, pp. 49–55. Chicago: Aldine Publishing Company.

———. 1980. Hunter-Gatherers Today and Reconstructing the Past. In E. Gellner, ed., *Soviet and Western Anthropology*, pp. 94–118. New York: Columbia University Press.

Wright, Walter. n.d. Wars of Medeek as Told by Nisdaxo'ok, Walter Wright. Unpublished manuscript, Kitsumkalum Treaty Office.

Yen, D. 1989. The Domestication of Environment. In D. Harris and G. Hillman, eds., *Foraging and Farming: The Evolution of Plant Exploitation*, pp. 55–78. London: Unwin Hyman.

Yerkes, R. 2000. Middle Woodland Settlements and Social Organizations in the Central Ohio Valley: Were the Hopewell Really Farmers? Paper presented in the Perspectives on Middle Woodland at the Millennium Conference. Center for American Archaeology, Grafton, IL, July 19–21, 2000.

Yesner, D. R. 1994. Seasonality and Resource "Stress" among Hunter-Gatherers: Archaeological Signatures. In E. S. Burch, Jr. and L. J. Ellanna, eds., *Key Issues in Hunter-Gatherer Research*, pp. 152–68. Oxford, UK: Berg Publishers.

Yoffee, N. 1993. Too Many Chiefs? (or, Safe Texts for the '90s). In N. Yoffee and A Sherratt, eds., *Archaeological Theory: Who Sets the Agenda*, pp. 60–78. Cambridge, UK: Cambridge University Press.

Zeder, M. 1994. After the Revolution: Post-Neolithic Subsistence Strategies in Northern Mesopotamia. *American Anthropologist* 96: 97–126.

———. 1999. Animal Domestication in the Zagros: A Review of Past and Current Research. *Paléorient* 25: 11–25.

Zeder, M., N. Cleghorn, and H. Lapham. 1995. A Reconsideration of the Evidence for Animal Domestication in the Zagros from the Perspective of the Upper Paleolithic in Highland Iran. Paper presented in the Fryxell Symposium on Interdisciplinary Studies in Archaeology, Annual Meeting of the Society for American Archaeology, Minneapolis, MN.

Zeder, M., and B. Hesse. 2000. The Initial Domestication of Goats (*Capra hircus*) in the Zagros Mountains 10,000 years ago. *Science* 287: 2254–57.

Zenk, H. B. 1976. Contributions to Tualatin Ethnography: Subsistence and Ethnobiology. Unpublished M.A. Thesis. Portland, OR: Department of Anthropology, Portland State University.

———. 1994. Tualitin Kalapuyan Villages: The Ethnographic Record. In P. W. Baxter, ed., *Contributions to the Archaeology of Oregon, 1989–1994*, pp. 147–66. Association of Oregon Archaeologists, Occasional Papers No. 5. Eugene: Oregon State Museum of Natural History.

Zvelebil, M. 1986a. Mesolithic Prelude and Neolithic Revolution. In M. Zvelebil, ed., *Hunters in Transition*, pp. 5–15. Cambridge, UK: Cambridge University Press.

————. 1986b. Mesolithic Societies and the Transition to Farming: Problems of Time, Scale and Organization. In M. Zvelebil, ed., *Hunters in Transition: Mesolithic Societies of Temperate Eurasia and Their Transition to Farming*, pp. 167–88. Cambridge, UK: Cambridge University Press.

————, ed. 1986c. *Hunters in Transition: Mesolithic Societies of Temperate Eurasia and Their Transition to Farming*. Cambridge, UK: Cambridge University Press.

————. 1993. Hunters or Farmers: The Neolithic and Bronze Age Societies of North-east Europe. In J. Chapman and P. Dolvkhanov, eds., *Cultural Transformations and Interactions in Eastern Europe*, pp. 146–62. Avebury, UK: Aldershot.

————. 1994. Plant Use in the Mesolithic and Its Role in the Transition to Farming. *Proceedings of the Prehistoric Society* 60: 35–74.

————. 1996. The Agricultural Frontier and the Transition to Farming in the Circum-Baltic Region. In D. Harris, ed., *The Origins and Spread of Agriculture and Pastoralism in Eurasia*, pp. 323–45. Washington, DC: Smithsonian Institution.

Contributors

DR. KENNETH AMES, Department of Anthropology, Portland State University, Portland, Oregon

DR. RICHARD ATLEO, *Umeek*, Ahousat First Nation, First Nations Studies, Malaspina University College, Nanaimo, British Columbia

MELISSA DARBY, Lower Columbia Research & Archaeology, Portland, Oregon

DR. DOUGLAS DEUR, Pacific Northwest Cooperative Ecosystem Studies Unit, University of Washington, Seattle, Washington

DR. DOUGLAS HALLETT, Center for Environmental Sciences & Quaternary Sciences Program, Northern Arizona University, Flagstaff, Arizona

JAMES T. JONES, Bamfield, British Columbia, School of Environmental Studies (graduate), University of Victoria, Victoria, British Columbia

DR. DANA LEPOFSKY, Department of Archaeology, Simon Fraser University, Burnaby, British Columbia

DR. KEN LERTZMAN, Department of Resource and Environmental Management, Simon Fraser University, Burnaby, British Columbia

DR. ROLF MATHEWES, Department of Biological Sciences, Simon Fraser University, Burnaby, British Columbia

DR. JAMES McDONALD, Department of Anthropology and First Nations Studies, University of Northern British Columbia, Prince George, British Columbia

ALBERT (SONNY) McHALSIE, Sto:lo First Nation, Chilliwack, British Columbia

DR. MADONNA MOSS, Department of Anthropology, University of Oregon, Eugene, Oregon

DR. SANDRA PEACOCK, Department of Anthropology, Okanagan University College, Kelowna, British Columbia

DR. BRUCE SMITH, National Museum of Natural History, Smithsonian Institution, Washington, D.C.

ROBIN SMITH, Indigenous Peoples' Health Research Centre, First Nations University of Canada, Northern Campus, Prince Albert, Saskatchewan

DR. WAYNE SUTTLES, Department of Anthropology, Portland State University, Portland, Oregon

DR. NANCY TURNER, School of Environmental Studies, University of Victoria, Victoria, British Columbia

KEVIN WASHBROOK, Sto:lo First Nation, Chilliwack, British Columbia

Index

bimodal model of food production, 39–42
Binford, L. R., 70
biodiversity, 275
biogeoclimatic zones, Northwest Coast, 9–10, 103, 246–47
biogeographic diversity, 11–12
biological conservation, 339–40
biomass, distribution of, 11
Bishop, Charles, 201
bitter cherry (*Prunus emarginata*), 108, 123, 136, 137
black currant, trailing (*Ribes laxiflorum*), 280
black huckleberry (*Vaccinium membranaceum*), 106, 137
black seaweed (*Porphyra* sp.), 279
blankets of muskrat pelt, 213
Blechnum spicant (deer fern), 107, 110, 137
blue camas. *See* camas, blue (*Camassia quamash; C. leichtlinii*)
blue-leaved huckleberry (Cascade bilberry) (*Vaccinium deliciosum*), 138, 224
blueberries. *See also* entries at huckleberry
blueberry, Alaska (*Vaccinium alaskaense*), 138, 224, 281
blueberry, bog or mountain (*Vaccinium uliginosum*), 139, 281
blueberry, Cascade. *See* Cascade bilberry (*Vaccinium deliciosum*)
blueberry, dwarf (*Vaccinium caespitosum*), 281
blueberry, oval-leaved (*Vaccinium ovalifolium*), 225, 281
  in anthropogenic landscapes, 137, 138, 141
  burning and, 223–27
  climate shifts and, 239
  management practices, 106
boa kelp (*Egregia menziesii*), 107
Boas, Franz
  on bark removal, 123
  on berry-picking rights, 165–66
  on digging sticks, 316
  dismissal of cultivation by, 24–26, 319–20
  on estuarine gardens, 296, 297, 325n13
  on harvesting practices, 317–18
  on large vs. small roots, 303
  on "Myth of the Salmon," 205–6
  on ownership, 304–5

on "prayers," 133
on soil and rock work in estuarine gardens, 311
theoretical and ideological agendas of, 25
on tobacco, 258, 259–61
on Tsimshian, 241–42, 243
Bob, George, 187
bog blueberry (*Vaccinium uliginosum*), 139, 281
bog cranberry. *See* cranberry, bog (*Vaccinium oxycoccos* or *Oxycoccus oxycoccos*)
bogs, anthropogenic plant communities in, 139, 143–44
Bogucki, Peter, 42
Boit, John, 201
Bolton, Alex, 247, 251
Bolton, Mark, 247, 250, 262
Bolton, Rebecca, 247
Boserup, Ester, 75–76
bottleneck problems, in food production, 76–77
Bouchard, Randall, 144, 297
boundaries, territorial, 166, 167, 172, 339
boundaries of gardens, marking of, 164, 310, 312, 326n16
boundary conditions and definitions, 44–46, 61, 71
Boyd, Robert, 210–11
bracken fern. *See* fern, bracken (*Pteridium aquilinum*)
bracket fungus (*Fomitopsis officinalis*), 111
bread-making, 192n6
British Columbia Treaty Process, 176. *See also* Douglas Treaties
broad-leaf maple (*Acer macrophyllum*), 108, 136
broad-leaved plantain (*Plantago major*), 110, 139, 150n5
Broughton, J. M., 75
Brown, Beatrice, 118
Brown, Bessie, 118
Brown, J., 41, 73–74
Brown, Robert, 183–84, 188, 193n9
buckbean (*Menyanthes trifoliata*), 279
bull kelp (*Nereocystis luetkeana*), 108
bulrush, American ("three-square") (*Schoenoplectus olneyi*), 126
bunchberry. *See* wild lily-of-the-valley (*Maianthemum dilatatum*)
Bunnell, Fred, 340

complexity, social
  as anomaly among hunter-gatherers, 101–2
  and hunting-gathering vs. agriculture, 41
  intensification and, 68–69, 72–74
  salmon intensification and, 222
Compton, Brian, 243, 272n12
Conioselinum pacificum. See carrot, wild construction, wood for, 108
Cook, James, 4, 161
Copper River railroad construction, 263
coppicing, 112, 120–25
Coptis trifolia (goldthread), 277
Corylus cornuta var. *californica* (hazelnut), 250–51, 251, 272n14
cosmology, 303
cost of failure, in intensification models, 81–84, 96
cottonwood (*Populus balsamifera* ssp. *trichocarpa*), 107–10, 138
Cove, J. J., 171
cow-parsnip (*Heracleum lanatum*), 113
  in anthropogenic landscapes, 136, 137
  management practices, 107, 120
  names for, 271n3
  time of harvesting, 113
  Tlingit use of, 278
  tobacco and, 261
Cowlitz, 203
Cox, Ross, 204
crabapple, Pacific (*Pyrus fusca* or *Malus fusca*), 13
  in anthropogenic landscapes, 138, 139, 143
  digging sticks made of, 316
  fish camps associated with, 293
  genetic alteration of, 147
  licorice fern harvested from, 115
  management practices, 106, 110, 117, 121
  multiple uses of, 13
  ownership of trees, 130, 131, 159
  as prestigious food, 175
  settlement, and concentration of, 272n13
  Sproat on, 178n11
  Tlingit use of, 279
  Tsimshian management of, 250
cranberry, bog (*Vaccinium oxycoccos* or *Oxycoccus oxycoccos*)
  in anthropogenic landscapes, 139, 143–44

Katzie ownership concepts, 155–56
  management practices, 106
  Tlingit use of, 281
cranberry, highbush (*Viburnum edule*), 165
  in anthropogenic landscapes, 138, 139
  management practices, 106, 125
  ownership of patches, 130, 131–32
  as prestigious food, 175
  Tlingit use of, 281
cranberry, lowbush (*Vaccinium vitis-idaea*), 281
creation, worldviews on, viii–x
creation story, Kwakwaka'wakw, 166
creeping raspberry (*Rubus pedatus*), 280
cress, rock (*Arabis* spp.), 277
crows, scaring from plant resource sites, 287
cultigens, 55
cultivating ecosystem type, 56
cultivation, as term, 55–56, 332–33. *See also* "middle ground" food production
cultivation on Northwest Coast, dismissal and misrepresentation of, 3–8, 24–26, 319–20
cultivation on Northwest Coast, reevaluation and reconceptualization of, 5, 14–16
cultural complexity. *See* complexity, social
cultural diversity, 103
cultural domesticates, 60–61
cultural restoration, 339–40
culturally modified trees (CMTs), 18, 123–25, 142
currant, gray or stink (*Ribes bracteosum*), 106, 120, 138, 279
currant, swamp (swamp gooseberry) (*Ribes lacustre*), 279
currant, trailing black (*Ribes laxiflorum*), 280
Curtis, E. S.
  Bark Gatherer, The (photo), 6
  Berry Picker, The (photo), 162
  Gatherer on the Beach (photo), 337
  Root Digger, The (photo), 299
  Tule Gatherer, The (photo), 12
Cyperus (nutgrass), 65

danger, in intensification models, 81
Darby, Melissa, 194–217
Darling, J., 249–50, 256–57

dating
  accelerator mass spectrometry
    (AMS), 235
  of fish weirs, 78
  oxidizable carbon ratio (OCR), 319
  radiocarbon, 230
Davidson, Florence, *124*
Dawson, G. M., 167
de Laguna, Frederica, 276–81, 282–83,
  288, 295n1
deer, origins of, viii–ix
deer browsing, impacts of, 150n6
deer fern (*Blechnum spicant*), 107, 110, 137
deerberry. *See* wild lily-of-the-valley
  (*Maianthemum dilatatum*)
Delgamuukw v. British Columbia, 171
Delgamuukw v. the Queen, x
DeLillo, Don, 37
demographic collapse, 20, 21, 327n26.
  See also population
depletion of resources, intensification
  and, 84, 95
determinism, 25–26
Deur, Douglas
  as contributor, 3–34, 296–327, 331–42
  on nettles, 295n3
  on population concentrations, 178n12
  on pruning, 125
developmental models, deterministic,
  24–26
devil's club (*Oplopanax horridum*), *122*
  in anthropogenic landscapes, 137, 138
  management practices, 110, 122
  spirit in, 335
  sqaw'a ms ("where grows devil's
    club"), 268
  Tlingit use of, 279
Dick, Chief Adam (Kwaxsistala), *32, 341*
  on bog meadows, 143
  on cultivation of plots, 101
  on digging sticks, 316–17
  on domestication, 325n12
  on harvesting practices, 318
  on ownership, 130, 131, 164
  on prevalence of gardens, 321–22
  on pruning, 120–21
  on replanting, 117, 148–49, 308
  on size of gardens, 306
  on soil modification, 310, 313–15
  on terms for "garden," 315
  training of, 324n9
  on transplanting, 309
  on weeding, 306–7

diet breadth models, 79–80, 82, 86. *See
  also* intensification of food production
diffusionism, 24
digging. *See also* harvesting
  ecological effects of, 112
  in estuarine gardens, 316–18
  Nuu-chah-nulth terms for, 326n19
  productivity of, 192n5
  tools and techniques, 115, *116*, *299*, 316
digging-houses at estuarine gardens, 318,
  327n24
digging sticks, 115, *116*, *299*, 316
discourse, worldview and, vii–xi
diseases, post-contact, 20, 213
dispossession, rationalizations for,
  27–29, 336. *See also* colonialism
  and colonization
diversity of Northwest Coast region, 103
division of labor, 77, 264–65
DNA sequencing, and domestication, 51
dock, western (*Rumex occidentalis*), 111,
  139, 271n3, 280
domestication
  cultivation vs., 15
  in estuarine gardens, 325n12
  of livestock, 50
  "middle ground" and, 45–52
  morphological vs. nonmorphological
    domesticates, 60–61
  phenotypic change and causal chain
    in, 48–52
dominion, biblical injunction of, vii–viii
Donald, L., 77
Douglas, James, 172, 336, 338, 342n5
Douglas, William, 4
Douglas-fir (*Pseudotsuga menziesii*)
  in anthropogenic landscapes, 136
  burning and, 19, 140
  in Coastal Douglas-fir zone, 9
  management practices, 108, 109, 110
Douglas-fir forest, 136–37, 140
Douglas-fir zone, Coastal, 9, 275
Douglas Treaties, 172–73, 338–39, 342n3,
  342n5
Downie, Major, 242
Drucker, Philip, 161, 302, 335
Dryopteris dilatata (wood fern), 278
Dryopteris expansa (spiny wood fern),
  106, 137, 277
dualistic assumption about agriculture,
  39–42, 70–71, 333
ducks, 197. *See also* waterfowl
Duff, Wilson, 142, 170, 172

dulce ribbon seaweed (*Palmaria palmata*), 279
Dunn, John, 254
dwarf blueberry (*Vaccinium caespitosum*), 281
dye materials, 109

Early Modern economies, and intensification, 78
East Hunter Creek, 232–35
Echinodontium tinctorium (Indian paint fungus), 109, 137
ecological diversity, 103
ecological effects of horticultural methods, 112, 134, 135, 237–38
ecological mechanisms, in burning, 226–27
ecology, evolutionary, 79, 80
economic determinism, 26
edible seaweed (red laver) (*Porphyra abbottiae*), 107, 114, 279
efficiency and intensification, 75–76, 79–80, 86–87, 88
Egregia menziesii (boa kelp), 107
elderberry, red (*Sambucus racemosa*), 125, 280
Eleocharis (spikerush), 65
Ellen, Roy, 40, 41
Elmendorf, William, 326n20
emergency foods, 107. *See also* risk and risk reduction
Emmons, George, 276–81, 288, 295n1
environmental determinism, 25
environmental relations, and Western vs. indigenous worldview, vii–xi
environmental zones. *See* biogeoclimatic zones, Northwest Coast
Epilobium angustifolium. See fireweed
Equisetum telmateia (giant horsetail or scouring rush), 107, 109, 120, 138
"era of incipient agriculture," 58
Erythronium revolutum (pink fawn lily), 137
estuarine gardens. *See also* salt marshes; wetlands, tidal
  in antiquity, 318–19
  cultivation practices, 304
  in economic, ceremonial, and cosmological life, 300–304
  expansion of gardens in marshes, 128
  harvest, 316–18
  maps of, 296, 298

overview, 296–300
ownership, 304–6
resources available in, 12–13
site preparation and soil modification, 310–15
straight vs. gnarled roots, 314
transplanting, replanting, and selective harvesting, 307–9
weeding and garden hunting, 306–7
Ethnographic Atlas (Murdock), 39–40
ethnography. *See* anthropology
eulachon grease, berries preserved with, 252
Europe, 42, 255–56
European clothing, 258
European colonialism. *See* colonialism and colonization
European model of agriculture
  Aboriginal land rights and, 171–74
  bias toward, 5, 14–15, 22–23
  introduction and imposition of (Tsimshian), 261–63
evergreen huckleberry (*Vaccinium ovatum*), 162
evolutionary development models
  Boas and, 24–26
  dualistic assumption, 39–42
  Harris's models, 46–47
  indigenous worldview vs., viii–x
evolutionary ecology models of intensification, 79, 80
exchanges, 334. *See also* trade in plant resources
explorers, misrepresentation of cultivation by, 3–5

failure, cost of, 81–84, 96
false (Indian) hellebore (*Veratrum viride*), 111, 138, 281, 289
false lily-of-the-valley. *See* wild lily-of-the-valley (*Maianthemum dilatatum*)
false Solomon's seal (*Smilacina racemosa*), 280
Feak, Eddie, 246
Feak Creek, 266
feasts, 301–2, 323n5
fens, 139, 143–44
fern, bracken (*Pteridium aquilinum*)
  in anthropogenic landscapes, 136–37, 141–42
  Coast Salish use of, 185
  management practices, 108

Tlingit use of, 279
Tsimshian management of, 249
fern, deer (*Blechnum spicant*), 107, 110, 137
fern, lady (*Athyrium filix-femina*), 277
fern, licorice (*Polypodium glycyrrhiza*), 115, 279
fern, maidenhair (*Adiantum pedatum*), 277
fern, oak (*Gymnocarpium dryopteris*), 278
fern, spiny wood (*Dryopteris expansa*), 106, 137, 277
fern, sword (*Polystichum munitum*), 108, 109, 136, 137, 181–82
fern, wood (*Dryopteris dilatata*), 278
fertilization, 117–19, 265, 289
fiber materials, 108–9
fir, Douglas. *See* Douglas-fir (*Pseudotsuga menziesii*)
fir, grand (*Abies grandis*), 109, 110, 136, 181–82
fir, silver (*Abies amabilis*), 107, 109, 137
fire management. *See* burning
fireweed (*Epilobium angustifolium*)
  in anthropogenic landscapes, 137, 138
  management practices, 107, 120
  Tlingit use of, 278
First-Fruits Ceremony, 132, 156
"First Nations," as term, 33n2
fish camps, 293
fish weirs, dating of, 78
Five Mile Rapids, 83
Fladmark, Knut, 68
flags of red-cedar bark, 310
flavorings, 107
floodplains, tidal, 139. *See also* estuarine gardens; wetlands, tidal
folklore, estuarine roots in, 303
Fomitopsis officinalis (bracket fungus), 111
"food production," as term, 59–61. *See also* "middle ground" food production
food storage. *See* storage of food
foraging, contemporary importance of, 270
foraging strategies, overview of, 105. *See also* harvesting
Ford, Richard I., 43
  on domestication, 46, 48
  on low-level food production economies, 61–62
  on terminology, 55, 59–60
Forde, Darryl, 297
Forestry Act (British Columbia, 1924), 268

forests
  age of, 230
  coastal rainforest, 137, 140–42
  montane, 138, 142
  rain-shadow Douglas-fir forest, 136–37, 140
Fort Langley Journal, 150n3, 185
Fort Simpson (Fort Naas), 252–54, 255, 261
Fort Victoria (Douglas) Treaties, 172–73, 338–39, 342n3, 342n5
Fowler, John, 260
Fowler, K. (Catherine S.), 66
Fragaria spp. (wild strawberry), 106, 136, 141, 278
Franchere, Gabriele, 204
Frank, Henry, 262
Fraser Valley, 143, 150n3, 185. *See also* Katzie
Fraser Valley Fire Period, 238
Fraser Valley prescribed burning case study
  data collection and analyses, 232–35
  ecological mechanisms, 226–27
  evidence, indirect and theoretical, 218–20
  methods, 221
  natural vs. cultural fires, predictions on distinguishing, 227–31
  precontact resource management, 222–23
  results and discussion, 236–39
  study sites, 231–32
  traditional practices, 223–26
freckle pelt (lichen) (*Peltigera aphthosa*), 279
French, David, 216n2
freshwater wetlands, 138–39, 143–44. *See also* marshes
Fritillaria camschatcensis. *See* rice-root lily, northern
Fritillaria lanceolata (chocolate lily), 136
Frozen Lakes, 232–36, 233, 234
fuel materials, 108
funeral rites, sacrificial burnings in, 259
fungus, bracket (*Fomitopsis officinalis*), 111
fungus, Indian paint (*Echinodontium tinctorium*), 109, 137
fur traders, 4, 204

gagemp ("grandfather"), 309
galts'ap (tribal population), 241

Gam-gak-muk, Chief, 170
gardening, as term, 56–57
Garfield, Viola
    on collective management by
        Tsimshian, 248–49
    on Port Simpson people, 242
    on tobacco, 258–59
    on Tsimshian plant resources, 243
Garry oak (*Quercus garryana*), 126, 219
Garry oak parkland, 140, 219
Gatherer on the Beach (Curtis), 337
Gaultheria shallon. See salal and salal
    berries
gender and labor. *See* division of labor
genetic change, 48–52, 147–48
gentian, mountain (*Gentiana
    platypetala*), 278
gentian, swamp ("land otter medicine")
    (*Gentiana douglasiana*), 278
Gentiana douglasiana (swamp gentian or
    "land otter medicine"), 278
Gentiana platypetala (mountain gentian),
    278
geographical diversity, 103
geography, human, 10
geophytes
    archaeological evidence for intensifi-
        cation, 93–97
    intensification models, 88–93
George, Agnes, 181
George, Chief Earl Maquinna, 151,
    159–60
George, Josephine, 186
George, Mary, 181
George, Ralph, 225–26
Geum macrophyllum (large-leaved
    avens), 278
giant horsetail (scouring rush)
    (*Equisetum telmateia*), 107, 109,
    120, 138
giant kelp (*Macrocystis integrifolia*), 107
Gibbs, George
    on bread-making, 192n6
    on camas cultivation, 187, 192n4
    on gardens, 297
    on resource ownership, 206
gifts, 334
ginger, wild (*Asarum caudatum*), 109, 136
Gitanyow (village), 170
Gitga'at, 129, 131–32
Gitxandakhl (village), 262, 272n19
Gitxsan (Gitksan)
    burning by, 247
    House groups, 249

ownership concepts, 167–71
    plant resources used by, 244
goatsbeard (*Aruncus dioicus*), 277
Goland, C., 79, 80. *See also*
    Winterhalder-Goland model
Gold, Captain, 145
goldthread (*Coptis trifolia*), 277
goose-tongue (*Plantago maritima*), 279
gooseberry, swamp (*Ribes lacustre*), 279
Gottesfeld, Leslie M. J. (Johnson),
    169–70, 246–50, 268
Gould, R., 73–74, 76
Grant, W. C., 127
grasses, 63–64, 108. *See also* bear-grass
    (*Xerophyllum tenax*)
Gray, Robert, 286
gray (stink) currant (*Ribes bracteosum*),
    106, 120, 138, 279
Great Basin Shoshoni groups, 66
green vegetables, management strategies
    overview, 107
growth cycles. *See* life cycles of plants
Gulf of Georgia period, 98
Gulf of Georgia region, 97–98
Gunther, Erna, 153
Gymnocarpium dryopteris (oak fern), 278

habitat diversity, 9, 103, 135
habitat improvements, in low-level food
    production economies, 65–66
habitat preferences in harvesting, 114–15
hahuulhi ("sovereign wealth," Nuu-
    chah-nulth)
    authority of, 159
    ownership responsibilities and, 131,
        163
    "subsistence" vs., x
    territories and, 151
Haida Gwaii (Queen Charlotte Islands),
    166, 257–58
Haida Nation
    estuarine gardens of, 297
    ownership concepts, 166–67
    plant resources used by, 244
    potato cultivation and trading, 253–55
    tobacco cultivation, 4, 257–58, 284,
        285
Haida tobacco. *See* tobacco (*Nicotiana
    spp.*)
hairy rock cress (*Arabis hirsuta*), 277
Haisla, 130, 159
Haiyupis, Roy, 145
Hajda, Yvonne P., 210–11

knowledge of plant gathering, women's monopoly on, 264

Komkanetkwa, 94

Kroeker, D. W., 212

kuh, 282. *See also* rice-root lily, northern (*Fritillaria camschatcensis*)

Kumeyaay, 63

Kwakwaka'wakw
  burning of berry bushes, 127
  ceremonial feasts, 301–2
  clans in, 163–66
  cosmology, 303
  creation story, 166
  domestication, potential, 325n12
  estuarine gardens of, 296, 297, 309
  harvesting practices, 316–18
  larger roots as chiefly tribute, 303
  ownership, concepts of, 130, 163–66, 304–5
  transplanting by, 309

Kwak'wala (language), 309, 314–15, 317

labor, division of. *See* division of labor

labor, increased, 74–75

labor, land as subject of vs. land as instrument of, 245

Labrador tea (*Ledum groenlandicum*), 107, 139, *250*

lady fern (*Athyrium filix-femina*), 277

lakeshores, anthropogenic plant communities in, 143

Lampman, Ben Hur, 195

land as subject of labor vs. as instrument of labor, 245

"land otter medicine" (swamp gentian) (*Gentiana douglasiana*), 278

land tenure. *See* ownership concepts and land tenure

landscapes, anthropogenic. *See* specific landscapes

languages, indigenous
  Chinook Jargon, 207, 216n2
  diversity of, 103, 150n1
  gardening words in, 147, 191, 317
  soil modification reflected in, 314–15
  vegetative propagation, linguistic evidence for, 309

large-leaved, avens (*Geum macrophyllum*), 278

Lathyrus spp. (wild pea), 278

laver, red (edible seaweed) (*Porphyra abbottiae*), 107, 114, 279

Lawrence, Scott, 255

laxabales or *lhaxabális* (large roots), 302–3

laxyupp (House lands and estates), 242–43, 266, 267

Ledum groenlandicum (Labrador tea), 107, 139, *250*

Ledum spp. (Hudson's Bay tea), 278

Lepofsky, Dana, 218–39, 323n5

Lertzman, Ken, 218–39

Lewis, Meriwether, 195–96, 202–4, 208, 213

lhagwa'nawe (chief, as "the thick root of the tribe"), 303

lichen, old man's beard (*Alectoria sarmentosa*), 109, 137

licorice fern (*Polypodium glycyrrhiza*), 115, 279

life-cycle intervention, 63

life cycles of plants, 112–13, 129

life force of plants, 176

lilies. *See also* camas, blue (*Camassia quamash; C. leichtlinii*); rice-root lily, northern (*Fritillaria camschatcensis*); yellow pond-lily (*Nuphar polysepalum*)
  chocolate lily (*Fritillaria lanceolata*), 136
  pink fawn lily (*Erythronium revolutum*), 137
  tiger lily (*Lilium columbianum*), 136

Lilium columbianum (tiger lily), 136

Liljeblad, Sven, 66

lily-of-the-valley, wild or false. *See* wild lily-of-the-valley (*Maianthemum dilatatum*)

lineage groups. *See* House groups

linear vs. simultaneous tasks, 77

linguistic diversity, 103, 105n1. *See also* languages, indigenous

livestock, 50, 213–14

local vs. regional resources, 98–99

Locarno Beach phase, 97–98

Locke, John, vii, x–xi, 173, 338

lodgepole pine (*Pinus contorta*), 109, 110, 139, 220

Loiseleuria procumbens (alpine-azalea), 278

low-level food production economies, 60, 61–66

lowbush cranberry (*Vaccinium vitisidaea*), 281

Lower Columbia River valley. *See also* Chinookan wapato intensification geography and people of, 198–99

post-contact disease, 201, 213
settlement patterns, 210–11
trade and exchange in, 208
wapato decimation, effect of, 214
as "Wapato Valley," 195–98
Lummi. *See* Salish, Coast
lupine, Nootka (*Lupinus nootkatensis*)
in anthropogenic landscapes, 139
in estuarine gardens, 296
management practices, 106
Tlingit use of, 278, 289
Lupinus nootkatensis. See lupine, Nootka
Lushootseed. *See* Salish, Coast
Lyons, Annie, 186
Lysichiton americanum. See skunk-
cabbage

MacNeish, Richard S., 58
Macrocystis integrifolia (giant kelp), 107
madrone, Pacific. *See Arbutus menziesii*
(arbutus or Pacific madrone)
Mahonia aquifolium (Oregon-grape),
109, 136
Maianthemum dilatatum. See wild lily-
of-the-valley
maidenhair fern (*Adiantum pedatum*),
277
maize agriculture, Mesoamerican shift
to, 53
malaria, 213
Malus fusca. See crabapple, Pacific
management practices, summary of, 103,
106–11, 135, 146
Manby, Thomas, 201
maple, broad-leaf (*Acer macrophyllum*),
108, 136
marine algae, 107
Marpole phase, 97–98
marshes. *See also* Chinookan wapato
intensification; estuarine gardens;
wapato (*Sagittaria latifolia*); wetlands
anthropogenic plant communities in,
138, 143
expansion of gardens in, 128
labor input on, 311–12
muskrats in, 212
salt marshes, 139, 312–13, *313*, 319
Maschner, H. D. G., 101
materials plants, 108–9
Mathewes, Rolf, 218–39
matrilineal descent, 167, 170, 191. *See also*
clans; kin groups

Matson, R. G., 222–23
maturity of harvested plants, 114
McDonald, James, 240–73, 269
McHalsie, Albert (Sonny), 218–39
McKenna-McBride Royal Commission,
28, 261–62
McNeary, Stephen, 247, 248, 269
meadows, in freshwater bogs, 143–44
meadows, low elevation, 136
Meares, John, 4
measurement of intensification, 75, 77–78
medicinal plants
management strategies overview,
110–11
places harvested, 115
Tlingit use of, 277–81
Tsimshian use of, 244
Meillasoux, C., 245
Menyanthes trifoliata (buckbean), 279
Menzies, Archibald, 258, 297–98
Mesoamerican agricultural
development, 53–54, 58
methods, horticultural. *See* technology
and tools
"middle ground" food production
domestication and, 45
dualistic assumption and, 39–42
food production, as term, 59–61
low-level food production
economies, 60, 61–66
models and boundaries of, 42–48
overview, 37–39
size of, 52–54
as stable state, 333
terminology issues, 43, 54–59
migration between resource sites
harvest scheduling and, 129–30
root-digging plots and, 317–18
settlement patterns and, 12
wapato intensification and, 210–11
miner's lettuce, Siberian. *See* Siberian
spring beauty (*Claytonia sibirica*)
misrepresentation and dismissal of cultivation, 3–8, 24–26, 319–20
modeling of intensification. *See* intensification of food production
modification, anthropogenic, in domestication, 48–52
monocultural plots, 308–9
montane forest, 138, 142
Moon, Johnny, *341*
Moran, E., 63
morphological change in domestication,
48–52

storage of food
   intensification and, 69, 87, 89
   risk buffered by, 81–82
   root crops, 288
   Tsimshian, 252
   wooden boxes for, 87
"Story of the Town Chief Peace," 259–60
Straits Salish (Nation), 120
strawberry, wild (*Fragaria* spp.), 106,
   136, *141*, 278
Streptopus amplexifolius (clasping
   twisted stalk), 280, 289
Supreme Court of British Columbia,
   166, 171
sustainable harvesting, 17, 114, 133–34
Suttles, Wayne
   as contributor, 181–93
   on ownership, 156, 207
   on potatoes, 256
   on social organization, 68
swamp currant (*Ribes lacustre*), 279
swamp gentian ("land otter medicine")
   (*Gentiana douglasiana*), 278
swamp gooseberry (*Ribes lacustre*), 279
swamps, anthropogenic plant communities in, 138, 143
Swan, James, 287
Swan, Luke, 160
Swaniset, George, 186–87
swans, 197, 200. *See also* waterfowl
sweet-cicely, mountain (*Osmorhiza
   chilensis*), 279
sweet potato, wild. *See* silverweed,
   Pacific (*Potentilla anserina* ssp.
   pacifica)
sweeteners, 107
sword fern (*Polystichum munitum*), 108,
   109, *136*, 137, 181–82
Sye, Jessie, *162*
Symphoricarpos albus (waxberry), 11, 137

t'aki'lakw (place of human-manufactured soil), 315
Tallio, Edward, 123
taste variation by environmental zone,
   246–47
Taxus brevifolia (Pacific or western yew),
   108, 114, 137, 316
Taylor, R. L., 284, 286
tea, Hudson's Bay (*Ledum* spp.), 278
tea, Labrador (*Ledum groenlandicum*),
   107, 139, *250*

technology and tools. *See also* burning;
   specific techniques
   for children, 308
   digging or tilling, 115, *116*
   disruption by colonization, and
      restoration of, 264
   ecological effects of, 112
   for estuarine gardens, 306–15
   intensification and technological
      change, 78, 89, 208–10
   potato cultivation, 288
   production as technological process,
      76–77
   pruning or coppicing, 120–25
   replanting, 117
   transplanting, 125–26
   Tsimshian, 263–66
   weeding, clearing, and fertilizing,
      117–20
Temple, Miriam
   on European-style farming, 262
   on laxyupp (House lands and
      estates), 242–43, 249
   on legislated restrictions, 268
   on tending, 265
tending
   Coast Salish practices, 187–88
   ecological effects of, 112
   in low-level food production
      economies, 62
   Tsimshian practices, 264–65
territories, tribal
   boundaries of, 10, 166, 167, 172
   ownership concepts and, 155
   Tsimshian resource properties,
      names of, 267–68
theft, 305–6. *See also* ownership
   concepts and land tenure
thimbleberry (*Rubus parviflorus*), *133*
   in anthropogenic landscapes, 137, 139
   First-Fruits Ceremony and, 132
   management practices, 107
   pruning of, 120
   Tlingit use of, 280
thistle (*Cirsium brevistylum*), 109
Thoms, A. V., 79. *See also* Thoms model
   of intensification
Thoms model of intensification, 79,
   88–92, 93, 96, 217n3
three-square (*Schoenoplectus olneyi*), 126
threshold of agriculture, 71. *See also* agricultural origins and development
Thuja plicata. *See* cedar, western red
   (*Thuja plicata*)

tidal floodplains. *See* floodplains, tidal
tidal wetlands and tidal flats. *See* estuarine gardens; wetlands, tidal
tiger lily (Lilium columbianum), 136
tilling, 62, 112, 115
Tlingit plant use and horticulture
  archaeological evidence of gardens, 289–93
  assessment of, 293–94
  environmental background, 274–75
  "horticulture," usage of term, 276
  initial European contact, 283–84
  potatoes, 286–89, 294
  tobacco, 283–86
  use of indigenous plants, 276–83
tobacco (*Nicotiana* spp.)
  in anthropogenic landscapes, 139
  chewing of, 260, 286
  cultivation abandoned for trade in, 286–87
  as "exception" to noncultivation, 17–18, 22
  "Haida tobacco," 285
  introduction of, 284
  management practices, 107, 128
  potato cultivation and, 286
  Tlingit cultivation of, 283–86
  Tsimshian and, 257–61
Tom, Virginia, 6, *299*, *337*
tools. *See* technology and tools
totem poles, and land ownership, 170
trade in plant resources
  estuarine root vegetables, 302
  exchange, role of, 334
  ownership and, 175
  potatoes, 252–54
  Tsimshian, 252–53, 265
  wapato intensification and, 207–8
tragedy of the commons (Hardin), 176, 270
trailing black currant (*Ribes laxiflorum*), 280
trails, 245–46
transhumance. *See* migration between resource sites
transplanting. *See also* replanting
  ecological effects of, 112
  in estuarine gardens, 307–9, 309, 325n13
  in low-level food production economies, 62
  of salmon eggs, 193n10
  sustainable harvesting and, 17

  by Tlingit, 283
  by Tsimshian, 245
transportation and intensification, 89, 92
trapping of waterfowl, 307
travel. *See* migration between resource sites
treaties, 172–73, 338–39, 342n3, 342n5
tree inner bark, 107
tree roots, 251
trees, culturally modified (CMTs), *18*, 123–25, 142
Trientalis arctica (northern starflower), 280
Trifolium spp. (clovers), 323n4
Trifolium wormskjoldii. See clover, springbank
Triglochin maritimum (arrow-grass), 139
tset ("wild sweet potato"), 282. *See also* silverweed, Pacific (*Potentilla anserina* ssp. *pacifica*)
Tsimshian plant use and horticulture
  bark and tree roots, 251
  berries, 245–49
  division of labor, 264–65
  ethnographic records on plant resources, 242–44
  and foraging vs. cultivation, 245
  fruit trees, 250–51
  nineteenth-century gardens, 261–63
  overview, 240–41
  ownership and colonization, 266–70
  ownership concepts, 167–71
  potatoes, 252–57
  root crops, 249–50
  seaweed, 251–52
  social organization, 241–42
  storage, 252
  technology, 263–65
  tobacco, 257–61
  trade and consumption, 252–54, 265–66
ts'isakis (place with soil), 315, 319
Tsuga heterophylla. See hemlock, western
Tsuga mertensiana (mountain hemlock), 10
Tualatin Kalapuyans, 196–97, 204–5, 211
tule (*Schoenoplectus acutus* or *Scirpus acutus*)
  in anthropogenic landscapes, 138, 143
  estuarine gardens, harvesting near, 13
  management practices, 109
  mats, amount used for, 14, 143
Tule Gatherer, The (Curtis: photo), 12

Milton Keynes UK
Ingram Content Group UK Ltd.
UKHW020338250924
448802UK00005B/117

9 780295 985657